AF794324

Helicopter Test
and Evaluation

HELICOPTER TEST AND EVALUATION

Alastair K. Cooke
BTech, BEng, MSc, PhD, CEng, MRAeS, MIMechE

Eric W. H. Fitzpatrick
MRAeS, Test Pilot

Blackwell
Science

QinetiQ

© QinetiQ Limited 2002

Blackwell Science Ltd, a Blackwell Publishing Company
Editorial Offices:
Osney Mead, Oxford OX2 0EL, UK
Tel: +44 (0)1865 206206
Blackwell Science, Inc., 350 Main Street, Malden, MA 02148-5018, USA
Tel: +1 781 388 8250
Blackwell Science Asia Pty Ltd, 54 University Street, Carlton South, Melbourne, Victoria 3053, Australia
Tel: +61 (0)3 9347 0300
Blackwell Wissenschafts Verlag, Kurfürstendamm 57, 10707 Berlin, Germany
Tel: +49 (0)30 32 79 060

The right of the Author to be identified as the Author of this Work has been asserted in accordance with the Copyright, Designs and Patents Act 1988.

All rights reserved. No part of this publication may be reproduced, stored in a retrieval system, or transmitted, in any form or by any means, electronic, mechanical, photocopying, recording or otherwise, except as permitted by the UK Copyright, Designs and Patents Act 1988, without the prior permission of the publisher.

First published 2002 by Blackwell Science Ltd

Library of Congress
Cataloging-in-Publication Data
Cooke, Alastair K.
Helicopter test and evaluation / Alastair K. Cooke, Eric W.H. Fitzpatrick.
p. cm.
Includes bibliographical references and index.
ISBN 0-632-05247-3 (Hardcover)
1. Helicopter – Flight testing.
2. Helicopters – Aerodynamics.
I. Fitzpatrick, Eric W.H. II. Title.
TL716.5 .C66 2002
629.133′352′0287 – dc21
2002010325

ISBN 13: 978-0-6320-5247-9

A catalogue record for this title is available from the British Library

For further information on Blackwell Science, visit our website:
www.blackwell-science.com

Contents

Preface

The authors of this book have been employed at the Empire Test Pilots' School (ETPS), part of QinetiQ, for the past ten years instructing student test pilots and flight test engineers on helicopter testing. Alastair Cooke has a masters degree in flight dynamics from Cranfield University and graduated as a flight test engineer from ETPS in 1989. Eric Fitzpatrick is a former military helicopter pilot and instructor who graduated as a test pilot from ETPS in 1986.

This book has been produced using the experience of the authors in flight test and flight test training gained over a combined period of 25 years in the field. Much of the material has its origin in the training notes produced by ETPS. These have been developed over a period of over forty years since the start of specific training for helicopter test pilots. The book has been designed to appeal to professionals working in the area of rotorcraft test and evaluation but it is hoped that it will also prove useful to a wider audience. In our experience, we have found that helicopter pilots are generally not well informed about the process that has led to their aircraft entering service, nor about why it has certain limitations imposed on it. We hope this book will provide pilots with this information as well as being a useful text for practising engineers and technologists. The rotor theory presented is more extensive than is found in most aeronautical degree courses and so the book should prove useful to graduates specializing in rotorcraft technology. Perhaps uniquely, this work approaches this important subject from both the theoretical and practical viewpoints.

For each topic the theory is explained briefly and is then followed by details of the practical aspects of testing a helicopter. These details include the safety considerations related to the anticipated tests, planning the tests themselves, and, where appropriate, the most efficient way to conduct individual flights. Following a description of each type of test, typical results are examined and an explanation given as to why they would be important to the clearance process. Whenever possible examples of actual test results have been presented and used in the subsequent discussion. The book is split into four main sections:

- Introduction: covering a methodology for testing and general aspects of test programmes.
- Performance: in this section level flight, vertical and climb/descent performance is addressed. The planning of performance trials is covered together with the methods for airborne data gathering and analysis of results.
- Handling qualities: this is a major section and covers the basics of helicopter stability and control testing. Also included are frequency domain methods and the use of mission task elements.
- Systems: in this section the major systems required to enable a helicopter to fly are

covered. This includes assessment of the cockpit, air data systems, engine control systems, and automatic flight control systems. Also addressed in this section is the testing of system failures.

Although highly specialized, the topic of helicopter testing is still vast and no single text could hope to cover everything. The authors have attempted, therefore, to concentrate on the most important aspects using their own knowledge of the subject as a guide. Inevitably, a number of important areas have not been covered; for example, it has not been possible to include information on specialist areas such as underslung load trials, deck operation trials or armament testing. The amount of information on systems testing to include in the book was a difficult decision due to the plethora of systems that can be fitted to a modern rotorcraft. Consequently, it was decided to detail the general methodology used in this type of testing and then to concentrate on systems which are intrinsic to the operation of all helicopters. Thus, it has not been possible to include the testing of hydraulics, electrics and lubrication systems. Similarly the testing of a number of cockpit systems such as piloting vision aids, navigation systems, weapons, and mission displays were considered to be beyond the scope of this book. Finally, it was not possible to include some types of testing which often play a large part in the life of a test pilot such as environmental trials, notably cold weather and icing. Despite these omissions, it is believed that the book covers all the essential areas of rotorcraft testing and will prove useful to a large part of the helicopter community. As the authors' background has been in military test and evaluation this has clearly influenced the subject matter that has been presented. However, most of the information given in the book can be applied, with only minor modification, to the testing of civil rotorcraft.

Before closing we would like to acknowledge the help of Mr Mike Cook of Cranfield University, Mr Mark Roots of QinetiQ Ltd and Miss Julia Burden of Blackwell Publishing in helping us develop our manuscript for publication.

Alastair Cooke and Eric Fitzpatrick
ETPS, QinetiQ Ltd
MOD Boscombe Down

Copyright acknowledgements

The following organizations are gratefully acknowledged for granting permissions for the use of copyright material.

The UK MOD for the cover photograph. NASA for Figure 3.37 and the AHS for Figures 3.18, 3.20 and 3.22.

The example helicopter

On several occasions throughout this book quantitative calculations have been made to support the theoretical trends being discussed. Where possible the calculations have been made using the same baseline data, referred to as the *example helicopter*. The details of this helicopter, which is loosely based on the Westland Lynx, are summarized below:

Main rotor

Radius	6.5 m	Blade chord	0.4 m
Standard rotor speed	35 rad/s	Number of blades	4
Lift curve slope	0.1/°	Twist	10°
Profile drag coefficient	0.008 or 0.01		

Tail rotor

Tail arm	7.5 m	Radius	1.1 m
Blade chord	0.2 m	Standard rotor speed	165 rad/s
Number of blades	4	Profile drag coefficient	0.010

Fuselage frontal drag area	2 m^2
Fuselage vertical drag area	8 m^2
Mass (unless otherwise indicated)	5000 kg

Abbreviations

AA	aircraft allowance
ACAH	attitude command/attitude hold
ACT	active control technology
ADI	attitude direction indicator
ADS	Aeronautical Design Standard
AFCS	automatic flight control system
AGL	above ground level
AI	attitude indicator
AltR	altimeter reading
AOA, AoA	angle of attack
AOB, AoB	angle of bank
APU	auxiliary power unit
ASE	automatic stabilization equipment
ASI	airspeed indicator
ASIR	airspeed indicator reading
ASW	anti-submarine warfare
AUM	all-up-mass
BCV	bleed control valve
CAA	Civil Aviation Authority
CAS	calibrated airspeed
CBEM	combined blade element and momentum (theory)
CFSS	collective fixed static stability
CG	centre of gravity
CP	collective pitch
CR Point	control reference point
CR Posn	control reference position
CWP	centralized warning panel
DA	density altitude
DEP	design eye position
DH	decision height
DVE	degraded visual environment
EAS	equivalent airspeed
ECS	engine control system
EFIS	electronic flight information system
EGT	exhaust gas temperature
EOL	engine-off landing
ERGA	engine and rotor governing assessment
ETPS	Empire Test Pilots' School
FADEC	full authority digital engine control
FAR	Federal Aviation Regulations
FBW	fly-by-wire
FCMC	flight control mechanical characteristics
FCS	flight control system
FFR	fuel flow rate
FIG	flight idle glide
FLIR	forward looking infra-red
FMECA	failure modes effect and criticality analysis

FOD	foreign object debris
FOR	field of regard
FOV	field of view
FPT	free power turbine
FTE	flight test engineer
FW	fixed wing
GCA	ground controlled approach
GPS	global positioning system
GSDI	groundspeed and drift indicator
GVE	good visual environment
GW	gross weight
HDD	head down display
HIGE	inside ground effect hover
HMD	helmet mounted display
HOGE	outside ground effect hover
HQR	handling qualities rating
HSI	horizontal situation indicator
HTA	helicopter type allowance
HUD	head-up display
IAS	indicated airspeed
IAT	indicated air temperature
ICAO	International Civil Aviation Organization
IFR	instrument flight rules
IGE	inside ground effect
IGV	inlet guide vanes
ILS	instrument landing system
IMC	instrument meteorological conditions
INS	inertial navigation system
IRE	identified risk element
ISA	ICAO standard atmosphere
JAR	Joint Airworthiness Regulations
KCAS	knots calibrated airspeed
KEAS	knots equivalent airspeed
KIAS	knots indicated airspeed
KTAS	knots true airspeed
LCD	liquid crystal display
LDO	lateral/directional oscillation
LDP	landing decision point
LED	light-emitting diode
M	Mach number
MAR	military aircraft release
MAUM	maximum all-up-mass
MAUW	maximum all-up-weight
MFD	multi-function display
MPOG	minimum pitch on ground
MPV	maximum power vertical
MSL	mean sea level
MTBF	mean time between failures
NDB	non-directional beacon
NOE	nap of the earth
NTPS	National Test Pilots' School

NVG	night vision goggles
OAT	outside air temperature
ODM	operating data manual
OEI	one engine inoperative
OFE	operational flight envelope
OGE	outside ground effect
PE	pressure error
PEC	pressure error correction
PFL	practice forced landing
PFLF	power for level flight
PH	(hover) position hold
PIO	pilot-induced oscillation
PTIT	power turbine inlet temperature
PUPO	pull-up/push-over
RAD ALT	radar altimeter
RCAH	rate command/attitude hold
RCDH	rate command/direction hold
RCDI	rate of climb and descent indicator
RCHH	rate command/height hold
REP	reference eye position
RH	relative humidity
ROC, RoC	rate of climb
ROD, RoD	rate of descent
RPM	revolutions per minute
RPV	reduced power vertical
RRM	risk reduction measure
RRPM	rotor revolutions per minute
RVR	runway visual range
RW	rotary wing
SAR	search and rescue; specific air range
SAS	stability augmentation system
SFC	specific fuel consumption
SHSS	steady heading sideslip
SSC	side stick controller
TAS	true airspeed
TCDB	trim control displacement band
TCWP	trim changes with power
TDP	take-off decision point
TE	tip effects
TFCP	trimmed flight control position
TGT	turbine gas temperature
TI	trials instruction
TO1C	turn on one control
TPS	trailing pitot-static
VFR	visual flight rules
VMC	visual meteorological conditions
VNE	maximum never to exceed speed
VNO	maximum normal operational speed
VOR	VHF omnidirectional range
VSI	vertical speed indicator
VSS	variable stability system

Notation

Arabic symbols

A	disk area
a	lift curve slope
A_1	lateral cyclic pitch
a_1	longitudinal cyclic flap
A_R	aspect ratio
B	tip loss factor
b	number of blades in main rotor or tail rotor
B_1	longitudinal cyclic pitch
b_1	lateral cyclic flap
c	rotor blade chord
C_D	drag coefficient
C_L	lift coefficient
$\tilde{C}_{L\,max}$	maximum mean lift coefficient
$\tilde{C}_{L\,nom}$	nominal mean lift coefficient
C_P	power coefficient
C_Q	torque coefficient
C_T	thrust coefficient
D	drag; rotor diameter
f	drag area
g	acceleration due to gravity
H_D	density altitude
H_P, h_P	pressure altitude
L	lift
l_{TR}	tail rotor moment arm
m	mass
N	engine RPM
N_c	compressor RPM
N_F	free turbine RPM
N_G	gas generator RPM
N_h	high speed spool RPM
N_l	low speed spool RPM
N_R	rotor speed RPM
P, P_s	static pressure
p	change in roll rate from trim
P_i	induced power
P_P	total or pitot pressure
P_{par}	parasite power
P_{pr}	profile power
P_∞	ambient pressure
q	change in pitch rate from trim
Q_{MR}	main rotor torque
R	rotor radius
r	change in yaw rate from trim

s	solidity
S, S_f, S_v	blade area, frontal drag area, vertical drag area
T	thrust; engine temperature
T_2, $T_{1/2}$	time to double or half bank angle
u	change in speed along X-axis
V	true airspeed
v	change in speed along Y-axis
V_c	calibrated airspeed; climb speed
V_e	equivalent airspeed
V_f	forward (horizontal) velocity component
V_H	maximum speed in level flight
V_i, v_i	indicated airspeed corrected for instrument error, induced velocity
V_{imp}	indicated minimum power speed
V_{ME}	maximum endurance airspeed
V_{MP}	minimum power airspeed
V_{MR}	maximum range airspeed
V_{toss}	take-off safety speed
V_T	tip speed
V_v	vertical velocity component
V_Y, v_y	speed for maximum rate of climb
W	weight
w	change in speed along Z-axis
Z	tapeline height

Greek symbols

α	angle of attack
β	sideslip angle; flap angle
γ	Lock number; flight path angle; disk tilt
δ	relative pressure; angle between rotor blade hinge lines
ε	rotor frequency parameter
ζ	relative damping
η	vertical velocity ratio (V_v/v_{ih}); efficiency
θ	relative temperature; pitch attitude; rotor feather angle
θ_0, θ_1	collective pitch and twist
θ_{TR}	tail rotor pitch
κ	non-uniform inflow factor
λ	inflow ratio
μ	advance ratio (V/V_T); forward velocity ratio (V_f/v_{ih})
ν	induced velocity ratio (v_i/v_{ih})
ξ	lead/lag angle
ρ	air density
σ	relative density
τ	time constant
ϕ	bank angle; inflow angle
χ	rotor wake skew angle
ψ	azimuth angle; yaw angle
Ω, ω	rotor speed; angular velocity; frequency

Chapter 1
The Flight Test Process

1.1 INTRODUCTION

Flight test is an expensive activity, which by its very nature attracts levels of risk higher than normal operations. To ensure that all trials are conducted as efficiently and safely as possible, a flight test process has been developed over the years. This process is used, with only minor variations, in nearly all test organisations throughout the world whether they be military, civilian or based at a manufacturing facility. Many of the steps in this process have evolved as the result of painful lessons and, therefore, the authors consider them vital to the overall test activity.

It is perhaps true to say that at the beginning of their careers, flight test personnel show greater interest in the test methods that they need to apply than in the overall system used to approve, authorize and regulate test flying. Consequently it was decided to make the test process the first chapter of this book to demonstrate the importance that the authors attach to this subject. It is our belief that without an understanding of the process the knowledge contained within the rest of the book cannot be applied effectively.

Brief details of the flight test process are given below. For more details the reader is referred to the AGARD flight test techniques and instrumentation series [1.1 and 1.2]. Generally the flight test process can be broken down into three major areas:

- Flight test planning.
- Conducting the test.
- Post-test actions.

1.2 FLIGHT TEST PLANNING

Thorough planning is vital for all flight trials to ensure that they are conducted safely, efficiently and that the trials objective is met. When a trial is first proposed to a test establishment a management plan is produced which includes a work breakdown structure defining the individual elements of the trial. For each element the department or departments which are to undertake the work are determined together with costs and timescales. Part of the management plan is to define the exact project technical objectives and it is at this level that this book will examine the planning process.

1.2.1 Technical objectives

To define the project technical objectives the exact requirements of the customer calling for the trial have to be understood clearly. Once this has been done then the next step is to decide on the assessment philosophy, in other words 'What are we going to do? – How much of it are we going to do?' Deciding on what approach to take depends on a number of factors such as the level of expertize within the test organization, the resources available, and if the trials results need to be compared with earlier results using a particular test method. The approach to the trial and the scope of the work both depend on the requirement to collect evidence and to identify where that evidence is located.

Determining what evidence is needed to be able to make well-supported recommendations relies upon the professional knowledge of the test team, although previous trials may provide a guide. The evidence itself does not necessarily need to come from a flight conducted as part of the trial, indeed because flight testing is such an expensive activity other sources of evidence are invariably considered first. These sources may be the aircraft manufacturer, who may have conducted the required test, earlier test results held at the test establishment, read across from other similar aircraft, another test organization, or possibly an operator who already has experience of the test article. Modelling and simulation are also used extensively. In each case the evidence and the source are considered carefully to determine to what extent it can be relied upon. If the existing evidence is determined to be reliable then it may be used without further testing or, more likely, a limited number of tests would be planned to 'spot check' the evidence to increase confidence in its validity.

Once the existing evidence has been established then the difference between this and the evidence required is what the flight trial must address, this defines the scope of the test. The team then decides how much testing is required to gather the missing evidence to allow the test objective to be met. From this is defined the detailed work breakdown structure including test techniques, required environmental conditions, trials location, the order and interdependency of tests, and the allocation of tasks to parts of the organization. Having defined the tests to be made, the facilities required to conduct them have to be identified: these usually include the aircraft build standard, ranges, test instrumentation and equipment, the test crew and their training standard, and data replay and analysis equipment.

1.2.2 Assessing and managing risk

Test flying by its very nature involves a degree of risk above that which is routinely accepted during normal operations. To ensure that tests are conducted as safely as possible a process of identifying and reducing risk is employed. The aim of this process is to determine what the risks are, decide on the best way to reduce them and then to decide if the residual risk is acceptable.

1.2.2.1 Conducting a hazard analysis

To identify risk areas the programme or flight is divided into sections and the potential dangers considered in each one: this will involve asking the question at each stage 'What could possibly present a hazard?' and 'What would be the possible consequences?' This procedure relies to a large extent on the experience of the test pilot or FTE, however, a thorough and logical assessment of each phase of flight should lead to the identification of individual risks. It is normal to involve a number of people by calling a review meeting where the group examine each aspect of the flight. Any databases available are also examined to ensure that the lessons learned on previous trials of a similar type are incorporated. On completion of this process a list of Identified Risk Elements (IRE) can be produced for each section of the programme or flight.

1.2.2.2 *Risk reduction measures*

Having identified the risks the next stage is to decide how they might be mitigated as much as possible. The measures which can be taken are known as risk reduction measures (RRM) and are typically divided into five main categories:

(1) *Training*: Correct and comprehensive training of the test team is vital to reducing the risk on any test programme. This training will be required on the aircraft itself and in the case of initial trials on a new aircraft encompasses the ground instruction, simulator training, conversion course and continuation training. On other aircraft it may only involve currency training on type. The training may also involve the practise and perfection of the test techniques to be used or the flight skills to be employed during the programme, such as night flying, NVG instrument flying. Safety and survival training for all crew members is also conducted.

(2) *Supervision*: The close supervision of the test programme from the initial inception of the requirement, through the raising of the trials instruction (TI), the test plan and the individual flight authorization is possibly the single most important RRM. It is often the greater experience and the objectivity of those in a supervisory position which is able to identify possible risks that may not be apparent to those who are intimately involved in the programme itself. The individuals with supervisory responsibility may also be able to identify any inadequacies with the performance of the trials personnel and take action accordingly.

(3) *Procedures*: The application of logical procedures enables any potential risks to be reduced. These include the method to be used for setting up test points, gathering data and recovering to a normal flight condition. In addition procedures are developed to cover emergencies and other unexpected events. These procedures are laid out in detail in the test plan for the flight(s) and are reviewed during the pre-flight briefing prior to authorization. Procedures are also developed for the dissemination of results and information within the test team and within the test establishment; this is particularly important where changes may be made to the test aircraft during the test programme.

(4) *Limitations*: Every flight and test programme is subject to a variety of limitations some of which are imposed by the manufacturer; the others are imposed by the test establishment. The limitations may be imposed for structural considerations, to prevent damage to systems, for handling reasons, or to enhance flight safety. In every case it is essential that all participants are aware of every limitation and that they are respected during all flights. The limitations are detailed in the test plan and revised in the pre-flight brief.

(5) *Monitoring*: The monitoring of a test programme or flight can take a variety of forms all of which may be employed if appropriate. This monitoring usually involves the comparison of the actual results with the anticipated results to ensure that the programme is progressing in the expected manner: a lack of correlation between these results may indicate that not all factors have been

addressed correctly during the planning stage and therefore the plan should be reassessed. The comparison of actual with anticipated results is also conducted for individual flights. In addition the actual results in flight are assessed for each test point prior to moving on to the next increment; this may require the use of telemetry to a ground monitor or the use of in-flight plots to predict the likely results from the next point. External monitoring of the flight may also be employed through the use of chase aircraft and ground observers.

1.2.2.3 Quantifying residual risk

Once the trials risks have been identified and the RRMs have been employed there will still be a certain amount of residual risk which is then quantified. This will then lead to the trial being categorized as high, medium, or low risk. The main reason for allocating a category of risk is to alert the responsible senior officer to those trials which require particularly careful supervision. One definition of risk categories used at the UK test establishment at Boscombe Down is as follows:

- *Low risk*: the residual risk is no higher than normal peacetime flying.
- *Medium risk*: there is an increased risk of injury to crews or damage to the aircraft over normal operations.
- *High risk*: there is an increased chance of loss of life or loss of an aircraft.

1.2.3 Trials documentation

Part of the planning process which is often not given sufficient attention, is the technical documentation and reporting requirement. All the documents needed to conduct the trial safely require definition. These include TIs, instrumentation schedules, limitations documents, and risk assessments. In most test establishments the management procedures call for these documents and define their content. The eventual trials report or reports also require identification early in the planning stage.

1.2.3.1 The trials instruction

The TI is the main output from the planning phase and is the document used to authorize and control the trial. In the TI is contained all the information that a senior person in the trials organization needs to be able to understand what is being planned and allows him or her to approve the trial. Thus the document needs to define the test methods, test procedures, and test conditions that will be used. The test methods and procedures are often standard and only require a reference to other documents which define them. The test conditions, on the other hand, always require careful thought as these vary with the trial being made. It is not simply a case of defining the target conditions for the trial but rather a range of permissible values of variables is defined so that the trial can proceed even if the exact desired conditions are not present. The TI also details the trials programme including such aspects as the size of increments in control parameters between successive tests and the criteria for progressing to the next test or phase. A key element of the TI is a section which details the safety precautions that will be taken. This safety section comes mainly from the risk assessment which is also included in the document so that the authorizing officer can ensure that the process has been conducted correctly.

1.2.3.2 *Sortie planning*

Each sortie in a test programme requires careful planning to ensure that the flight follows a safe progression and that the maximum amount of high quality data is gathered in the flight time available. Where possible tests are conducted concurrently, for example, information on flight control positions and engine bay temperatures can be gathered during level flight performance tests. This clearly requires careful liaison between the different parts of a flight test organization if they are all involved in the same trial. The individual test points are organized such that successive points can be achieved quickly. The principle is that no flight time is wasted. It is better to have too many test points in a sortie plan than too few but the plan still needs to be reasonable and achievable. The possible effects of weather are considered as well so that tests are organized into high level and low level sorties to ensure that time is not wasted if the cloud base is low. For each sortie the go/no go criteria is decided upon at the planning stage, for example, this might be a minimum light level or a maximum wind strength. Part of the sortie planning procedure is to produce a written flight brief, which will be used during the pre-flight briefing to ensure that all trials participants are fully prepared. Included in this brief will be the flight safety points, priorities, limitations, go/no go criteria, possible alternate tests, criteria for stopping the tests, and responsibilities of individual participants.

1.2.3.3 *Test cards*

A set of test cards is normally produced for each test flight. These cards serve two purposes: firstly they act as an aide-mémoire of the tests to be conducted and limitations that apply, and secondly they provide a means of recording manual data to back up flight test instrumentation data. It would be hard to overemphasize the importance of having carefully produced test cards. Each card defines the test being made with the conditions required and has sufficient space to record data if required. The limits for that particular test are written clearly together with any reminders. On each card the planned start and finish time for the test are included with the expected fuel state. It is often helpful to colour code the cards to indicate the priority of tests, which allows easy airborne re-planning in the event that the flight falls behind time. When producing the cards it is best to sit in the cockpit and note the most efficient order to gather information from the aircraft or test instruments. Once the cards have been made the crew again sit in the aircraft and go through a sortie rehearsal to ensure that the cards are satisfactory.

1.3 CONDUCTING THE TRIAL

1.3.1 Establishing the test conditions

For the results of a trial to have any meaning the conditions which pertained at the time of the test will be of critical importance. Thus, for example, if a performance trial is being conducted the test pilot will have to fly the aircraft at exactly the right speed, with the right value of sideslip, and at the correct altitude. The environmental conditions such as temperature, winds, turbulence levels, etc. will also have to be appropriate. In many trials a variety of conditions will be required and often at opposite extremes. For example, very still air conditions may be needed to obtain accurate results but high levels of turbulence may also be required to give the test pilot conditions representative of operational flying to allow him or her to make qualitative judgements of the handling qualities. In all cases the test conditions need to be recorded for inclusion in the flight report.

1.3.2 Moving into the unknown

Nearly all test flying involves going from a known condition into an unknown one and it is how this is done which is important. Although there will always be an element of risk in this situation, provided the correct methodology is applied nasty surprises can mostly be avoided. The main part of this methodology is to adopt the incremental approach, i.e. only small changes are made in a single parameter when moving from a known condition to the next test point. As an example if control response tests are being conducted by making step cyclic inputs into a control fixture the size of the inputs would be increased very gradually. After each input the change in aircraft response is compared with the previous results with the aim of determining the relationship between changes in the input parameter and the aircraft output. Once a relationship has been established this can be used to predict the output for the next test point to determine if the test should continue. Any change in the relationship can also serve as a warning that extra caution is needed. To establish this input/output relationship a 'how goes it plot' is often constructed which in its simplest form consists of a graph where the two parameters are cross-plotted. Also marked on the graph will be the pre-determined test limits. Each test point adds to the trend line and the line is extrapolated prior to each subsequent point to ensure that the predicted result lies inside the test limits. For this system to work it is vital that only one parameter is changed at a time. Thus an input size change should not be made at the same time as the aircraft speed is changed. In addition to establishing trend lines for the test results the team also need to have a sound idea of the expected results prior to the trial taking place. This information can come from a variety of sources, such as results of previous trials under similar circumstances or as a result of computer modelling. Again the predicted trend information is compared with the actual results and major differences serve as a warning that all factors have not been accounted for.

1.4 POST-TRIALS ACTIONS

1.4.1 Trials reporting

The deliverable from most trials is a written report which addresses the trials objective and records the test data. Careful planning of the report takes place before the trial is conducted to ensure that the final product is of good quality. This planning will scope the document, produce a skeleton structure, and define responsibilities. One of the most important planning stage actions, particularly for large, multi-discipline reports, is to nominate a single person to be in overall charge of the report. He or she can then ensure that the style and content are consistent across the authoring departments and eliminate duplication, or worse, contradictions. Each report has an introductory section that contains all the information relating to how the trial was conducted. This includes who performed the trial, when it was performed, the state of the aircraft or equipment tested, the instrumentation used, and the tests that were performed. The next section documents the results of the trial, including the conclusion of the test team and any recommendations made. In many test establishments this is written in a set layout known as the seven-part format. The aim of this format is to ensure that a person reading the report understands exactly what was done, what the results were, what the data mean, what implications a deficiency would have in operations, what the conclusion is, what the recommendations are, and whether or not the article under test meets the relevant specification or not. The seven-part format is shown below:

- Establish the test conditions;
- Present the data;
- Analyze and discuss the data;
- Relate to the operational role;
- Conclude;
- Recommend;
- Specification compliance.

Most reports have a summary and a section that groups together the recommendations. In all cases a successful report will make it clear what was done during the trial and will provide a convincing argument to implement the test team's recommendations.

1.4.2 Learning from the past

Human nature being what it is, there is a tendency to believe that a trial is complete once the report has been delivered and accepted by the customer. However it is important to have a trial's closure procedure which ensures that all the data gathered is retained. This data may be required to provide baseline information for comparison purposes when the aircraft is modified in the future. Given the long periods that rotorcraft are kept in service the data may have to be retained for a considerable number of years. As well as keeping copies of the customer reports all the other trials documentation such as trials instructions, risk assessments, etc. are retained to serve as templates for the future. A 'lessons learned' system is also maintained where information concerning mistakes, problems, solutions and good ideas can be retained even though the trials participants may move on.

Chapter 2
Performance Theory

2.1 INTRODUCTION

Unlike a conventional fixed wing aircraft, which has separate means of generating forces for lift and forward propulsion, the helicopter uses only the thrust from a rotor to meet these two essential requirements for sustained flight. This limitation does, however, afford the helicopter a unique advantage over most fixed wing aircraft: the ability to generate a lift force even when the vehicle is stationary. Understanding the basic manner in which a rotor system develops thrust is fundamental to any study of the measurement of rotorcraft performance. It is also required in any introduction to the stability and control attributes of a typical helicopter. There are two basic theoretical approaches to understanding the generation of thrust from a rotor system: momentum theory; and blade element theory. Other more complex theories tend to build on the fundamentals introduced by these two approaches. Since many standard texts on helicopter performance cover momentum theory and blade element theory [2.1 to 2.7] it is only necessary for us to review the key points here. Forward flight produces asymmetric flow across a rotor disk and thus it is desirable for us to start by restricting our discussion to purely axial flight.

2.2 AXIAL FLIGHT: MOMENTUM THEORY

Momentum theory was initially developed by Rankine and Froude from their study of ship propellers or water screws. It later found application in the design of airscrews [2.3] and is used here to represent a climbing rotor. The theory makes certain assumptions:

- Air is an inviscid and incompressible fluid.
- The rotor acts as a uniformly loaded or 'actuator' disk with a infinite number of blades so that there is no periodicity in the wake.
- The flow both upstream and downstream of the disk is uniform, occurs at constant energy and is contained within a streamtube.
- No rotation is imparted on the fluid by the action of the rotor.

These assumptions necessarily limit the accuracy of the Froude theory. The momentum approach has, however, been extended to cover the more realistic case of a compressible fluid [2.7]. Other real effects such as swirl and unsteady flow can be accommodated by appropriate empirical factors [2.4, 2.6 and 2.8].

2.2.1 Froude theory

The Froude theory postulates that the flow above and below a climbing rotor can be considered as constant energy with energy being imparted only by the actuator disk. This energy is added to the airflow in the form of an increase in static pressure. The theory then draws some conclusions about the streamtube that the disk influences. Far above the rotor, the velocity of flow in the streamtube is equal to the freestream that is dependent on the rate of climb of the rotor itself. As the rotor draws air through its disk the velocity just above the disk is greater than the freestream and as a consequence of the continuity equation the streamtube will have contracted; also by virtue of Bernouilli's relationship the pressure just above the rotor will be less than ambient. Immediately below the disk the pressure will be greater than ambient because of the energy added by the rotor however, the velocity and streamtube area will be the same as just above the rotor. Far below the disk in what is termed the *ultimate wake*, the flow will achieve a pressure equal to ambient but the velocity will exceed the freestream again because of the energy imparted by the rotor. The continuity equation and Bernoulli relationship require that the cross-sectional area of the ultimate wake be less than the disk area, see Fig. 2.1. Adoption of the concept of an actuator disk leads to a very simple relationship for the thrust developed by a rotor:

$$T = A(P_2 - P_1)$$

where A is the disk area.

The pressure difference $(P_2 - P_1)$ generated by the disk can be related to the vertical velocity by considering the changes in pressure and velocity occurring in the streamtube. Consider a helicopter climbing vertically at speed V_c and assume that the

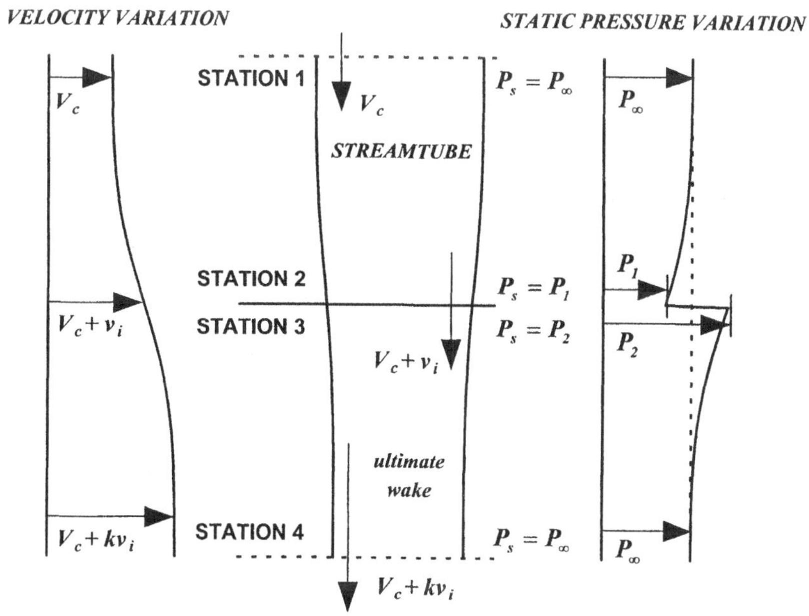

Fig. 2.1 Flow through the actuator disk.

acceleration of flow caused by the action of the rotor results in an increase in the flow velocity of v_i (Fig. 2.1). Likewise, assume that the continued acceleration of flow below the rotor leads to a total velocity rise of kv_i at the ultimate wake. Now Bernoulli states that:

$$P_\infty + \frac{1}{2}\rho V_c^2 = P_1 + \frac{1}{2}\rho(V_c + v_i)^2$$

$$P_2 + \frac{1}{2}\rho(V_c + v_i)^2 = P_\infty + \frac{1}{2}\rho(V_c + kv_i)^2$$

Thus:

$$T = A(P_2 - P_1) = \frac{1}{2}\rho A(2V_c + kv_i)kv_i$$

This relationship requires a value for k before it can be used to estimate the thrust required for a given rate of climb. If the momentum change caused by the disk is considered, an alternative expression for thrust can be developed. Recalling that force is equal to rate of change of momentum or massflow multiplied by a change in velocity gives [2.2]:

$$T = \rho A(V_c + v_i)kv_i$$

Hence $k = 2$ and:

$$T = 2\rho A(V_c + v_i)v_i \tag{2.1}$$

This fundamental equation predicts that the induced velocity at the rotor disk is equal to half the total increase in flow velocity required to match the thrust requirement of the rotor. The maximum increase in velocity occurs at the ultimate wake, which is usually taken as one rotor diameter downstream of the disk. This *momentum disk* model can be applied to any working state of the rotor in which a continuous streamtube is formed.

2.2.2 Power considerations

The total change in energy per unit mass (ΔE) along the streamtube is given by:

$$\Delta E = 2(V_c + v_i)v_i$$

As power can be defined as massflow multiplied by change in energy:

$$P = \dot{m}\Delta E = \rho A(V_c + v_i) \times 2(V_c + v_i)v_i$$

$$P = T(V_c + v_i) = TV_c + Tv_i \tag{2.2}$$

So the power required to drive a climbing rotor can be seen to come from two sources: the power required to generate the rate of climb (the *useful* power $= TV_c$) and the power required to generate the thrust (the *induced* power $= Tv_i$). The power calculated using Equation (2.2) represents an ideal minimum value because this simple theory neglects all forms of losses.

It has been seen that power estimations require a knowledge of thrust, rate of climb and induced velocity. Whilst the thrust can be related to the weight of the helicopter and the rate of climb is an easily specified variable the induced velocity is more difficult to determine. However, Equation (2.1) can be re-written as:

$$v_i^2 + v_i V_c - \frac{T}{2\rho A} = 0$$

Only the positive root of this quadratic has any meaning in this case, so:

$$v_i = -\frac{V_c}{2} + \sqrt{\frac{V_c^2}{4} + \frac{T}{2\rho A}} \tag{2.3}$$

Note that Equation (2.3) can be used to predict the induced velocity in the hover ($V_c = 0$):

$$v_{ih} = \sqrt{\frac{T}{2\rho A}} \tag{2.4}$$

2.3 AXIAL FLIGHT: BLADE ELEMENT THEORY

Simple momentum theory treats the rotor as an actuator disk through which a uniform flow passes. Unfortunately this theory tells us little about the flow around the individual blades that make up the rotor system. Momentum theory cannot, therefore, be used to predict the magnitude of any losses associated with realistic flow around rotor blades. Blade element theory overcomes some of the restrictions inherent in momentum theory as it is based upon the idea that the rotor blades function as high aspect ratio wings constrained to rotate around a central mast as the rotor system advances through the air. As before our study of blade element theory begins with purely axial flight.

2.3.1 Elemental forces, thrust, torque and power

Consider a rotor consisting of b blades climbing at speed V_c. The blades are each of length R, and turn at a rotational speed of Ω rad/s. If we now examine the forces generated on a small element, δr, of a blade located at r from the hub we can gain insight into how a complete rotor system generates thrust and drag. Figure 2.2 shows a blade element and Fig. 2.3 depicts the forces acting on such an element. Each element of the blade is assumed to develop the full aerodynamic forces and moments as it would in two-dimensional flow at the same conditions that occur at its radial station. No allowance is made at this stage for finite span or blade wake effects.

Firstly, we must determine the total flow through the disk. As we have seen for a rotor in a climb this is composed of the climb velocity, V_c, and the induced velocity v_i. The resultant velocity at a blade element therefore has a vertical component, $V_c + v_i$, and a horizontal component Ωr. From Fig. 2.3 it can be seen that the resultant velocity, V, is given by:

$$V = \sqrt{(V_c + v_i)^2 + \Omega^2 r^2}$$

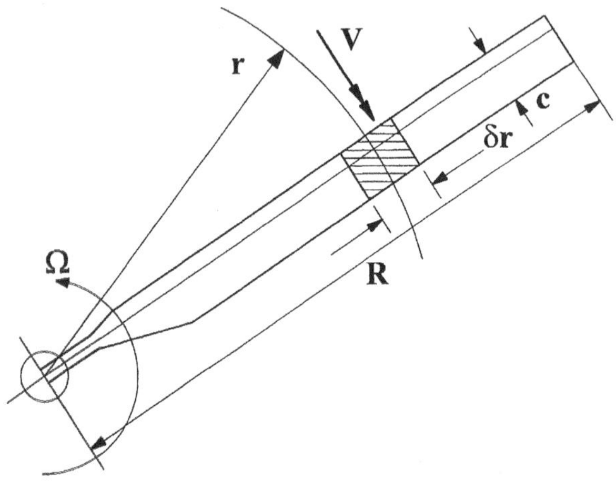

Fig. 2.2 Definition of the blade element.

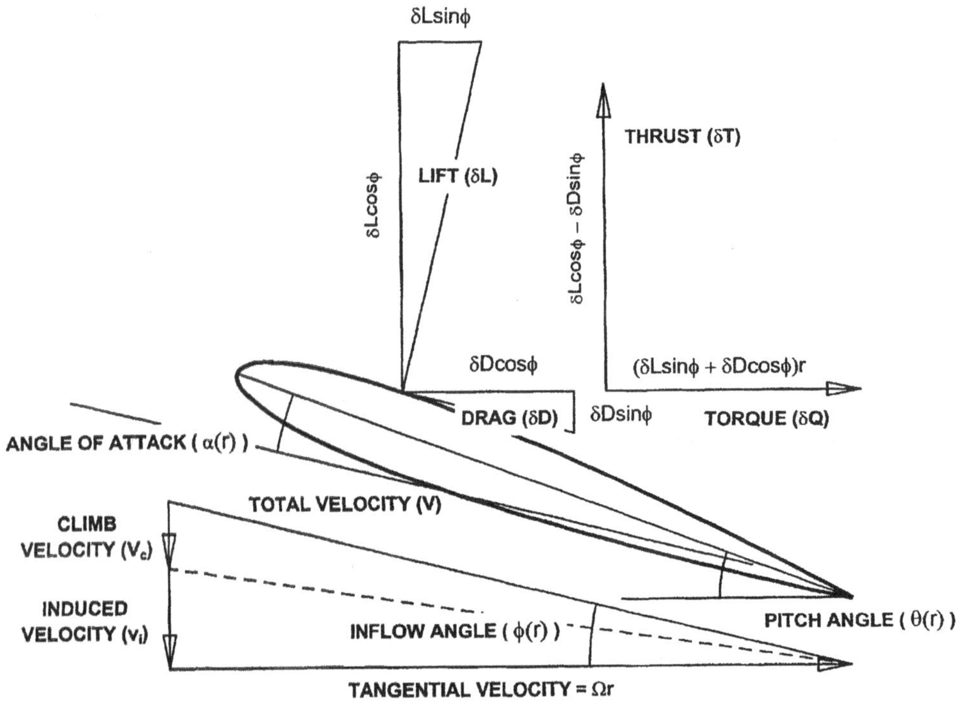

Fig. 2.3 Conditions at the blade element.

This is commonly approximated to $V = \Omega r$. Justification for this approximation can be seen by considering the example helicopter in a rapid vertical climb (2000 ft/min or 10.2 m/s). Using Equation (2.4), $v_{ih} = 12.3$ m/s and later work will show that under these conditions $v_i = 0.7v_{ih} = 8.6$ m/s. If the blade tip is considered then $V = \Omega R = 227.5$ m/s whereas including the vertical component gives $V = 228.3$ m/s. Thus in this case the standard approximation underestimates the total velocity by only 0.3%. Now the blade section incidence, α, will depend on its radial position, r, and from Fig. 2.3 for any spanwise position r:

$$\alpha(r) = \theta(r) - \phi(r)$$

Using small angle approximations:

$$\alpha(r) = \theta(r) - \left(\frac{V_c + v_i}{\Omega r}\right) = \theta(r) - \lambda(r)$$

This local incidence will lead to elemental lift and drag:

$$\delta L = \frac{1}{2}\rho V^2 S C_L = \frac{1}{2}\rho\Omega^2 r^2 c(r) a\alpha(r)\delta r$$

$$\delta D = \frac{1}{2}\rho V^2 S C_D = \frac{1}{2}\rho\Omega^2 r^2 c(r) C_D(r)\delta r$$

which, from Fig. 2.3, can be combined to give elemental thrust and torque. Assuming the inflow angle is small and the lift/drag ratio is large leads to:

$$\delta T = b\delta L$$

$$\delta Q = b(\phi\delta L + \delta D)r$$

The elementary power, δP, required to produce the elementary thrust can be found from the elementary torque:

$$\delta P = \Omega\delta Q = b\Omega(\phi\delta L + \delta D)r = \Omega r(\phi\delta T + b\delta D) = (V_c + v_i)\delta T + \Omega r b\delta D$$

If this elemental equation is integrated along the rotor blade:

$$P = (V_c + v_i)T + \frac{1}{2}\rho\Omega^3 b\int c(r) r^3 C_D \mathrm{d}r \tag{2.5}$$

Equation (2.5) is similar to the result obtained by adopting momentum theory except that we now have an extra term in the expression that represents the *profile power* required to keep the rotor turning against the torque produced by the profile drag. The total rotor thrust, torque and power can be obtained by integrating analytically the expressions for δT, δQ and δP over the span of the blade. For a rectangular blade at constant pitch, where C_D is now the mean profile drag coefficient for whole rotor, then:

$$\left.\begin{aligned} T &= \frac{1}{2}\rho abc V_T^2 R\left(\frac{\theta}{3} - \frac{\lambda}{2}\right) \\ Q &= T\frac{(V_c - v_i)}{\Omega} + \frac{1}{8}\rho bc V_T^2 R^2 C_D \\ P &= T(V_c - v_i) + \frac{1}{8}\rho bc V_T^3 R C_D \end{aligned}\right\} \tag{2.6}$$

2.3.2 Constant chord blades with linear twist

Most modern blades feature a degree of negative twist, decreasing the pitch angle towards the tip, as means of optimizing the blade loading distribution. This can be expressed using the following equation:

$$\theta(r) = \theta_0 = \frac{r}{R}\theta_1$$

where θ_0 is the collective pitch applied at the hub; θ_1 is the total twist (normally negative) and r/R factors the total twist according to radial position. Now:

$$\alpha(r) = \theta(r) - \left(\frac{V_c + v_i}{\Omega r}\right) = \theta_0 + \frac{r}{R}\theta_1 - \left(\frac{V_c + v_i}{\Omega r}\right)$$

Therefore, for a rectangular blade, the elemental thrust equation becomes:

$$\delta T = \frac{1}{2}\rho abc\Omega^2\left[\theta_0 r^2 + \frac{r^3}{R}\theta_1 - \left(\frac{V_c + v_i}{\Omega r}\right)r\right]\delta r$$

So:

$$T = \frac{1}{2}\rho abc\Omega^2\left[\frac{\theta_0}{3}R^3 + \frac{\theta_1}{4}R^3 - \frac{1}{2}\left(\frac{V_c + v_i}{\Omega}\right)R^2\right]\delta r$$

$$T = \frac{1}{2}\rho abc V_T^2 R\left(\frac{\theta_0}{3} + \frac{\theta_1}{4} - \frac{\lambda}{2}\right) \qquad (2.7)$$

2.4 NON-DIMENSIONAL COEFFICIENTS

Before examining how the realism of the theories introduced above might be improved, it is necessary to discuss the concept of non-dimensional coefficients. These coefficients are analogous to the lift and drag coefficients that are a common feature of aerodynamics. Rather than lift and drag, in the case of rotorcraft, coefficients of thrust (C_T), torque (C_Q) and power (C_P) are used. These coefficients are defined as:

$$C_T = \frac{T}{\rho A\Omega^2 R^2} = \frac{T}{\rho A V_T^2}$$

$$C_Q = \frac{Q}{\rho A\Omega^2 R^3} = \frac{Q}{\rho A V_T^2 R}$$

$$C_P = \frac{P}{\rho A\Omega^3 R^3} = \frac{Q\Omega}{\rho A V_T^2 \Omega R} = C_Q$$

Note that power coefficient is numerically equal to torque coefficient (though, of course, they are different physically). Now converting Equation (2.7) into coefficient terms gives:

$$C_T = \frac{1}{2}a\frac{bcR}{A}\left(\frac{1}{3}\theta_0 + \frac{1}{4}\theta_1 - \frac{1}{2}\lambda\right)$$

$$C_T = \frac{1}{2}as\left(\frac{1}{3}\theta_0 + \frac{1}{4}\theta_1 - \frac{1}{2}\lambda\right)$$

where s is defined as the *solidity* of the disk that is the ratio of total blade area to disk area. Hence for a rotor with rectangular blades $s = bcR/A$. The relationship for thrust coefficient can be simplified by altering the definition of blade pitch from the blade root to a position at three-quarter radius ($\theta_{0.75}$), since:

$$C_T = \frac{1}{2} sa \left(\frac{1}{3}\theta_0 + \frac{1}{4}\theta_1 - \frac{1}{2}\lambda \right)$$

$$= \frac{1}{2} sa \left[\frac{1}{3}\left(\theta_0 + \frac{3}{4}\theta_1 \right) - \frac{1}{2}\lambda \right]$$

$$C_T = \frac{1}{2} sa \left(\frac{1}{3}\theta_{0.75} - \frac{1}{2}\lambda \right)$$

Hence a rotor with zero twist will generate the same thrust coefficient as one with linear twist, provided the pitch setting of the untwisted blade is equal to that at the three-quarter radius on the twisted blade. Also from Equation (2.6):

$$P = T(V_c - v_i) + \frac{1}{8}\rho bc V_T^3 R C_D$$

Thus:

$$C_P = \frac{T}{\rho A V_T^3}(V_c + v_i) + \frac{1}{8}\frac{\rho bc R V_T^3}{\rho A V_T^3} C_D$$

$$= C_T \frac{V_c + v_i}{V_T} + \frac{1}{8} s C_D$$

$$C_P = C_T \frac{V_c}{V_T} + C_T \frac{v_i}{V_T} + \frac{1}{8} s C_D$$

This equation is frequently used in performance analysis. It shows that for a climbing rotor the power required can be sub-divided into three parts: the first term represents the useful power; the second the induced power; and the third the profile power. If the particular case of *hovering* flight is considered ($V_c = 0$) then from momentum theory the induced velocity can be related to the thrust coefficient:

$$v_{ih} = \sqrt{\frac{T}{2\rho A}} = \sqrt{\frac{T V_T^2}{2\rho A V_T^2}} = \sqrt{\frac{C_T V_T^2}{2}} = V_T \sqrt{\frac{C_T}{2}}$$

In addition, the power coefficient becomes:

$$C_P = C_T \sqrt{\frac{C_T}{2}} + \frac{1}{8} s C_D = \frac{1}{\sqrt{2}} C_T^{3/2} + \frac{1}{8} s C_D$$

Now for a given rotor with fixed solidity, provided the profile drag coefficient remains constant, then:

$$C_P = k_1 C_T^{3/2} + k_2 \qquad (2.8)$$

Equation (2.8) is a very important result since it suggests that for a hovering rotor the power coefficient is directly proportional to the thrust coefficient (or aircraft weight) *provided* the drag coefficient remains unchanged. Likewise if the helicopter mass is fixed and the rotor profile drag coefficient is constant then the power required to drive the rotor will vary as Ω^3 so that C_P will remain unchanged. This simple rule forms the basis of hover performance testing and is illustrated in Fig. 2.4.

2.5 AXIAL FLIGHT: IMPROVED THEORETICAL ESTIMATES

Although the blade element theory introduced above has allowed the real effect of rotor driving torque arising from profile drag to be included, it has still relied upon the assumption that the inflow is constant along the rotor radius. Likewise, no allowance has been made for three-dimensional flow effects such as tip vortices. Minor modifications to the blade element theory are used to account for these important effects.

2.5.1 Variations in induced velocity

So far, it has been assumed that the induced velocity is constant across the disk. In addition, no description has been given of the precise relationship between blade pitch and induced velocity. In fact flow is induced downwards through the disk as a *consequence* of the aerofoil's *inclination* to the direction of rotation. As such it is often necessary to develop a relationship between blade pitch and induced velocity for any radial station. Combining the momentum and blade element theories introduced earlier leads to:

$$\delta T = \frac{1}{2}\rho abc\Omega^2 r^2\left[\theta(r) - \left(\frac{V_c + v_i}{\Omega r}\right)\right]\delta r = 2\rho(V_c + v_i)v_i 2\pi r\delta r$$

Therefore:

$$\frac{1}{2}\rho abc\Omega r[\Omega r\theta(r) - (V_c + v_i)] = 4\rho(V_c v_i + v_i^2)\pi r$$

or:

$$\left[\frac{v_{ih}}{V_T}\right]_r = -\frac{as}{16} + \sqrt{\left(\frac{as}{16}\right)^2 + \left(\frac{as}{8}\right)\frac{r}{R}\theta(r)} \qquad (2.9)$$

where $[v_{ih}/V_T]_r$ represents the ratio of the induced velocity in the hover at radius r to the tip speed, and $\theta(r)$ is the pitch at radius r. Figure 2.5 shows the variation of induced velocity for a hovering rotor calculated using Equation (2.9).

2.5.2 Fuselage download

The slipstream from the rotor exerts a downward force on the helicopter fuselage, which is in addition to any vertical drag associated with axial flight. This means that in the hover the main rotor must generate sufficient thrust to support not only the weight of the helicopter but balance this download. The rotor wake contracts from the diameter of the rotor to its ultimate wake size in about a quarter of rotor radius

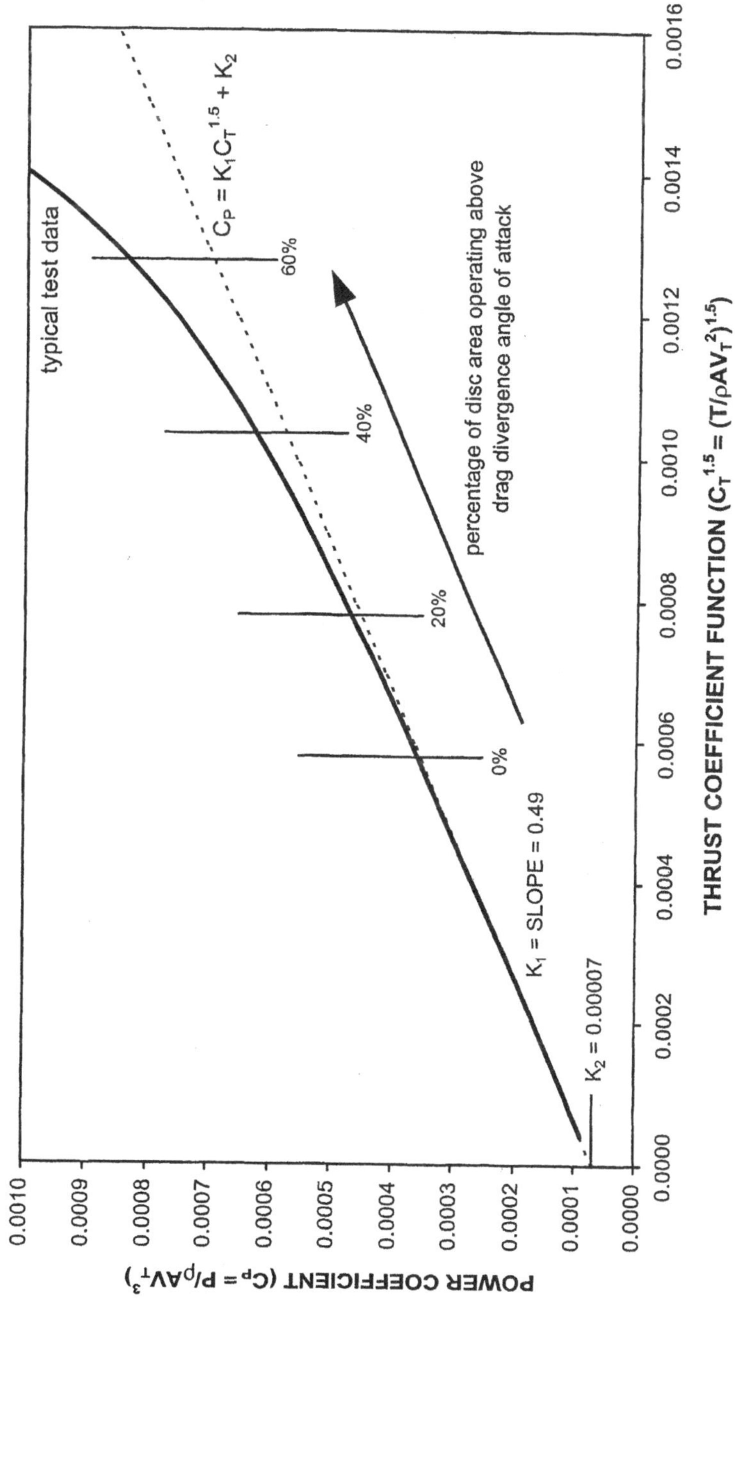

Fig. 2.4 Typical hover performance results.

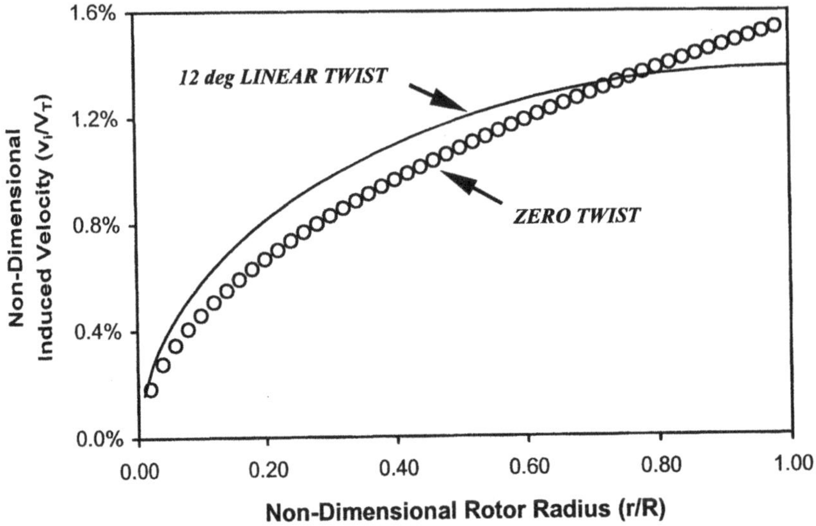

Fig. 2.5 Variation of inflow with rotor radius (CBEM).

[2.6]. Even so it is often assumed that the fuselage is immersed in the ultimate wake and receives the full effect of the downwash. Simple estimates of download use a projected area and assume a drag coefficient of 0.3 [2.6]. It is often convenient to express the vertical drag as a percentage of gross weight (GW) and typical values range between 1% and 4% of GW.

2.5.3 Ground effect

Proximity to the ground results in a decrease in the induced power to produce a given thrust. The presence of the ground is effectively the same as that of having a mirror image rotor below ground. The slipstream of the image rotor induces an upward velocity at the real rotor so that the inflow velocity will be reduced and the incidence of the real rotor increased at constant pitch or feather angle. Hence, the lift developed by the real rotor is increased and the induced power required to maintain a hover is reduced.

Figure 2.6 shows the general trend in the form of a graph of the ratio of the induced power required in and out of ground effect against the height of the rotor above ground (Z), non-dimensionalized by the rotor diameter, *D*. We can see that ground effect decreases with increasing height, effectively disappearing at about one rotor diameter from the ground. Empirical relationships exist which describe the variation of power, in-ground-effect, with hover height [2.4]. One example is:

$$P_{\mathrm{i}} = P_{\mathrm{iOGE}}\left[-0.1276\left(\frac{Z}{D}\right)^4 + 0.7080\left(\frac{Z}{D}\right)^3 - 1.4569\left(\frac{Z}{D}\right)^2 + 1.3432\left(\frac{Z}{D}\right) + 0.5147\right]$$

2.5.4 Tip loss and non-uniform flow

A characteristic of the actuator disk theory is that the linear theory of lift is maintained right out to the edge of the disk. In reality, because the rotor consists of a finite

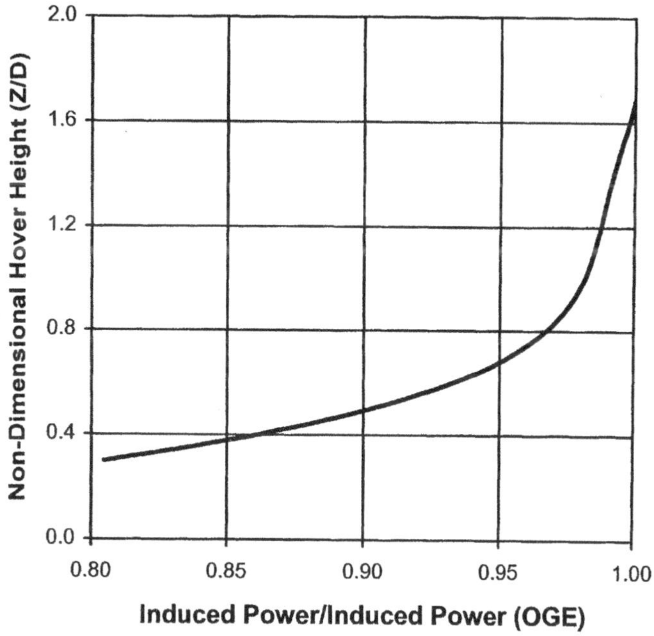

Fig. 2.6 Ground effect trends.

number of blades, some air is able to escape outwards, from the streamtube, by the action of the tip vortices. The total induced flow is thus less than the actuator disk theory would prescribe. This deficiency is called *tip loss* and is shown by a rapid reduction in lift over the last few per cent of the span near the tip. It is common practice to account for this discrepancy by assuming that beyond a certain outboard station the blade sections develop drag but no lift. Thus the thrust integral becomes:

$$C_T = \frac{1}{2} sa \int_0^{BR} (\theta r^2 - \lambda r) \mathrm{d}r$$

This increases the hover power to:

$$C_P = \frac{1}{B\sqrt{2}} C_T^{3/2} + \frac{s}{8} C_{D_0} \quad \text{or} \quad C_P = \frac{\kappa}{\sqrt{2}} C_T^{3/2} + \frac{s}{8} C_{D_0}$$

In 1927, Prandtl and Betz approximated the tip loss factor, B, to:

$$B = 1 - \frac{\sqrt{2C_T}}{b}$$

Goldstein and Lock showed satisfactory correlation with this approximation for lightly loaded rotors [2.9]. The value of κ can be increased slightly to account for the losses incurred due to the non-uniform nature of the swirling flow from a real rotor. Typical values for κ vary between 1.13 [2.4] and 1.15 [2.10].

2.6 AXIAL FLIGHT: VERTICAL DESCENTS AND THE VORTEX-RING STATE

Momentum theory and simple blade element theory are based on constant energy flow above and below the rotor. They both also assume that the slipstream has a finite velocity either side of the rotor. When these conditions are not satisfied these theories can no longer be used to make theoretically accurate predictions. Provided the rotor ascends, continuous flow through the rotor exists and use of the momentum theory is valid. The hover therefore represents the lowest vertical climb speed for which momentum theory is sound. Consider a vertical descent at low rates of descent: the rotor is now working to impart a downwards velocity on the fluid whilst the general flow along the streamtube is upwards. Clearly smooth flow is not possible in this situation and thus a vortex will form in the region of the rotor. A limiting condition arises when the rate of descent matches the induced velocity necessary to generate the thrust. Here the flow stagnates in the disk and *fully developed vortex flow* has been established. Such a situation leads to a rapid descent with uncontrolled pitching and rolling motion being caused by a violent flapping motion of the rotor blades. At rates of descent above this limiting value a steady smooth flow upward through the rotor is established and the momentum theory can be applied. It is possible to determine the value of V_D $(-V_c)$ equating to fully developed vortex ring flight. It is often surmised that provided V_D exceeds the induced velocity required for hovering flight then the rotor is extracting sufficient energy as it descends through the atmosphere to support the weight of the helicopter. Thus momentum theory:

- applies in *vertical ascents* and in the *hover* (when $V_c \geqslant 0$);
- is invalid in the *vortex-ring* and *turbulent windmill-brake* states (when $0 < V_D < 2v_{ih}$);
- applies in the *windmill-brake* state (when $V_D > 2v_{ih}$).

The variation of induced velocity through all phases of axial flight can be found using a combination of momentum theory for the vertical ascent and windmill-brake states and empirical relationships for the vortex-ring and turbulent windmill-brake states. An expression developed by Glauert [2.3] has found favour:

$$\tilde{F} = \frac{3\tilde{f}^2}{(1 - 2\tilde{f})^2}$$

where:

$$\tilde{F} = \left(\frac{v_{ih}}{U}\right)^2 \quad \tilde{f} = \left(\frac{v_{ih}}{V_c}\right)^2 \quad U = V_c \pm v_i$$

Plotting the variation of $1/\tilde{f}$ with $1/\tilde{F}$, see Fig. 2.7, is a convenient method of portraying the different flow states. As has already been mentioned momentum theory cannot be applied for rates of descents less than twice the induced velocity. It can be shown that this condition equates to a value of $\tilde{f} < 0.25$. The hover condition relates to a value of $\tilde{F}$ equal to zero. These two conditions form boundaries inside which non-momentum flow takes place.

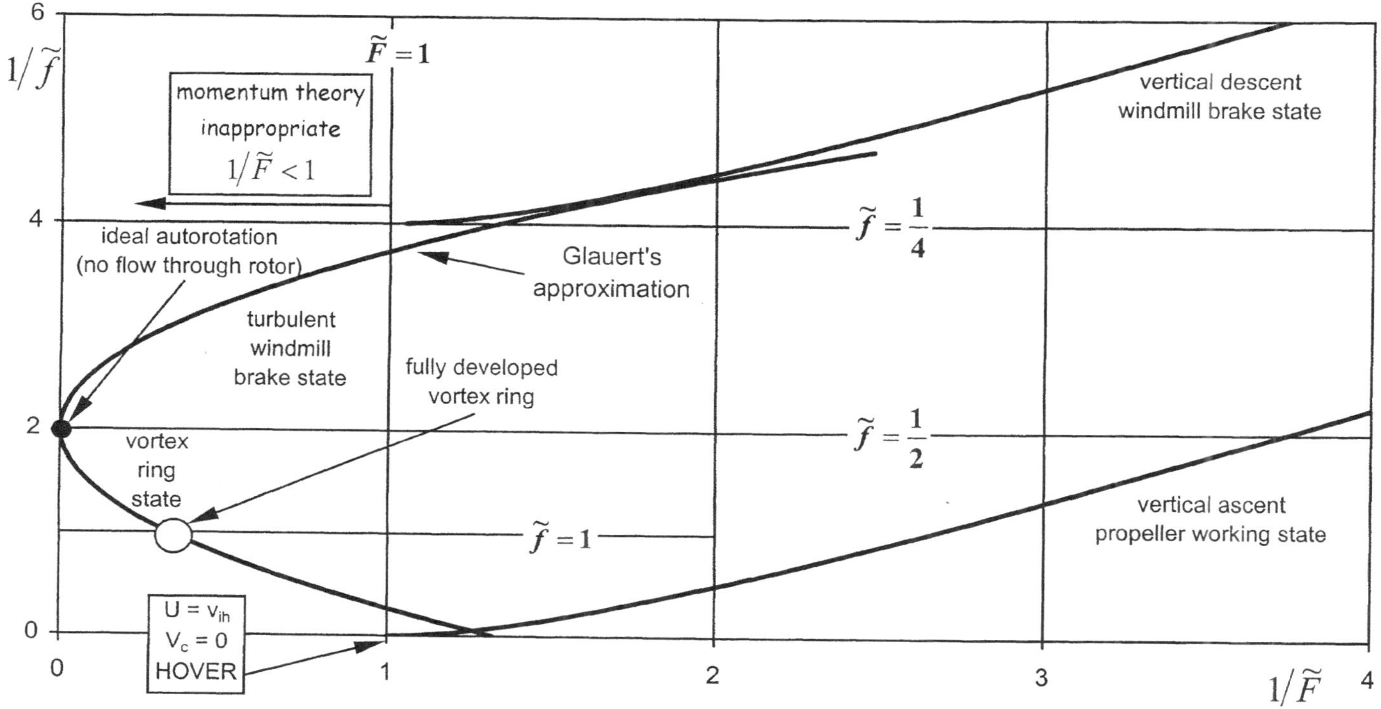

Fig. 2.7 Axial rotor working states.

2.6.1 The vortex-ring state

For low rates of descent, that is when the vertical velocity is less than the induced velocity ($\tilde{f} < 0.5$), the rotor is said to be operating in the *vortex-ring state*. As the rotor descends with power in this state, it is effectively pumping air from beneath the rotor by the action of the vortex ring located in-plane with the rotor. At high power settings, it is possible to pump so much air from beneath the rotor that the rate of descent increases thereby exacerbating the problem. The pilot may attempt to arrest the rate of descent by application of increased collective pitch but this tends to increase the rate of descent still further. If on the other hand the pilot lowers the collective lever the thrust is reduced and the helicopter also increases its rate of descent. This dangerous phenomenon is often called *power settling*. It can cause helicopters to develop significant rates of descent when operating below maximum gross weight even when full torque is applied.

Care must be taken to avoid *entry into vortex ring* particularly when hovering without external references at high altitude and when making steep approaches. Operation of the rotor in a state of *fully developed vortex ring* can be prevented by avoiding vertical descents at rates between 70% and 150% of v_{ih} (1700 to 3600 ft/min for the example helicopter). However vertical descents at modest speeds may be required operationally in these circumstances: the onset of significant vibration; a sudden uncommanded increase in the rate of descent; or a tendency for the helicopter to gyrate in pitch and roll, are taken as indicators of the incipient stages of power settling or the vortex ring condition and recovery action is initiated.

Other causes of vortex ring are the application of a high collective setting in a flare manoeuvre or operating in an upflow (equal to v_i) close to a cliff. It should be noted that yaw manoeuvres can establish a vortex-ring state in the tail rotor.

2.6.2 The windmill-brake state

At rates of descent in excess of v_{ih} the vortex ring rises above the disk and the rotor is said to be operating in the *turbulent windmill-brake state*. Although less problematic it is still a working state for which momentum theory is inappropriate. For still higher rates of descent (*the windmill-brake state*) momentum flow is restored with the rotor extracting sufficient energy from the rising air to maintain the rotor speed and generate thrust. In theory, the rotor can operate satisfactorily up to very high rates of descent in the windmill-brake state. A rotor is said to be operating in an *ideal autorotative* state if there is no mean flow through the rotor. Consequently the induced power is zero and the helicopter is able to make a controlled descent with potential energy being used to meet the power requirements of the rotorcraft. Theoretically, autorotation is achieved in vertical flight at a rate of descent equal to $1.8v_{ih}$. In practice however the requirement to overcome the profile drag of the rotor blades changes the rate of descent to $1.7v_{ih}$.

2.7 HORIZONTAL FLIGHT: MOMENTUM THEORY

It has already been stated that the momentum theory is inaccurate for low rates of descent. It is also true that these inaccuracies will persist if the theory is applied at low forward speed. Glauert [2.11] proposed that if the forward speed is large compared with the induced velocity then the momentum theory could be applied successfully. He also suggested that it was possible to make a direct analogy between a rotor and an elliptically loaded wing.

2.7.1 Elliptically loaded wing analogy

The analogy with an elliptically loaded wing, suggested by Glauert, provides us with a relationship for thrust. From Fig. 2.8 it can be seen that a lifting wing imparts a downwash velocity, w, on the flow as it passes over the aerofoil. If the wing is elliptically loaded then its effect on the air will be the same as a rotor with uniform inflow, (Fig. 2.9). Induced power, P_i, equals Tv_i; since the rotor is assumed to act like a wing, P_i is considered to be the power required in overcoming the *induced drag*. Thus:

$$P_i = Tv_i = V'D_i = \frac{1}{2}\rho A(V')^3 C_{D_i}$$

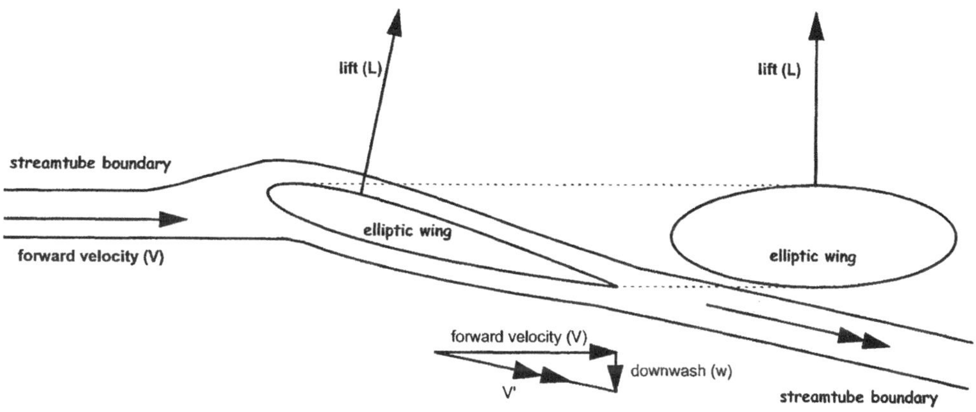

Fig. 2.8 Elliptic wing in level flight.

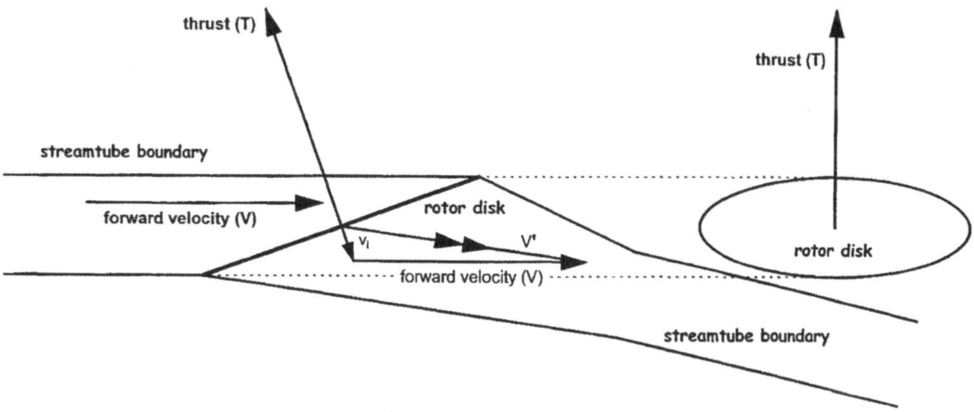

Fig. 2.9 Momentum disk in level flight.

But for an elliptically loaded wing (after McCormick [2.12]):

$$C_{D_i} = \frac{C_L^2}{\pi AR} \qquad AR = \text{aspect ratio} = \frac{\text{span}}{\text{chord}} = \frac{\text{span}^2}{\text{area}}$$

Now a wing of span b is the same as a rotor of diameter $2r$. Thus:

$$Tv_i = \frac{1}{2}\rho A(V')^3 C_{D_i} = \frac{1}{2}\rho A(V')^3 \frac{C_L^2}{\pi AR} = \frac{\rho A^2(V')^3}{8\pi r^2}\left[\frac{2T}{\rho(V')^2 A}\right]^2 = \frac{T^2}{2\rho AV'}$$

and so:

$$T = 2\rho AV'v_i \tag{2.10}$$

Equation (2.10) provides the means to calculate the induced velocity in horizontal flight since T and V' can be estimated from Fig. 2.8 and by considering the forces on the helicopter. The power required to drive the helicopter (P_{IDEAL}) can be determined, as before, from the product of the thrust and the axial component of the relative speed of flow through the disk. Thus, the ideal power is given by:

$$P_{IDEAL} = T(V \sin\gamma + v_i) \tag{2.11}$$

2.7.2 Estimating the induced velocity

If a helicopter in horizontal flight at speed V (see Fig. 2.10) is considered then in order to sustain level flight the rotor disk must be tilted forward. This is so that the thrust vector, T, can provide both the forward propulsive force ($T \sin\gamma$) and support the weight of the helicopter ($T \cos\gamma$). Now the rotor will induce a flow through itself at right angles to its plane of rotation. The combination of the forward speed, V, and this induced velocity, v_i, is the oblique flow across the disk, V'. From Equation (2.10):

$$v_i = \frac{T}{2\rho AV'}$$

and from Fig. 2.10:

$$V' = \sqrt{(V + v_i \sin\gamma)^2 + (v_i \cos\gamma)^2} = \sqrt{(V^2 + 2Vv_i \sin\gamma v_i^2}$$

Also:

$$T\cos\gamma = W \qquad T\sin\gamma = D \qquad \tan\gamma = \frac{D}{W}$$

Thus:

$$v_i = \frac{W}{2\rho A \cos\gamma\sqrt{V^2 + 2Vv_i \sin\gamma + v_i^2}} = \frac{v_{ih}^2}{\cos\gamma\sqrt{V^2 + 2Vv_i \sin\gamma + v_i^2}}$$

or:

$$\tilde{v}_i = \frac{1}{\cos\gamma\sqrt{\tilde{V}^2 + 2\tilde{V}\tilde{v}_i \sin\gamma + \tilde{v}_i^2}} \tag{2.12}$$

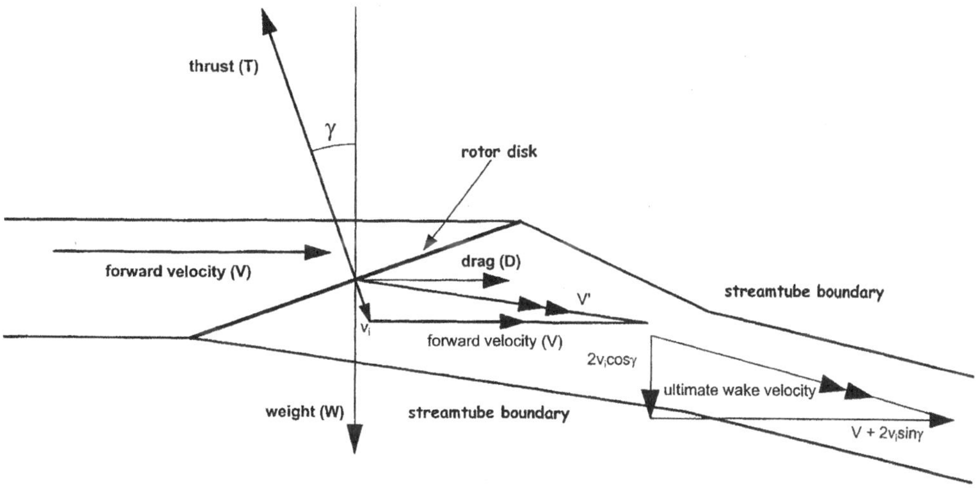

Fig. 2.10 Velocity changes through actuator disk in level flight.

where:

$$\tilde{v}_i = \frac{v_i}{v_{ih}} \qquad \tilde{V} = \frac{V}{v_{ih}}$$

Equation (2.12) suggests that as the horizontal speed is increased the induced velocity will reduce. Now typical disk tilt angles are small such that γ approximates to zero. Hence:

$$\tilde{v}_i = \frac{1}{\sqrt{\tilde{V}^2 + \tilde{v}_i^2}}$$

and using the quadratic solution of $\tilde{v}_i^4$ with a positive root:

$$\tilde{v}_i = \sqrt{-\frac{\tilde{V}^2}{2} + \sqrt{\frac{\tilde{V}^4}{4} + 1}}$$

which can be approximated at higher forwards speeds, when V is much greater than v_i, by:

$$v_i = \frac{W}{2\rho A V}$$

Figure 2.11 shows the predicted variation of induced velocity ratio with forward velocity ratio. At speeds in excess of twice the induced velocity in the hover the simple approximation can be used. (For the example helicopter this equates to a forward speed of around 30 m/s or 60 KTAS).

2.7.3 Estimating the ideal power:

From Fig. 2.10 it can be seen that by using Equation (2.11) it is possible to write the ideal power in coefficient terms:

$$C_P = \frac{P_{IDEAL}}{\rho A V_T^3} = \frac{T(V\sin\gamma + v_i)}{\rho A V_T^3} = \frac{DV + Tv_i}{\rho A V_T^3} = \frac{1}{2}\mu^3\left(\frac{f}{A}\right) + \lambda C_T$$

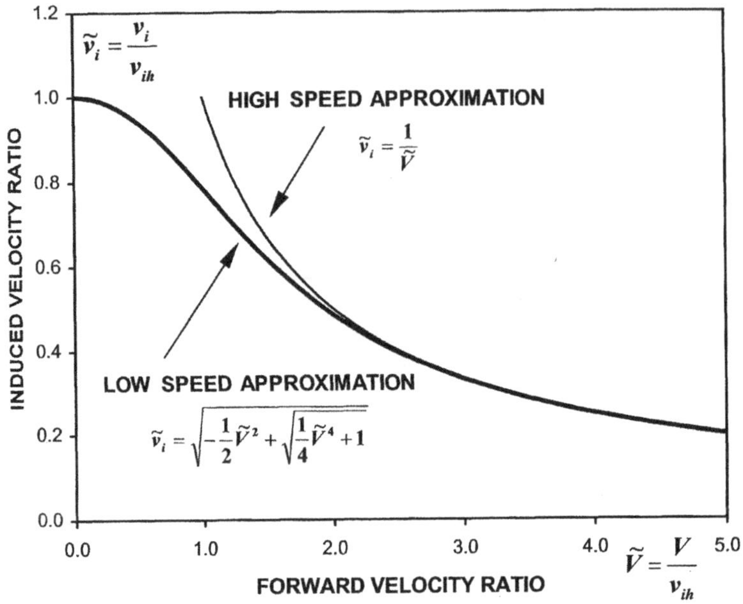

Fig. 2.11 Variation of induced velocity with forward speed.

where $f = SC_D$ = drag area (area of flat plate that produces equivalent drag to helicopter body) and μ = advance ratio (V/V_T). Thus the ideal power coefficient comprises two components: one proportional to the cube of the *advance ratio* – the *parasite power*; and one proportional to the *inflow ratio* (λ) – the *induced power*. Figure 2.12 shows the variation of ideal power coefficient for the example helicopter, note the presence of the familiar '*power bucket*'.

2.8 HORIZONTAL FLIGHT: BLADE ELEMENT THEORY

Although momentum theory can be used with some success to determine the power required by a helicopter in horizontal flight it does not gives us a complete analysis of the situation. As with axial flight we have to adopt a *blade element* approach if we are to understand the power requirements more fully. The elemental thrust, torque and power can be written as:

$$\delta T = \delta L \cos\psi - \delta D \sin\psi$$

$$\delta Q = (\delta L \sin\psi + \delta D \cos\psi) r$$

$$\delta P = (\delta L \sin\psi + \delta D \cos\psi) \Omega r$$

On integrating the torque equation, it was found to contain elements that matched those obtained using momentum theory. These elements can be traced to the presence of $\delta L \sin\psi$ in the elemental torque equation. It is therefore possible to simplify the mathematics associated with using blade element theory by only considering the drag component. The estimate of power required using this method is then added to that

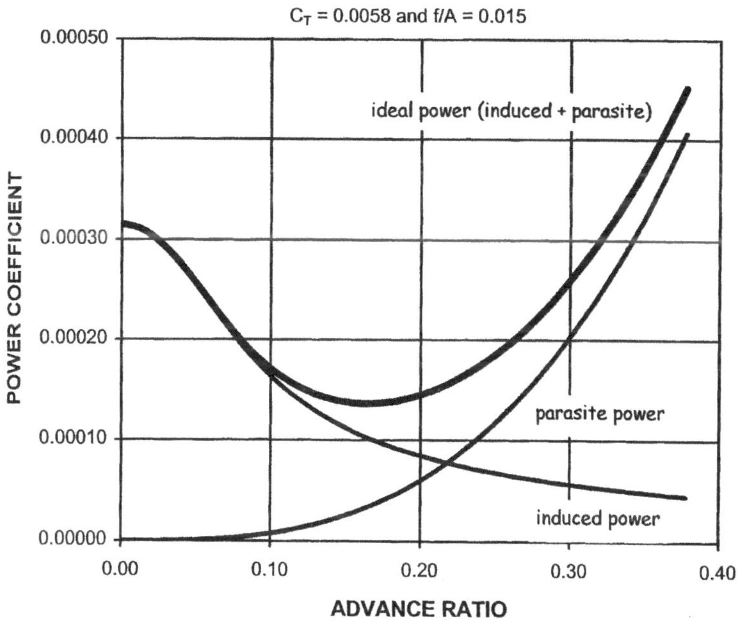

Fig. 2.12 Estimate of power required in level flight using momentum theory alone.

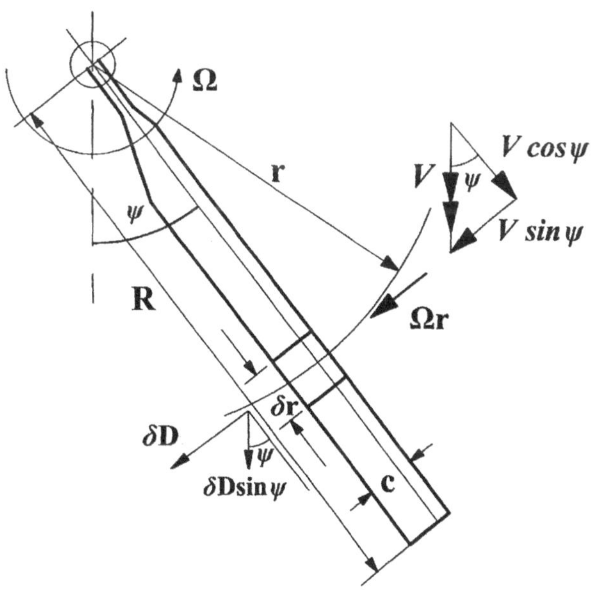

Fig. 2.13 Blade element in forward flight.

found using momentum theory to yield a more complete estimate. Figure 2.13 shows a blade element with its associated elemental drag force. It is now evident that the speed of flow over the blade element contains a term that varies with blade azimuth ($V\sin\psi$) and that consequently, the drag force will vary as the element moves around the hub. By integrating the drag force acting normal to the blade element a relationship for the profile power is obtained. Unfortunately, this is not the complete picture. Although the power required to drive the rotor around the hub has been found, in the face of a horizontal airstream (V), it has not accounted for the power required to drive the rotor system forward: the *rotor parasite power*. This additional power arises because there will be a hub force acting rearwards which must be matched by a component of the rotor thrust. The hub force appears because there is an asymmetry of drag acting on opposing blades. When a blade reaches the advancing side, it will see increased drag, a component of which will act rearwards. An opposing blade will see reduced drag as it enters the retreating side. Although a component of the drag on the retreating side acts forwards, it will be less than on the advancing side and a net rearward hub force will result.

2.8.1 Rotor profile power

From Fig. 2.13 the elemental rotor profile power is given by:

$$\delta P = \Omega\,\delta Q = \Omega r\,\delta D = \Omega r \frac{1}{2}\rho(\Omega r + V\sin\psi)^2 c\,\delta r\, C_D$$

The profile power can be obtained by multiplying δP by the number of blades, b, and integrating both along the radius and around the azimuth:

$$P_{pr} = \frac{1}{2\pi} b \int_0^{2\pi}\int_0^R \frac{1}{2}\rho\,\Omega r\, C_D(\Omega r + V\sin\psi)^2 c\,\mathrm{d}r\,\mathrm{d}\psi$$

Note that $1/2\pi$ is introduced because the average power required around the azimuth is needed. Now:

$$P_{pr} = \frac{1}{2\pi}\rho b c\Omega \frac{1}{2} C_D \int_0^{2\pi}\int_0^R r(\Omega^2 r^2 + 2Vr\sin\psi + V^2\sin^2\psi)\mathrm{d}r\,\mathrm{d}\psi$$

Thus from Layton [2.4]:

$$P_{pr} = \frac{1}{8}\rho b c V_T^3 R C_D(1 + \mu^2) \tag{2.13}$$

Comparison with Equation (2.6) shows that the profile power in horizontal flight can be found by calculating the profile power required in the hover and scaling it by a factor equal to $(1 + \mu^2)$.

2.8.2 Rotor parasite power

Likewise from Fig. 2.13 the elemental rotor parasite power is given by:

$$\delta P = V\delta H = V\delta D\sin\psi = V\frac{1}{2}\rho(\Omega r + V\sin\psi)^2 c\,\delta r\, C_D\sin\psi$$

Therefore:

$$P_{par} = \frac{1}{4}\rho b c V_T^3 R C_D \mu^2 \tag{2.14}$$

2.8.3 Power required for horizontal flight

We are now in a position to determine the power required for horizontal flight using Equations (2.11), (2.13) and (2.14). It is customary to add the contribution from rotor parasite power to the rotor profile power rather than the fuselage parasite power. Thus:

$$P = T(V\sin\gamma + v_i) + \frac{1}{8}\rho bcRV_T^3 C_D(1+\mu^2) + \frac{1}{4}\rho bcRV_T^3 C_D\mu^2$$

$$P = \frac{1}{2}\rho V^3 f + Tv_i + \frac{1}{8}\rho bcRV_T^3 C_D(1+3\mu^2)$$

and in coefficient terms:

$$C_P = \frac{1}{2}\mu^3\left(\frac{f}{A}\right) + \lambda C_T + \frac{1}{8}sC_D(1+3\mu^2) \qquad (2.15)$$

Equation (2.15) is another important result since it suggests that for a helicopter (fixed f/A) at a given weight (fixed C_T) the power coefficient will depend solely on the advance ratio (μ). Also if the advance ratio is fixed and the rotor profile drag coefficient is constant then the power required to drive the rotor will vary as Ω^3 so that C_P remains unchanged. This relationship gives us a mechanism for assessing the effect of changes in rotor drag on the level flight performance of a helicopter.

2.9 HORIZONTAL FLIGHT: IMPROVED THEORETICAL ESTIMATES

When blade element theory was applied to axial flight, it proved useful in identifying the profile power required. Unfortunately, the theory failed to capture the complexities of the real situation requiring both theoretical and empirical adjustments. Likewise the relationships so far developed for horizontal flight can be modified to account for:

- Spanwise flow;
- Reversed flow;
- Non-uniform inflow;
- Blade stall and drag divergence.

2.9.1 Spanwise flow

The method of accounting for the hub force generated by the asymmetry of drag around the azimuth (introducing $\mathrm{d}D\sin\psi$) fails to consider the flow of air along the blade, the so-called *spanwise* flow. This omission becomes clear if one considers a blade aligned with the longitudinal axis of the helicopter and situated over its nose. Here the freestream flow is along the blade from tip to root. Although the drag associated with this flow will generate a hub force the $\sin\psi$ term fails to account for it since $\sin 180^\circ = 0$. The practice of replacing the factor 3 with a larger empirical value k in Equation (2.15) rectifies this omission. Studies have suggested that appropriate values lie between 4.5 and 4.7 [2.13 and 2.14].

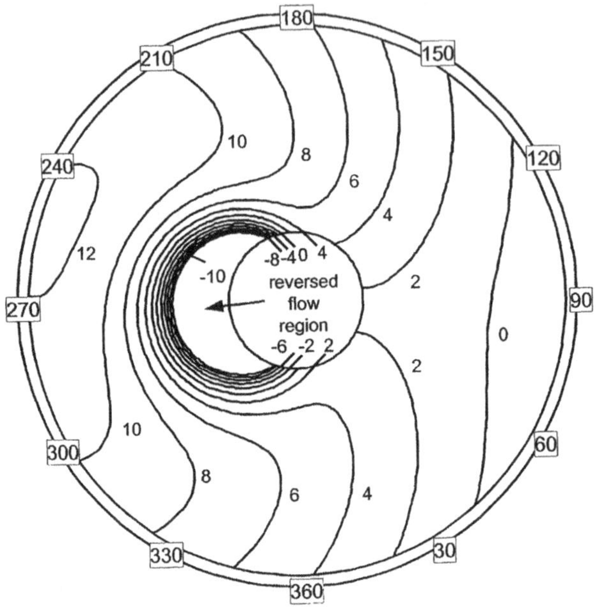

Fig. 2.14 Variation of angle of attack across rotor disk.

2.9.2 Reversed flow

It has been shown that the chordwise velocity of a blade element is given by:

$$U = \Omega r + V \sin \psi$$

On the retreating side the components of rotational velocity and horizontal velocity are subtractive and a region of reversed flow will exist. It can be shown that this region is circular in nature with a diameter equal to μR. At low forward speed this region is usually of little consequence as the aerodynamic section of the blade does not begin until some distance from the hub ($0.25R$ is typical). Since the flow in the reversed flow region is from trailing edge to leading edge it has the effect of pushing the rotor around the mast thereby extracting power from the airflow. Although the amount of negative power is usually quite small and often ignored, it may, however, be approximated by an 8% reduction in the forward speed dependent component of rotor power [2.4], thus, for example, reducing k from 4.65 to 4.30. Hence the power required in forward flight may be estimated using:

$$C_P = \frac{1}{2}\mu^3\left(\frac{f}{A}\right) + \lambda C_T + \frac{1}{8}sC_D(1 + 4.3\mu^2)$$

Figure 2.14 shows the variation of angle of attack across a rotor disk at a high advance ratio ($\mu = 0.34$). Note that the reversed flow region is clearly delineated.

2.9.3 Non-uniform inflow

Although Glauert [2.11] provides a simple method of calculating the mean induced velocity for a rotor in horizontal flight he, along with others who have studied lifting rotors [2.15 and 2.16], appreciated that the induced velocity over a rotor is far from uniform. Consequently they proposed additional formulae designed to generate an upwash ahead of the rotor disk and a steadily increasing inflow along the rotor diameter similar to that seen along the chord of an aerofoil. The formula proposed by Glauert is sufficiently accurate for our purposes:

$$v_i = v_{i0}\left[1 + K\left(\frac{r}{R}\right)\cos\psi\right] \tag{2.16}$$

Note that if K is set to a value greater than 1 (typically 1.2 [2.17]) an upwash results and a linear fore/aft variation in induced velocity is predicted. Lateral variations, although present in reality, have been ignored at this level of simplification. The term v_{i0} represents the mean induced velocity found using Equation (2.12). Before determining the effect of non-uniform induced velocity on the induced power required in forward flight, it is necessary to consider the *skew angle* of the rotor wake. This skew angle (χ) is measured from the vertical and indicates the direction taken by the ultimate wake from the rotor disk. Once again several formulae exist to predict χ [2.6, 2.15 and 2.18]. Although some include wake contraction, the simplest assume a cylindrical wake. The following relationship from Prouty [2.6] is most often quoted:

$$\frac{1}{\tan\chi} = \frac{v_{i0}}{V}$$

Note that since a cylindrical wake is assumed, the induced velocity at the rotor hub is used rather than the ultimate wake value of $2v_{i0}$. Now from Equation (2.12):

$$\frac{1}{\tan\chi} = \sqrt{-\frac{1}{2} + \sqrt{\frac{1}{4} + \frac{1}{\tilde{V}^4}}}$$

Figure 2.15 shows a typical variation in skew angle with horizontal speed. As with tip losses in axial flight it can be shown [2.17] that the extra induced power required by non-uniform inflow can be taken into account by applying a factor κ to the estimate of induced power based on a uniform inflow. Therefore, for small disk tilt:

$$P_i = \kappa P_{i0} = \kappa T v_i = \kappa W v_i$$

Estimates for κ vary from 1.17 [2.17] to 1.2 [2.10].

2.9.4 Blade stall and drag divergence

The high subsonic Mach numbers and high angles of attack experienced by rotor blades cause reductions in lifting performance (*stall*) and increases in drag (*drag divergence*). As both these effects cause an increase in the power required by a rotor they are typically included as additions to the power coefficients already determined. Although the variation of angle of attack and Mach number around the azimuth of a rotor in horizontal flight is complex these additional coefficients are calculated by assuming that drag divergence is most significant at $\psi = 90°$ and blade stall at $\psi = 270°$. Estimation of the power increment due to compressibility (C_{Pm}) begins by determining the tip Mach number at $\psi = 90°$ (M_{90}). This Mach number is then compared with the

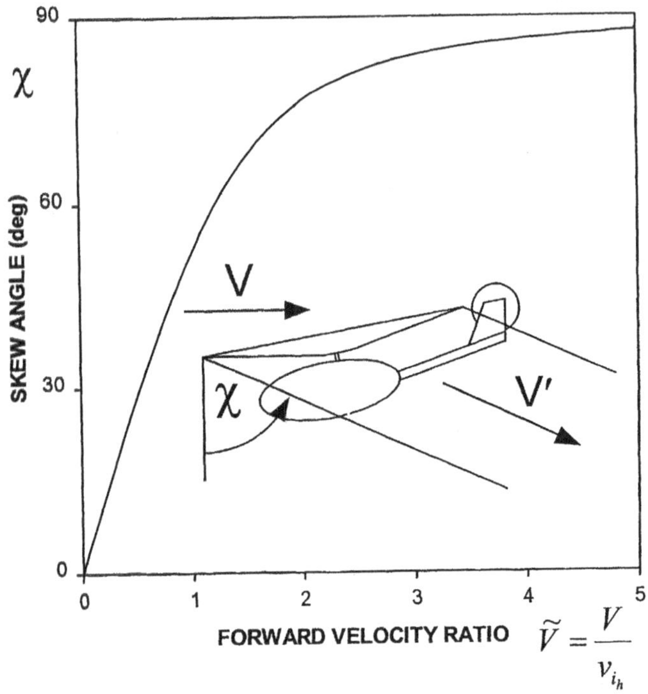

Fig. 2.15 Variation of main rotor skew angle with forward speed.

drag divergence Mach number that is itself dependent on the local angle of attack (α_{90}). Thus after Gessow and Crim [2.19]:

$$C_{Pm} = s[0.012\,\Delta M_d + 0.1(\Delta M_d)^3]$$

where:

$$\Delta M_d = M_{90} - M_d - 0.06$$

and, as an example, from Prouty [2.6] for NACA 0012:

$$M_d = 0.726 - 2.45\alpha_{90}$$

The local angle of attack is found by combining collective pitch (θ_0), blade twist (θ_T), longitudinal cyclic pitch (B_1) and an inflow angle obtained from the mean induced velocity (v_{i0}) [2.4]:

$$\alpha_{90} = \theta_0 + \theta_T + B_1 + \frac{\lambda}{1-\mu}$$

The increment in power coefficient due to retreating blade stall (C_{Ps}) is handled by assuming that the stalled region is diamond-shaped and centred about $\psi = 270°$. Within this region the section profile drag coefficient is taken to increase by 0.08 (for a NACA 0012 section). Hence, from Castles and New [2.20]:

$$C_{Ps} = \frac{s}{24\pi}(1-\mu^2)(1-X_s)\sqrt{1-X_s^2}$$

where X_s is the dimensionless radius beyond which blade stall is present. This radius can be found by setting the general equation for angle of attack along the radial at $\psi = 270°$ equal to α_{max} [2.4 and 2.21]:

$$\alpha_{max} = \theta_0 + X_s\theta_T - B_1 + \frac{\lambda}{X_s - \mu}$$

Although α_{max} varies with Mach number it is often sufficiently accurate to use the angle of attack above which low speed separation occurs (12.5° for a NACA 0012 section [2.21]). A correction to C_{Ps} is included if an unstalled section at the retreating tip bounds the stalled region.

2.9.5 Main rotor power

It is now possible to combine all these improvements into a single relationship for the power required by a single main rotor (including an estimate of the effect of tip losses on the induced power), for example:

$$\left.\begin{aligned} C_P &= 1.2\lambda C_T + \frac{1}{2}\mu^3\frac{f}{A} + \frac{1}{8}sC_D(1 + 4.3\mu^2) + C_{Pm} + C_{Ps} \\ C_P &= C_{Pi} + C_{Ppar} + C_{Ppr} + C_{Pm} + C_{Ps} \end{aligned}\right\} \quad (2.17)$$

where the subscripts represent: *i* – induced; *par* – parasite; *pr* – profile; *m* – Mach and *s* – stall.

Figure 2.16 shows a more realistic variation of power coefficient with advance ratio.

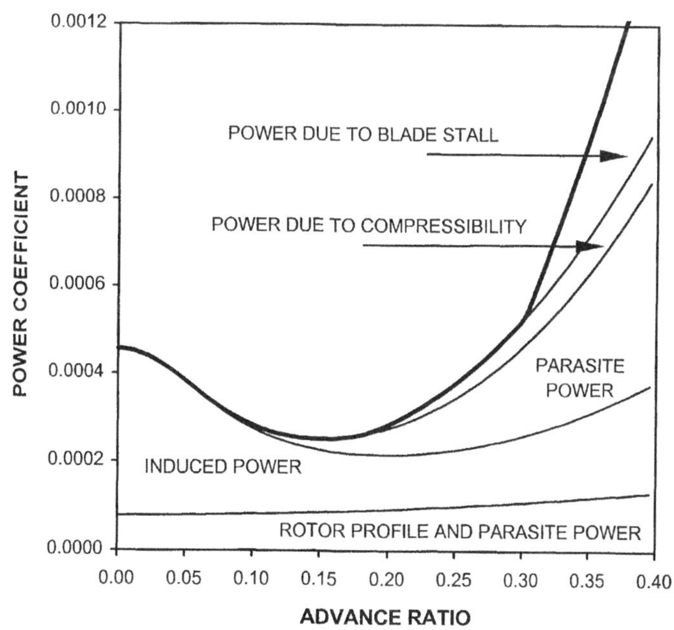

Fig. 2.16 Improved estimate of power required in level flight.

Note how the induced power rises at high forward speed as larger disk tilts and higher thrusts are required to balance the drag. Likewise, note the significance of compressibility and blade stall at high advance ratio. These two factors are often combined under a single term '*tip effects*'.

2.10 TAIL ROTOR POWER

The discussion so far has only been concerned with determining the power required by a single *main* rotor. Such a rotor system will generate a *torque reaction* in the fuselage and without some form of *anti-torque* device the fuselage will spin uncontrollably. The most common anti-torque system is a small rotor affixed to a tail boom and orientated so that it generates a moment as it rotates which acts in opposition to the torque reaction. Alternative systems based on ducted fans, or *fenestrons*, and pressurized air (the NOTAR) have been used with great success. It should be remembered that tail rotors are required mainly in low speed flight when there is insufficient airflow over the fin to generate a side force that can oppose the torque reaction.

The tail rotor power can be estimated by considering the torque it must oppose. The main rotor torque is responsible for the torque reaction and thus the tail rotor thrust can be found by a simple balance of moments. Therefore:

$$T_{TR} l_{TR} = Q_{MR} = \frac{P_{MR}}{\Omega_{MR}}$$

where, in forward flight:

$$P_{MR} = \rho A_{MR} (\Omega R)^3_{MR} \left[1.2\lambda C_T + \frac{1}{2}\mu^3 \frac{f}{A} + \frac{1}{8} s C_D (1 + 4.3\mu^2) + C_{Pm} + C_{Ps} \right]_{MR}$$

or, in axial flight:

$$P_{MR} = \rho A_{MR} (\Omega R)^3_{MR} \left[1.15 \left(\frac{V_c + v_i}{V_T} \right) C_T + \frac{1}{8} s C_D + C_{Pm} + C_{Ps} \right]_{MR}$$

2.10.1 Power required by the tail rotor in the hover

In the hover the tail rotor will be rotating about a fixed point in space and will not, therefore, be subjected to any form of oncoming airflow. Thus the tail rotor power will be given by:

$$P_{TR} = T_{TR} \sqrt{\frac{T_{TR}}{2\rho A_{TR}}} + \frac{1}{8} \rho s_{TR} A_{TR} (\Omega R)^3_{TR} C_D$$

2.10.2 Power required by the tail rotor in forward flight

In forward flight the tail rotor will be subjected to an oncoming air stream in much the same manner as the main rotor. If we ignore the effects of compressibility and

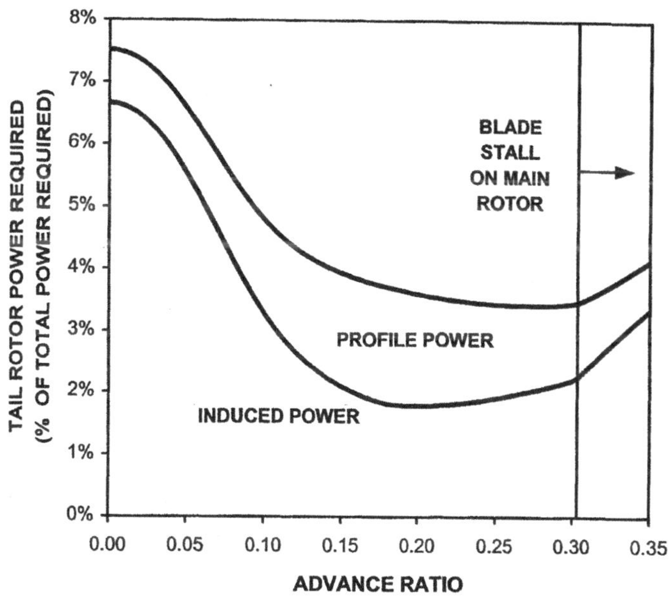

Fig. 2.17 Variation of tail rotor power in level flight.

blade stall then it is possible to determine the tail rotor power increment using a similar approach as that taken for the main rotor [2.4]. Thus the tail rotor profile power will be given by:

$$P_{\mathrm{prTR}} = \frac{1}{8}\rho s_{\mathrm{TR}} A_{\mathrm{TR}} (\Omega R)^3_{\mathrm{TR}} C_{\mathrm{D}} \left[1 + 4.3\mu^2 \frac{(\Omega R)^2}{(\Omega R)^2_{\mathrm{TR}}} \right]$$

and the total tail rotor power will be:

$$P_{\mathrm{TR}} = \frac{P_{\mathrm{MR}}}{l_{\mathrm{TR}}\Omega_{\mathrm{MR}}} v_{\mathrm{iTR}} + P_{\mathrm{prTR}}$$

where the induced velocity v_{iTR} is given by:

$$v_{\mathrm{iTR}} = \Omega R \sqrt{ -\frac{1}{2}\mu^2 + \sqrt{\frac{1}{4}\mu^4 + \left(\frac{AC_{\mathrm{P}}R}{2A_{\mathrm{TR}}l_{\mathrm{TR}}} \right)^2 } }$$

Figure 2.17 shows a typical variation of tail rotor power with forward speed. The power is expressed as the ratio of the tail rotor power to the total power required. Note the rapid reduction in power with forward speed and that although there is an increase in power after a minimum is reached it is not as steep as the main rotor power. This trend is due to the effects of forward speed on the induced velocity of the tail rotor.

2.10.3 Unloading the tail rotor

The variation of power depicted in Fig. 2.17 has been calculated assuming that the tail rotor provides all the anti-torque force required (no fin contributions). Typically as the forward velocity of a helicopter increases the vertical stabilizer or *fin* generates an aerodynamic force that acts in sympathy with the tail rotor thereby reducing its loading. In fact, for many helicopters operating close to their cruise speed, the tail rotor is completely unloaded with the fin providing all the anti-torque force required. Layton [2.4] quotes a recent military specification that requires that a helicopter be capable of an 80 knot run-on landing with no tail rotor. He concludes that the vertical stabilizer must generate sufficient sideforce at 80 knots to balance the main rotor suggesting that the tail rotor is fully unloaded at that speed and that above that speed reversed pedal is required.

2.10.4 Tail rotor stall

It has already been seen that the vortex-ring state can cause a lifting rotor to lose effectiveness in descending flight. Likewise vortices can lead to tail rotor stall and a consequent loss of yaw control. Any situation that causes the tail rotor to *pump* air into an oncoming airstream has the potential to generate a vortex ring state. Sideways flight, rapid spot turns or out-of-wind hovering are situations when the tail rotor is operating in a state equivalent to either a climbing or descending rotor. If flow into the rotor opposes the inflow (the descending case) then a vortex ring may form which reduces the ability of the tail rotor to generate an anti-torque thrust. Depending on the size and loading of the tail rotor it may be possible for the pilot to detect the incipient stages of this phenomena by observing unusual activity in yaw or by feeling irregular buffeting of the tail cone. Failing either of these cues the pilot may only become aware that something is amiss when full pedal deflection has no effect in arresting an unwanted yaw rate. In this situation the tail rotor is effectively *power settling* and the only course of action left is to reduce the anti-torque requirement by either lowering the collective lever or closing down the engine(s).

2.11 POWERED DESCENTS

Prior to analysing the performance characteristics of a helicopter in an autorotation it is necessary to consider the limits on stable descent. Two basic conditions exist: descent under power characterized by the rotor imparting a downwards induced velocity vector on the upward moving flow, and autorotation when smooth momentum flow takes place with the rotor blades generating thrust via a transfer of energy from the airstream to the rotor. Descent under power from the hover or at low forward speed can lead to the potentially hazardous flight regime of vortex ring therefore determining the combination of airspeed and RoD that leads to the vortex-ring state is important.

Earlier the vortex-ring state was mentioned in association with purely vertical flight when it was concluded that for rates of descent between zero and $2v_{ih}$, a vortex would exist somewhere in the streamtube passing through the rotor. This broad definition was further refined by the realization that stable rotor behaviour was only in question when a vortex existed close to the disk. Consequently the *danger area* was reduced to vertical rates of descent between $0.7v_{ih}$ and $1.5v_{ih}$. It is now necessary to determine the effect of a forward speed component on the rates of descent that can generate this potentially hazardous situation.

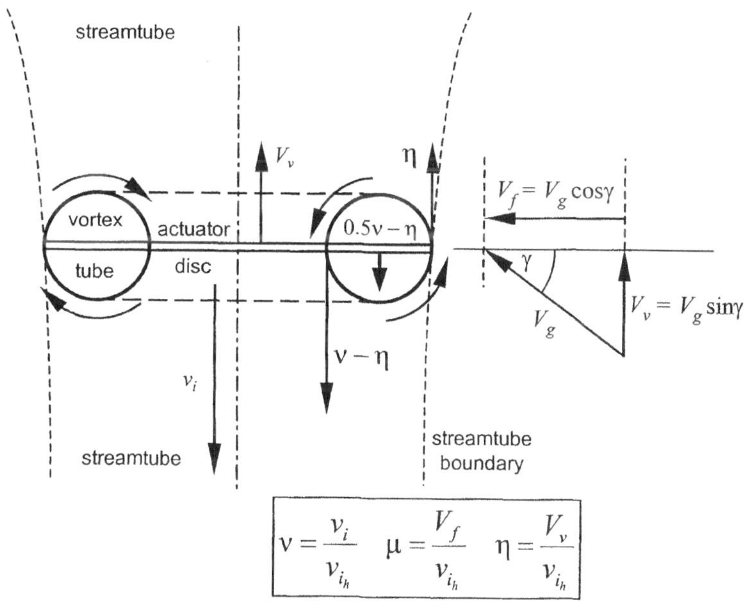

Fig. 2.18 Wolkovitch analysis of the vortex-ring state (adapted from [2.27]).

2.11.1 Predicting vortex-ring boundaries using dynamic inflow methods

Perhaps the most accurate predictions of the flight conditions that lead to vortex-ring can be obtained by considering so-called *dynamic inflow* [2.22]. Dynamic inflow attempts to account for the effect of the vortex-ring condition on the inflow characteristics through the whole rotor. Wolkovitch [2.23] analysed the vortex-ring state by considering a slipstream that is surrounded by a protective tube of vorticity, see Fig. 2.18.

It can be seen that the vortices descend at an average speed of $0.5\nu - \eta$ or $v_i/2 - V_g \sin\gamma$ relative to the rotor, whereas the rotor itself descends at a speed $V_g \sin\gamma$. Thus for low rates of descent the vortex-ring will move away from the rotor and steady flow can exist. If, however, the glide slope is too steep the relative velocity of the vortices will fall to zero and they will remain within the rotor disk resulting in the vortex-ring condition. This critical condition can be expressed in terms of a critical glide speed, V_{crit}:

$$V_{crit} = \frac{v_i}{2 \sin\gamma} \tag{2.18}$$

or normalizing using the induced velocity in the hover (v_{i_h}):

$$\frac{V_{crit} \sin\gamma}{v_{i_h}} = \frac{V_v}{v_{i_h}} = \frac{v_i}{2 v_{i_h}}$$

or:

$$\eta = \frac{\nu}{2} \tag{2.19}$$

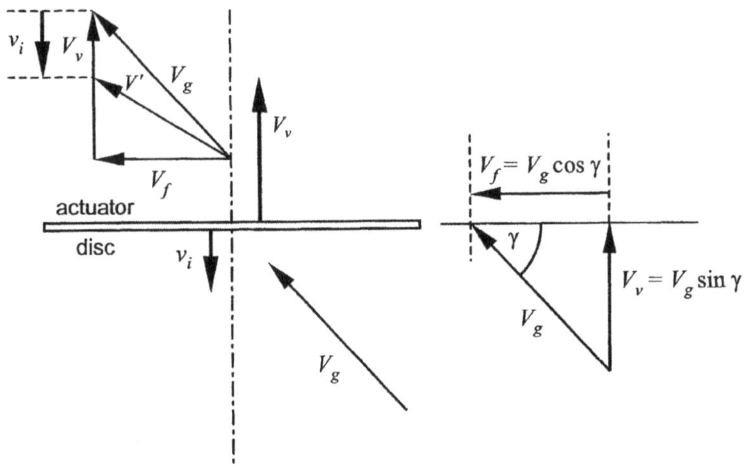

Fig. 2.19 A gliding rotor.

To express Equation (2.19) in a more convenient form it is necessary to rewrite it in terms of horizontal velocity ($V_f = V_g \cos\gamma$) and vertical velocity ($V_v = V_g \sin\gamma$). This is achieved by use of momentum theory, since:

$$T = 2\rho A v_i V'$$

Since noting that thrust and weight must balance both in a steady descent and in the hover (neglecting the effects of download and vertical drag), we can write:

$$v_i V' = \frac{T}{2\rho A} = v_{i_h}^2$$

Now from Fig. 2.19 it can be seen that for a gliding rotor:

$$V' = \sqrt{V_f^2 + (V_v - v_i)^2}$$

Thus:

$$v_{i_h}^2 = v_i \sqrt{V_f^2 + (V_v - v_i)^2}$$

$$v_{i_h}^2 = v_i \sqrt{V_f^2 + V_v^2 - 2V_v v_i + v_i^2}$$

$$v_{i_h}^4 = v_i^2 (V_f^2 + V_v^2 - 2V_v v_i + v_i^2)$$

Now from Equation (2.18) the vortex-ring condition is entered when: $2V_v = v_i$. Hence the lower vortex-ring boundary will be represented by:

$$v_{i_h}^4 = 4V_v^2 (V_f^2 + V_v^2)$$

$$1 = 4\eta^2\mu^2 + 4\eta^4$$

Thus:

$$2^2\eta^4 + 2^2\eta^2\mu^2 - 1 = 0 \tag{2.20}$$

If the rate of descent is allowed to increase *and* the blade pitch is reduced the helicopter will eventually enter the windmill-brake state. Wolkovitch [2.23] recommended that the upper vortex-ring boundary be represented by:

$$\eta = \frac{1.4v}{2} \tag{2.21}$$

Again we can recast this equation, since:

$$v_{i_h}^4 = v_i^2(V_f^2 + V_v^2 - 2V_v v_i + v_i^2)$$

or:

$$1 = v^2(\mu^2 + \eta^2 - 2\eta v + v^2)$$

Therefore substituting for η using Equation (2.21) and simplifying:

$$30^2\eta^4 + 70^2\eta^2\mu^2 - 49^2 = 0 \tag{2.22}$$

2.11.2 Improved predictions

The 'Wolkovitch' boundaries although based on a detailed representation of the vortex-ring condition fail to match empirical data. Experience gained from wind tunnel experiments and flight test suggests that the vortex-ring condition can be escaped by allowing a forward velocity component to develop. Thus the lower boundary should show that as the horizontal velocity (μ) is increased higher rates of descent (η) are required to enter the vortex-ring state. This can be achieved if the Wolkovitch approach is modified to account for the effect of forward speed on the wake geometry, see Fig. 2.20.

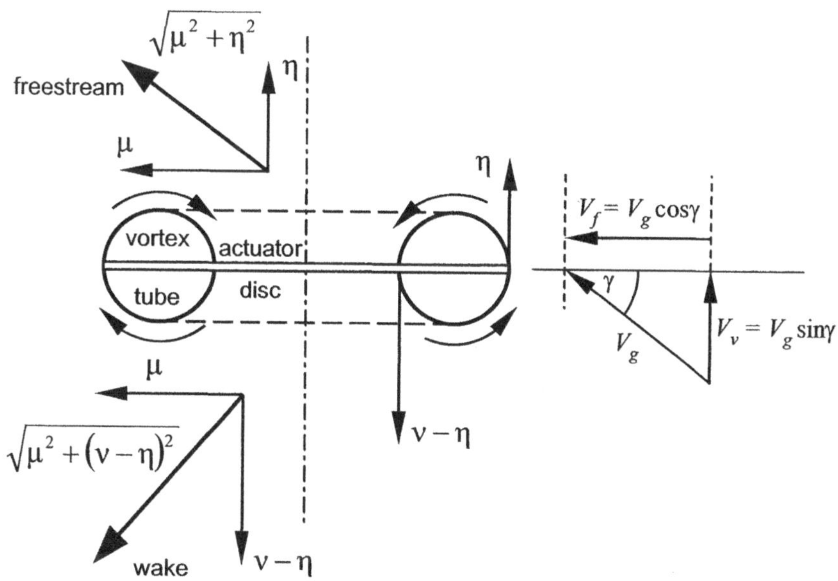

Fig. 2.20 Modified Wolkovitch analysis of the vortex-ring state (adapted from [2.27]).

Figure 2.20 shows that the magnitude and direction of both the freestream flow and the wake can be represented by vectors:

$$\mathbf{a} = \text{freestream flow} = \sqrt{\mu^2 + \eta^2}$$
$$\mathbf{b} = \text{wake} = \sqrt{\mu^2 + (v - \eta)^2}$$

Now the dot product ($\mathbf{b} \cdot \mathbf{a}$) can be thought of as the magnitude of one vector multiplied by the component of the other vector in the direction of the first. Thus in this context the dot product represents the magnitude of the wake velocity multiplied by the component of the freestream in the direction of the wake. Hence:

$$\frac{\mathbf{b} \cdot \mathbf{a}}{|\mathbf{b}|} = \text{component of freestream flow in direction of wake}$$

$$\frac{\mathbf{b} \cdot \mathbf{a}}{|\mathbf{b}|} = \frac{\mu \times \mu + (v - \eta) \times -\eta}{\sqrt{\mu^2 + (v - \eta)^2}} = -\frac{\mu^2 + \eta(\eta - v)}{\sqrt{\mu^2 + (v - \eta)^2}}$$

The Wolkovitch approach assumes that the vortex-ring state begins when the velocity of the vortex tube relative to the actuator disk is zero, that is:

$$\frac{\mathbf{b} \cdot \mathbf{a}}{|\mathbf{b}|} + |\mathbf{b}| = 0$$

$$\frac{\mu^2 + \eta(\eta - v)}{\sqrt{\mu^2 + (v - \eta)^2}} + \sqrt{\mu^2 + (v - \eta)^2} = 0$$

Therefore:

$$\mu^2 + \eta(\eta - v) + \mu^2 + (v - \eta)^2 = 0$$
$$2\mu^2 + 2\eta^2 + v^2 - 3\eta v = 0$$

or:

$$\mu^2 = \frac{3}{2}\eta v - \eta^2 - \frac{1}{2}v^2 \qquad (2.23)$$

As before from momentum theory, $1 = v^2(\mu^2 + \eta^2 - 2\eta v + v^2)$, thus:

$$1 = v^2\left(\frac{3}{2}\eta v - \eta^2 - \frac{1}{2}v^2 + \eta^2 - 2\eta v + v^2\right) = v^2\left(\frac{1}{2}v^2 - \frac{1}{2}\eta v\right)$$
$$2 = v^4 - \eta v^3$$
$$\eta = v - \frac{2}{v^3}$$

And, returning to Equation (2.23):

$$\mu^2 = \frac{3}{2}v\left(v - \frac{2}{v^3}\right) - \left(v - \frac{2}{v^3}\right)^2 - \frac{1}{2}v^2 = \frac{1}{v^2} - \frac{4}{v^6}$$

Therefore the lower vortex-ring boundary is given by:

$$\left.\begin{aligned}\eta &= v - \frac{2}{v^3}\\ \mu^2 &= \frac{1}{v^2} - \frac{4}{v^6}\end{aligned}\right\} \tag{2.24}$$

The upper vortex boundary is obtained in an analogous manner, by replacing v with kv in Equation (2.23). As mentioned above, Wolkovitch recommended $k = 1.4$ as denoting the edge of the fully developed vortex ring condition. Peters and Chen [2.22], on the other hand, prefer $k = 2$ as indicating the condition beyond which no vortex is present anywhere in the streamtube. Hence Equation (2.23) becomes:

$$\mu^2 = \frac{3}{2}\eta(2v) - \eta^2 - \frac{1}{2}(2v)^2 = 3\eta v - \eta^2 - 2v^2$$

Once again from momentum theory, $1 = v^2(\mu^2 + \eta^2 - 2\eta v + v^2)$, thus:

$$1 = v^2(3\eta v - \eta^2 + \eta^2 - 2v^2 - 2\eta v + v^2)$$

$$1 = v^2(\eta v - v^2) = \eta v^3 - v^4$$

$$\eta = v + \frac{1}{v^3}$$

Consequently:

$$\mu^2 = 3v\left(v + \frac{1}{v^3}\right) - \left(v + \frac{1}{v^3}\right)^2 - 2v^2 = \frac{1}{v^2} - \frac{1}{v^6}$$

Therefore the upper vortex-ring boundary is given by:

$$\left.\begin{aligned}\eta &= v + \frac{1}{v^3}\\ \mu^2 &= \frac{1}{v^2} - \frac{1}{v^6}\end{aligned}\right\} \tag{2.25}$$

These boundaries are shown in Fig. 2.21. Clearly the lower boundary approximates to the empirical trend whereby with forward speed higher rates of descent are required to enter the vortex-ring state. There are, however, two issues that remain unresolved. First the upper and lower boundaries should predict the same forward velocity, above which vortex-ring can be completely avoided. Secondly the upper boundary is based on entry to the windmill-brake state ($V_v = 2v_{ih}$ in the hover) whereas the lower indicates entry into fully developed vortex-ring ($V_v = 0.7v_{ih}$ in the hover). In order to resolve these problems the lower boundary must be further modified to start at $\mu = 0$, $\eta = 0$. This is achieved by noting that a vortex will appear in the streamtube when $(\mathbf{b}\cdot\mathbf{a})/|\mathbf{b}| < 0$. Thus the lower boundary should be given by:

$$\frac{\mu^2 + \eta(\eta - v)}{\sqrt{\mu^2 + (v - \eta)^2}} = 0$$

$$\mu^2 = \eta v - \eta^2$$

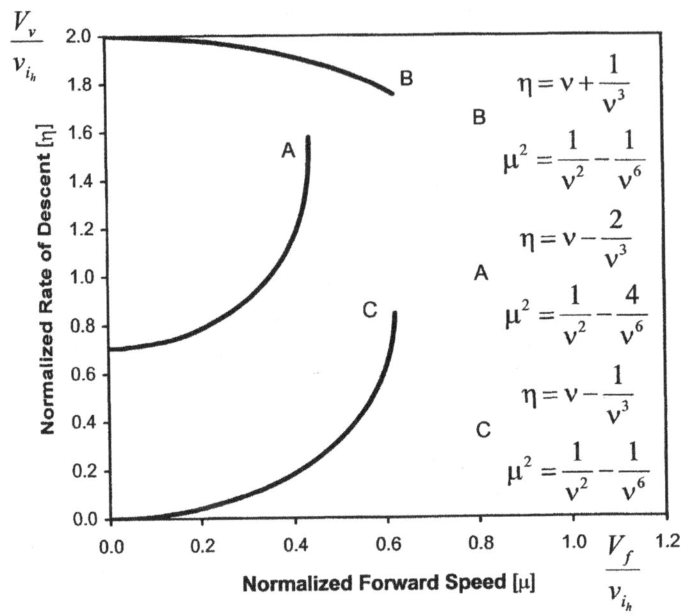

Fig. 2.21 Vortex-ring boundaries using modified Wolkovitch analysis.

Applying momentum theory gives:

$$1 = \nu^2(\eta\nu - \eta^2 + \eta^2 - 2\eta\nu + \nu^2)$$

$$1 = \nu^2(\nu^2 - \eta\nu)$$

$$\eta = \nu - \frac{1}{\nu^3}$$

Consequently:

$$\mu^2 = 3\nu\left(\nu - \frac{1}{\nu^3}\right) - \left(\nu - \frac{1}{\nu^3}\right)^2 - 2\nu^2 = \frac{1}{\nu^2} - \frac{1}{\nu^6}$$

Therefore the complete criteria, see Fig. 2.21, for the vortex-ring state are:

$$\eta = \nu \pm \frac{1}{\nu^3}$$

$$\mu^2 = \frac{1}{\nu^2} - \frac{1}{\nu^6}$$

These criteria can be used to predict a complete and coherent boundary for the onset of the vortex ring condition [2.22].

2.11.3 Incipient vortex-ring condition

Empirical measurements suggest that for a conventional helicopter significant penetration into the vortex ring condition can be made without hazard. It is possible therefore

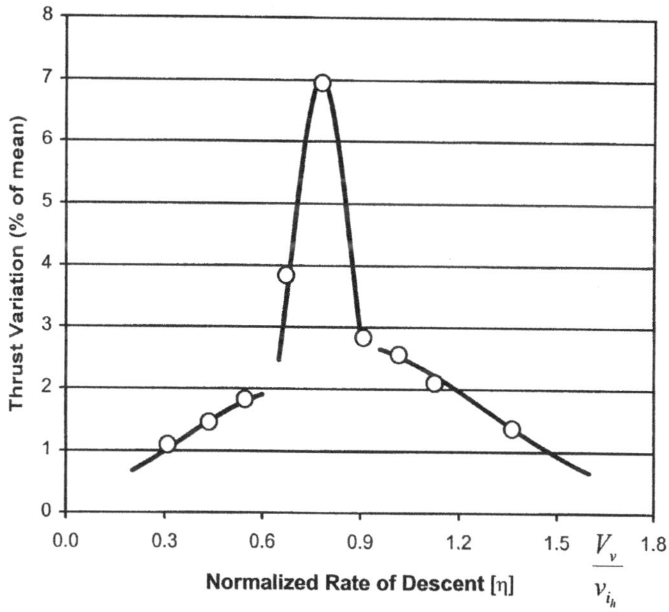

Fig. 2.22 Thrust fluctuations in powered descents (adapted from [2.27]).

to examine the *incipient stages* of the condition without compromising a safe recovery. Any theoretical vortex ring boundary should therefore reflect this reality by showing contours of increasing vortex strength. Model tests on a rotor, with $-8°$ of linear twist, operating at a range of collective pitch settings in a vertical descent highlight a correlation between vortex strength and thrust fluctuation [2.24], see Fig. 2.22.

The trend in thrust fluctuations implies that the worse case situation occurs with a RoD equal to approximately $0.8v_{ih}$, slightly lower than the theoretical value of v_{ih}. Figure 2.22 also shows that for rates of descent less than $0.3v_{ih}$ and greater than $1.5v_{ih}$ only 1% thrust fluctuation occurs which might be equated to the incipient stage of the vortex ring condition. Based on the vertical flight data shown in Fig. 2.22 and using the boundary shape suggested by Peters and Chen [2.22], a series of contours can be drawn (Fig. 2.23). Note that if the rate of descent is low (around $0.5v_{ih}$) glideslopes steeper than approximately 60° will cause the helicopter to enter the incipient stage of the vortex ring state. Also note that if an approach angle of less than 10° is used vortex ring can be almost completely avoided thus the simple rule-of-thumb of '*no more than 500 ft/min below 30 kts*' is sound since it gives an approach angle of 9°. Figure 2.24 presents the same information in a slightly different manner and also illustrates the 30 kts/500 ft/min rule of thumb for v_{ih} between 15 m/s and 50 m/s. Here normalized glide velocity (V_g/v_{ih}) and glide angle (γ) are used to show the vortex ring boundary. The figure clearly shows that for glide angles of less than 60° incipient vortex ring can be completely avoided. Also if a modest rate of descent (around $0.3v_{ih}$) is generated from the hover (glide angle equals 90°) then the vortex ring condition can be explored with a degree of safety provided the helicopter is sufficiently high to initiate recovery by generating forward speed. Note that for rotorcraft with high disk loading and consequently high values of v_{ih} the rule of thumb cannot be guaranteed to keep the helicopter clear of the vortex ring condition.

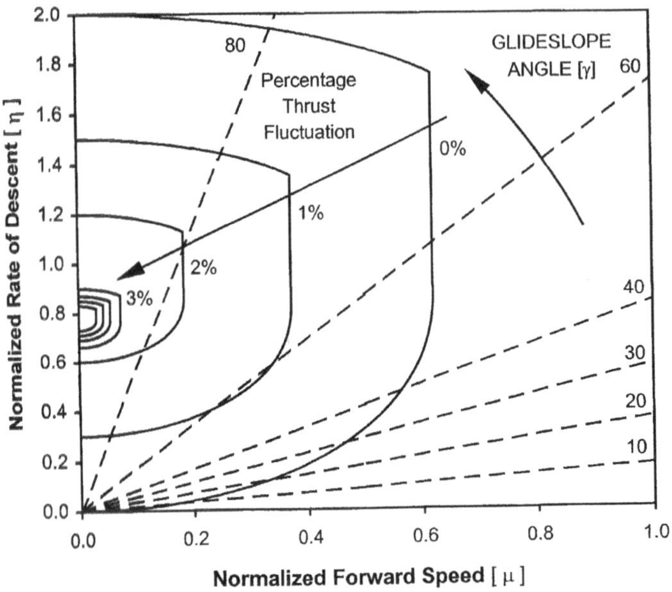

Fig. 2.23 Vortex-ring boundaries – forward speed versus vertical speed.

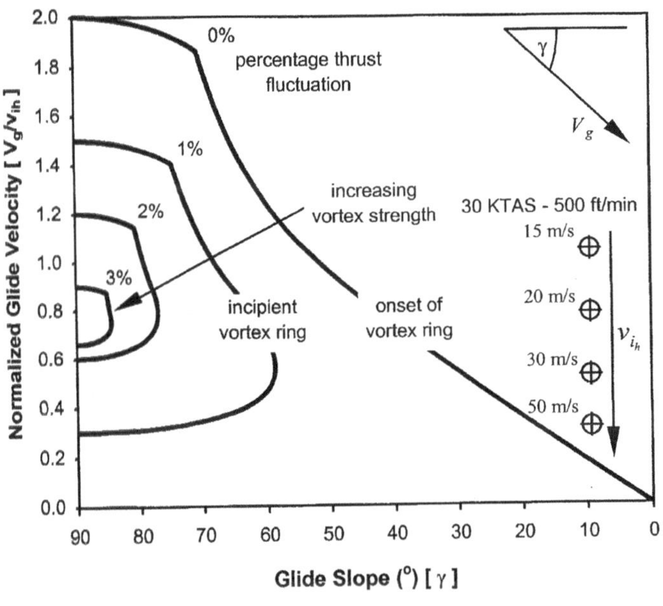

Fig. 2.24 Vortex-ring boundaries – glide slope angle versus glide speed.

2.12 AUTOROTATION

Assessment of rotorcraft performance following an engine failure is necessary to ensure its safe operation. For single-engined rotorcraft an engine failure will result in a forced, or *engine-off*, landing (EOL). For multi-engined helicopters, however, it is still necessary to assess the ability of the aircraft to continue flight particularly in critical areas of the flight envelope. In any event the pilot must react quickly to avoid loss of control and prevent the rotor speed from decaying below acceptable minimums. As the power fails directional control will go out of trim and the residual anti-torque moment will yaw the helicopter, generating additional sideslip that may possibly induce a roll if no corrective action is taken. (An undemanded yaw may be the first indication to the pilot that a power failure has occurred.) The rotor speed decay rate can be estimated by applying simple rotational mechanics to the problem. Now:

$$Q = I\dot{\omega}$$

Thus:

$$\dot{\omega} = \frac{\mathrm{d}\Omega}{\mathrm{d}t} = \text{rotor speed decay rate} = -\frac{Q}{I}$$

where Q = rotor torque and I = moment of inertia of rotor system.

Thus reducing the rotor torque will reduce the rate of rotor speed decay. This can be achieved by reducing the blade pitch in order to decrease the rotor drag. In steady autorotation a given value of collective pitch will cause the helicopter to settle on a unique descent speed and rotor speed combination. The pilot can therefore control the rotor RPM using collective pitch – the lower the pitch the higher the rotor speed – but in practice the usable range of RRPM is very restrictive. If the rotor speed is too low the blade will stall and lose lift; too high and there will be excessive loading on the rotor hub and blade roots. The safe range is typically within 80% and 120% of the nominal power-on speed for transient excursions and between 90% and 110% for stabilized conditions.

2.12.1 Factors affecting rotor speed decay

2.12.1.1 Minimum rotor speed and rotor inertia

It is vital that the RRPM does not decay too far following a loss of power as it may then be impossible to establish the aircraft in a stabilized autorotation. The rate of rotor speed decay is therefore a very important factor, and the minimum rotor speed below which recovery is impossible is of critical concern to the rotor system designer. It is rather difficult to determine the minimum rotor speed accurately but by making some reasonable assumptions it is possible to get a basic understanding of the issue. Suppose the minimum allowable rotor speed (Ω_{min}) corresponds to the maximum mean lift coefficient achievable by the rotor ($\tilde{C}_{\text{L max}}$). Now the rotor thrust, T, is given by [2.25]:

$$T = \frac{1}{2}\rho V^2 bcR\tilde{C}_{\text{L}}$$

If the thrust is assumed to remain constant during the initial phase of the power failure [2.25] and the inflow velocity is small compared with the rotor speed, then:

$$T = \frac{1}{2}\rho\Omega_{\text{nom}}^2 bcR^3\tilde{C}_{\text{L nom}} = \frac{1}{2}\rho\Omega_{\text{min}}^2 bcR^3\tilde{C}_{\text{L max}}$$

Thus:

$$\Omega_{\min} = \Omega_{\text{nom}} \sqrt{\frac{\tilde{C}_{\text{L nom}}}{\tilde{C}_{\text{L max}}}}$$

where Ω_{nom} = rotor speed at instant of power failure and $\tilde{C}_{\text{L nom}}$ = average lift coefficient at instant of power loss.

In addition if the torque coefficient remains unchanged [2.25], then:

$$C_Q = \frac{Q}{\rho A V_T^2 R} = \frac{Q}{\rho A \Omega^2 R^3} = \frac{Q_{\text{nom}}}{\rho A \Omega_{\text{nom}}^2 R^3} = \frac{Q_{\min}}{\rho A \Omega_{\min}^2 R^3}$$

where Q_{nom} = torque required at instant of power failure and $Q_{\min}$ = torque required at minimum rotor speed.

Hence the torque required to drive the rotor at any instant following the power failure will be given by:

$$Q = Q_{\text{nom}} \left(\frac{\Omega}{\Omega_{\text{nom}}} \right)^2$$

In the absence of any shaft power from the powerplants the torque requirement will tend to cause the rotor speed to decay. Thus:

$$\dot{\omega} = -\frac{Q_{\text{nom}}}{I} \left(\frac{\Omega}{\Omega_{\text{nom}}} \right)^2$$

Reorganizing, integrating and applying the initial condition that at $t = 0$, $\Omega = \Omega_{\text{nom}}$, gives:

$$t = \frac{I\Omega_{\text{nom}}^2}{Q_{\text{nom}}} \left(\frac{1}{\Omega} - \frac{1}{\Omega_{\text{nom}}} \right)$$

Therefore, the time taken for the rotor speed to reduce to the minimum permissible value can be obtained from:

$$t_{\min} = \frac{I\Omega_{\text{nom}}}{Q_{\text{nom}}} \left(\frac{\Omega_{\text{nom}}}{\Omega_{\min}} - 1 \right) = \frac{I\Omega_{\text{nom}}}{Q_{\text{nom}}} \left(\sqrt{\frac{\tilde{C}_{\text{L max}}}{\tilde{C}_{\text{L nom}}}} - 1 \right)$$

Thus we see that $t_{\min}$ is dependent on the rotor inertia, the rotor speed and torque requirement at the instant of power loss and the ratio of the thrust required to the maximum thrust the rotor can produce. Consider the effect of changing some of these variables on a 4-bladed rotor of 6.5 m radius and 0.4 m chord with a lift curve slope of 6 per radian operating in the hover, see Table 2.1. Note that from the simplistic analysis detailed above the minimum rotor speed for this example helicopter is 18 rad/s based on a nominal mass of 5000 kg. The data in Table 2.1 indicates that the decay time can be lengthened by increasing the rotor inertia, increasing the nominal rotor speed or by reducing the thrust that the rotor has to produce under normal power-on conditions.

Table 2.1 Examples of decay time (based on conditions at 75% rotor radius).

Rotor speed Ω_{nom}(rad/s)	Mass (kg)	Torque Q_{nom} (Nm)	Rotor inertia I (kg/m^2)	Lift coefficient $C_{L\,nom}$	Maximum lift coefficient $C_{L\,max}$	Time to minimum rotor speed t_{min} (s)
35.0	5000	23 087	6000	0.2640	1.0024	8.63
35.0	5000	23 087	5400	0.2640	1.0024	7.76
35.0	5000	23 087	6600	0.2640	1.0024	9.49
31.5	5000	22 923	6000	0.3258	1.0411	6.49
38.5	5000	23 712	6000	0.2183	0.9586	10.67
35.0	4500	21 036	6000	0.2377	1.0024	10.52
35.0	5500	25 242	6000	0.2904	1.0024	7.14

2.12.1.2 *Pilot action and rate of descent*

Whilst the simplistic analysis detailed above gives a basic guide to the factors affecting the time taken for the rotor speed to decay to the minimum permissible value it does not take account of important additional factors. These include the action of the pilot in attempting to contain the rotor speed within limits and the effect of rate of descent on the angle of attack of the rotor blades. To investigate these effects further it is necessary to construct a more detailed model of a rotor entering a vertical autorotation. As before consider the three-quarter radius as being representative of the conditions on the complete rotor blade. Now:

$$T = L\cos\phi - D\sin\phi$$

$$Q = 0.75R(D\cos\phi + L\sin\phi)$$

Using basic aerofoil theory it can be shown that:

$$L = \frac{1}{2}\rho V_{0.75}^2 abcR\alpha_{0.75}$$

$$D = \frac{1}{2}\rho V_{0.75}^2 bcRC_D$$

Now the angle of attack and airflow velocity will depend on the blade pitch, the vertical velocity and the rotational velocity at the three-quarter radius. Assuming a linear twist :

$$\theta_{0.75} = \theta_0 + 0.75\theta_1$$

$$\alpha_{0.75} = \theta_{0.75} - \lambda_{0.75} = \theta_{0.75} - a\tan\left(\frac{V_c + v_{ih}}{0.75R\Omega}\right)$$

and:

$$V_{0.75}^2 = (0.75R\Omega)^2 + (V_c + v_{ih})^2$$

$$\frac{v_{ih}}{\Omega R} = -\left(\frac{as}{16}\right) + \sqrt{\left(\frac{as}{16}\right)^2 + \left(\frac{as}{8}\right)0.75\theta_{0.75}}$$

The effect of changes in thrust developed and torque required on the vertical velocity and the rotor speed can be determined by assuming constant acceleration over a short time interval, δt:

$$[V_c]_{t+\delta t} = [V_c]_t + \frac{[T - mg]_t}{m}\delta t$$

$$[\Omega]_{t+\delta t} = [\Omega]_t - \frac{[Q]_t}{I}\delta t$$

A more detailed study of the situation following an engine failure (see Fig. 2.25) highlights the role of the rate of descent in changing the magnitude and direction of the forces acting on the rotor blades and thus the unbalanced torque causing the rotor speed to decay. It can be seen that the speed of a rotor initially operating at 35 rad/s, with an inertia of 6000 kg m^2, stabilizes at 21.5 rad/s once the vertical rate of descent has reached 1500 ft/min. It should be remembered that the analysis presented ignores the effect of entry into the vortex-ring condition. In reality the pilot will attempt to arrest the rotor speed decay and contain the N_R within narrower limits by rapid reductions in collective pitch. This has the effect of increasing the rate of descent and unloading the rotor blades thereby reducing, and ultimately reversing, the decelerating torque applied to the rotor. Since prompt action following an engine failure is often vital to success it is important to know if the pilot has sufficient time to identify a problem with the powerplant(s) and take corrective action. This is determined by measuring the *delay time.*

2.12.1.3 *Delay time*

Delay time is simply defined as the time between an engine failure and the pilot commencing corrective action by rapidly lowering the collective lever. Hence the *maximum* delay time is the delay that causes the rotor speed to reach the *minimum power-off transient value* before rising to achieve a stable autorotative condition. Since entry into power-off flight will inevitably involve some change in rotor speed it is common practice to relate key instances to the corresponding rotor speed, see Fig. 2.26.

Having worked-up to a repeatable collective lever lowering rate (by routinely acquiring zero '*g*' for example) and with reference to a sensitive rotor speed gauge a test pilot will incrementally reduce N_{R_1} until N_{R_3} equates to the transient power-off minimum rotor speed. Alternatively the test may be curtailed before the minimum is reached if a sufficiently long delay time has been demonstrated. The model developed in the previous section can be used to show the effect of increasing delay time, see Fig. 2.27. Note that a maximum delay of 1.5 s is possible with the example rotor. Figure 2.28 shows the rotor speeds N_{R_1} and N_{R_3} in the form of an in-flight cross-plot or '*how-goes-it*' chart. Test teams often use this style of chart to get some idea of the effect of increasing delay time on the minimum rotor speed. They can, therefore, ensure that the minimum rotor speed limit is approached in a safe manner by using the chart to select the next incremental reduction in N_{R_1}. It is worth reminding the reader that in flight the test team will use values of N_{R_1} to progress the evaluation. The delay times would be determined from appropriate post-flight analysis.

2.12.2 Windmilling rotors and autorotation

In power-off flight the rotor blades may be in a state of true autorotation or they may be 'windmilling'. In a true autorotation the blades are set at a pitch angle that

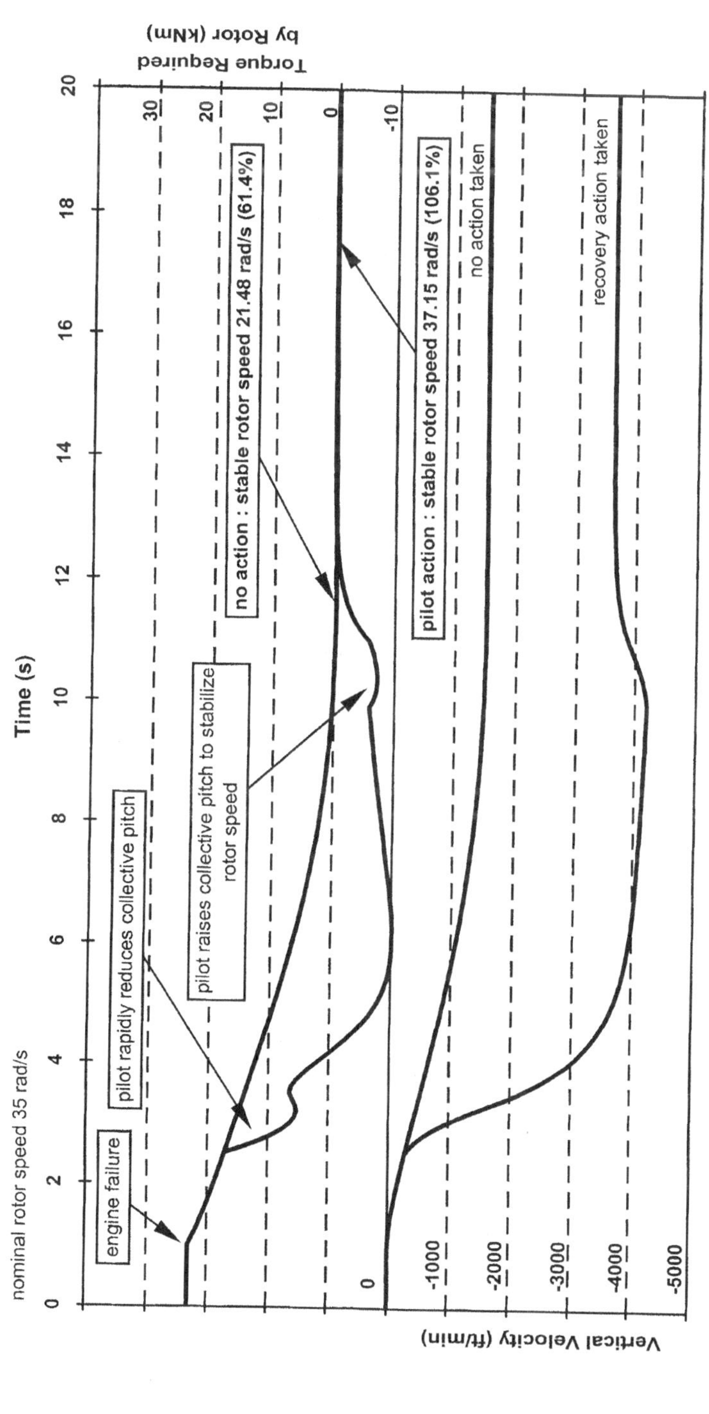

Fig. 2.25 Effect of pilot action on autorotative flight conditions.

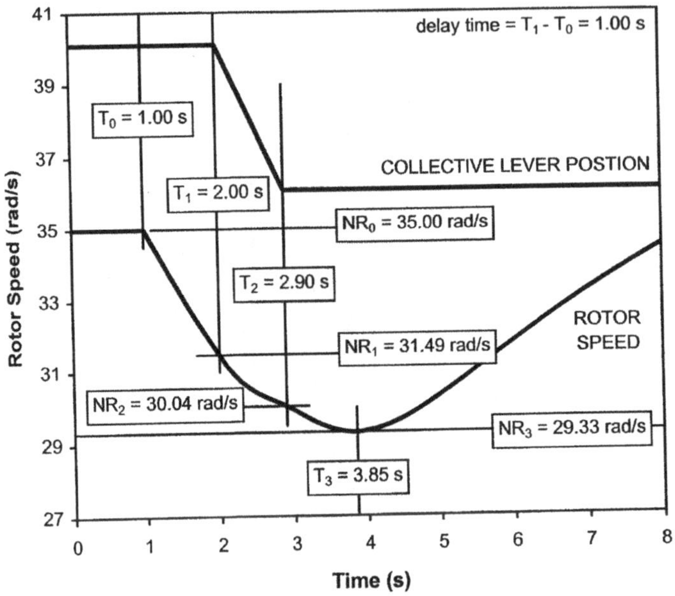

Fig. 2.26 Definition of key rotor speeds associated with delay time testing.

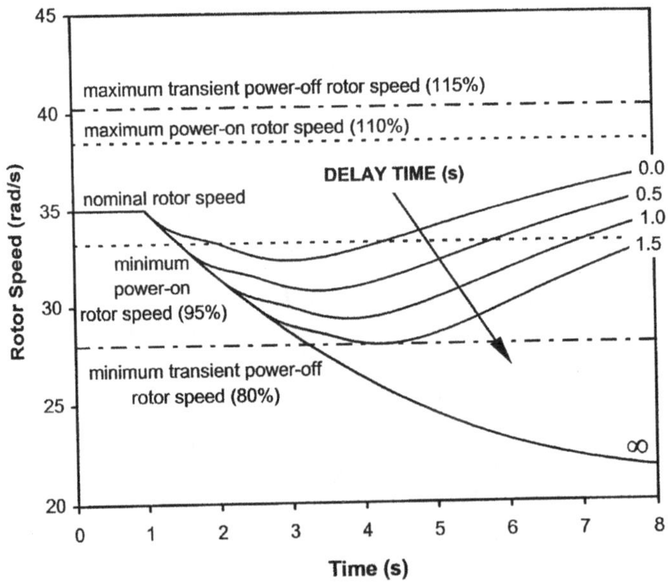

Fig. 2.27 Effect of increasing delay time on rotor speed decay.

combines with the rate of descent to produce no net torque and thereby stabilizes the rotor speed. From this condition the pilot can make minor changes in collective pitch to reduce or increase the rotor speed. If the balance of in-plane forces results in an accelerative condition the rotor blades are said to be *windmilling*. Figure 2.29 compares

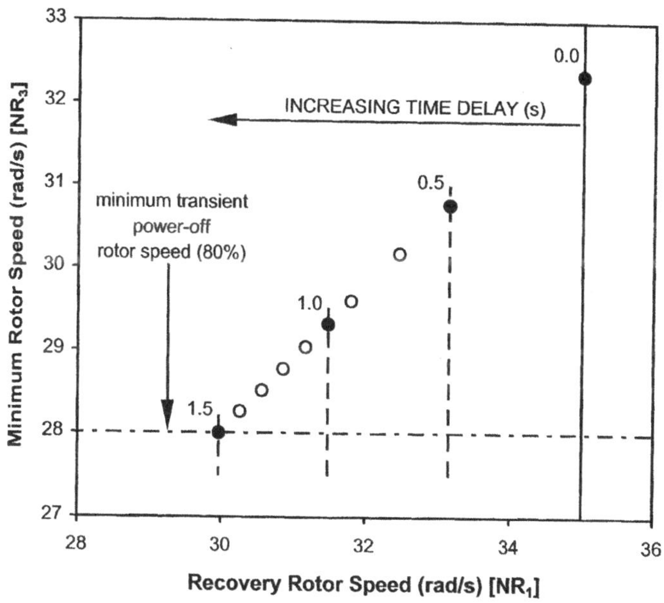

Fig. 2.28 In-flight cross-plot – recovery N_R versus minimum N_R.

the radial variation of elemental thrust and torque for the example rotor in a hover (at 5000 kg) and in a true vertical autorotation. Note how the requirement to balance the accelerative and decelerative in-plane forces along the blade length changes the lift distribution. Likewise the drag associated with the blade root and the tip-loss region results in the need for negative torque in the region $0.25 < r/R < 0.85$.

2.13 AUTOROTATIVE PERFORMANCE

As a consequence of the problems associated with the vortex-ring state and visually clearing the proposed landing sight, it is common practice for pilots to conduct autorotative descents in forward flight rather than in vertical flight. It is therefore often necessary to determine the performance of helicopters operating in a stable autorotation at some forward airspeed. The approach [2.26] is based on the familiar equations of forward flight and starts by calculating the vertical (F_Z) and horizontal (F_X) forces that must be balanced by rotor:

$$F_X = \frac{1}{2}\rho V^2 S_f$$

$$F_Z = mg$$

where S_f = frontal drag area.

The drag force, F_X, gives rise to a power requirement (the parasite power P_{par}) which is added to the rotor profile power to give the shaft power (P_S) that must be provided by the rate of descent:

$$P_S = P_{pr} + P_{par} = \frac{1}{8}\rho bc V_T^3 R C_D(1 + 4.3\mu^2) + F_X V$$

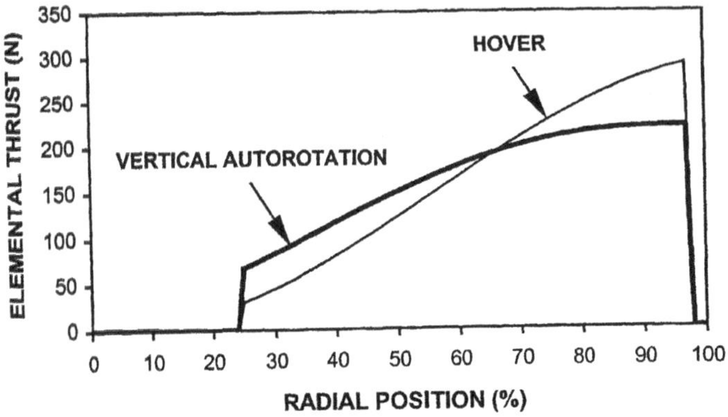

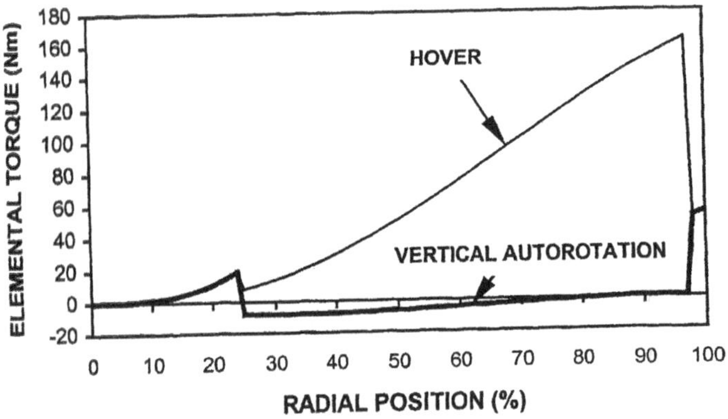

Fig. 2.29 Blade conditions – hover and vertical autorotation.

Because this shaft power must be extracted from the rotor, by a rate of descent, it is given a negative sign before being normalized using the hover power (P_H):

$$\frac{P_S}{P_H} = \frac{-(P_S)}{\sqrt{\frac{(mg)^3}{2\rho A} + \frac{1}{8}\rho bc V_T^3 R C_D}}$$

Now it can be shown that [2.26]:

$$\frac{P_S}{P_H} = \frac{V_f}{v_{i_h}}\tan\theta + \frac{w_0}{v_{i_h}\cos^2\theta} + \frac{V_v}{v_{i_h}} \qquad (2.26)$$

where θ is the disk tilt ($\tan\theta = F_X/F_Z$) and w_0 is the vertical component of the induced velocity.

A momentum analysis of a descending rotor in forward flight yields [2.26]:

$$(1+\tan^2\theta)\left[\frac{w_0}{v_{i_h}}\right]^4 + 2\left(\frac{V_v}{v_{i_h}} + \frac{V_f}{v_{i_h}}\tan\theta\right)\left[\frac{w_0}{v_{i_h}}\right]^3 + \left(\frac{V_v^2 + V_f^2}{v_{i_h}}\right)\left[\frac{w_0}{v_{i_h}}\right]^2 - 1 = 0 \qquad (2.27)$$

Equations (2.26) and (2.27) can be solved simultaneously for given values of P_S/P_H, V_f and v_{ih} to determine V_v (the autorotative rate of descent, RoD). These equations can also be used to predict the effect of aircraft configuration and flight condition on the autorotative performance, see Figs 2.30 to 2.32, which show the effect of aircraft mass, altitude and rotor speed.

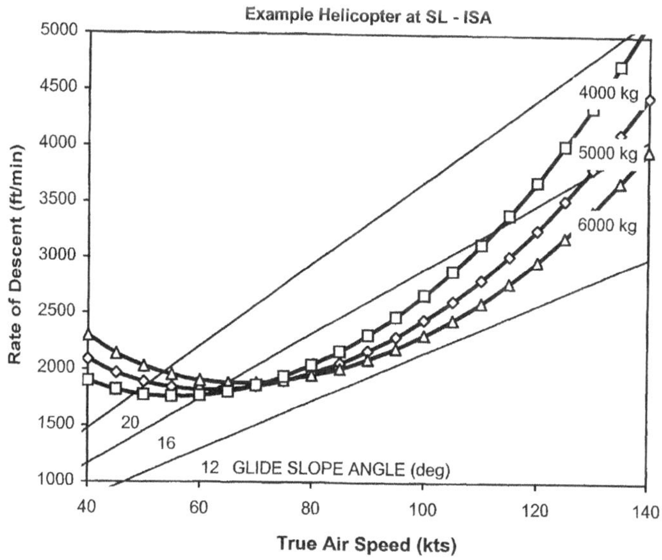

Fig. 2.30 Autorotative performance – effect of all-up-mass.

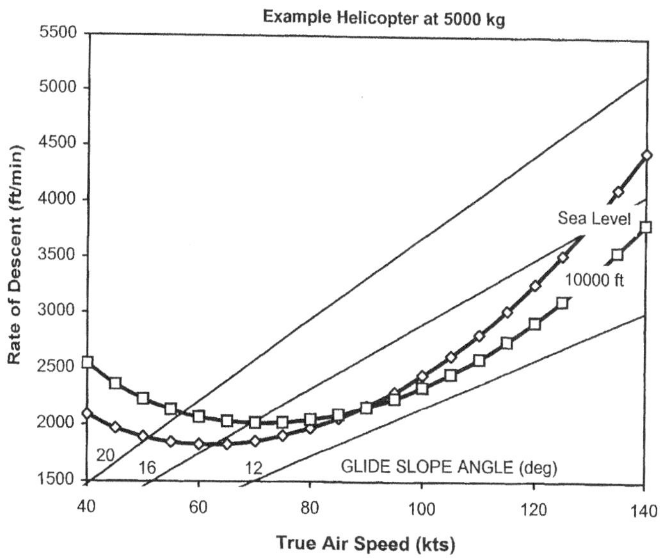

Fig. 2.31 Autorotative performance – effect of altitude.

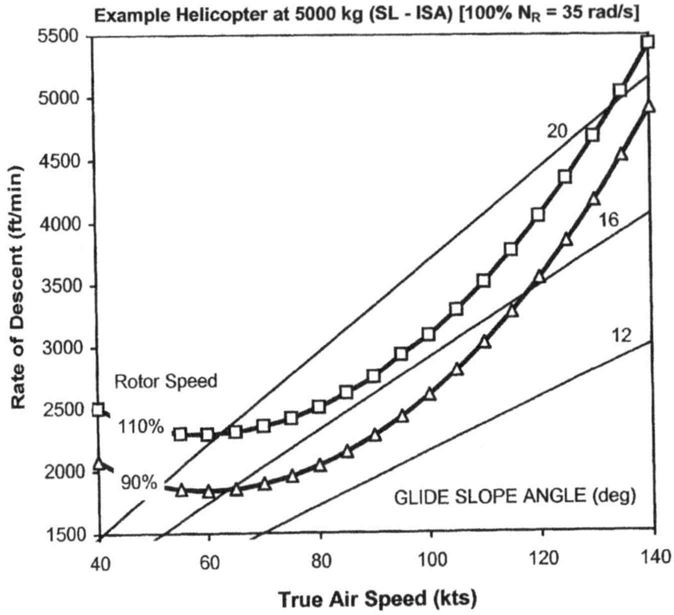

Fig. 2.32 Autorotative performance – effect of rotor speed.

2.13.1 Speeds for maximum range and endurance

Since the variation of RoD in autorotation with airspeed resembles a power curve it is possible to use it to identify speeds for maximum range or endurance. In an autorotative descent the helicopter will obviously remain aloft longer if it is flown at the speed that results in the lowest rate of descent. Therefore V_{ME} is the 'bucket speed' obtained from the performance curve, see Fig. 2.33. An autorotative performance curve typically presents RoD versus *indicated* airspeed. If the pitch attitude of the aircraft during the descent is not excessive and the pitot-static system is free of significant errors it can be assumed that the curve presents two velocities along orthogonal axes (forward advance and rate of descent). Consequently the slope of a line from the origin that intersects the performance curve represents the angle of glide and therefore the speed that gives the minimum angle (that is tangential to the performance curve) will be V_{MR}, the 'range speed' in autorotation, see Fig. 2.33.

2.14 FLARE CHARACTERISTICS AND ENGINE-OFF LANDINGS

By examining the autorotative performance of a typical helicopter and noting the conditions leading to entry into vortex ring it has been possible to show that engine-off flight can be conducted safely at all airspeeds provided the rotor speed required falls within the power-off range for the rotor. In fact stable autorotations are possible even in vertical flight but these often result in high rates of descent. High RoDs do not present a problem provided they allow sufficient time for the pilot to select a suitable landing site and compose himself for the ensuing engine-off landing. In theory,

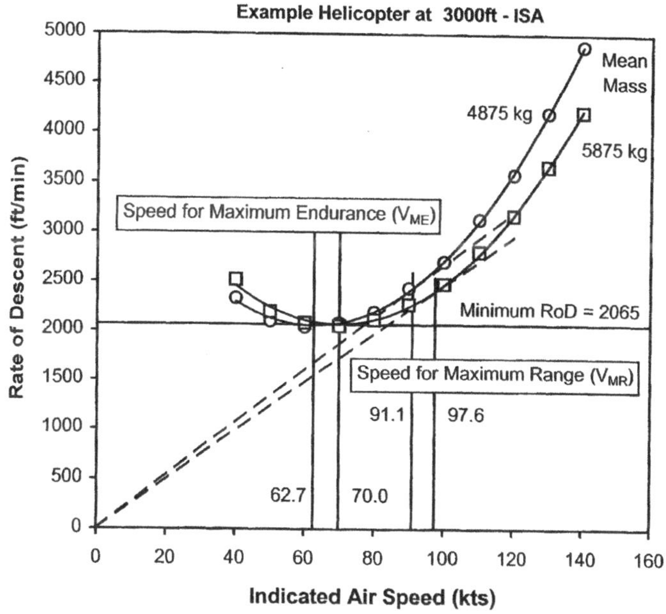

Fig. 2.33 Autorotative performance speeds.

if the problems of vortex ring are ignored, it is possible to autorotate vertically and by applying increased collective pitch, at the correct moment, arrest the sink rate to land with zero ground speed. However, from the typical autorotative performance detailed above it is more likely that the pilot will be descending at V_{ME}. Although he may have opted to accelerate to V_{MR} and/or droop the rotor, to increase his time aloft or cover ground more efficiently, he will inevitably reselect V_{ME} and raise the rotor speed before attempting a landing. The raised N_R stores more energy for use in the landing phase of the manoeuvre.

2.14.1 The flare, the check and the level manoeuvre

A successful touchdown occurs at the end of a series of manoeuvres designed to transfer the helicopter from a condition of moderate horizontal and vertical velocity to a condition of little or no velocity in either direction [2.6]. The following design features, along with an assessment of the acceptable level of damage, will affect the flare angle the pilot can employ and the aircraft attitude, sink rate and run-on speed allowed at ground contact:

- undercarriage type – wheeled or skidded;
- available stroke in undercarriage – main and tail/nose;
- crashworthiness of cockpit, fuselage floor or cabin and fuel system;
- tail boom strength;
- field of view;
- rotor inertia.

The idealized manoeuvre begins with a cyclic flare at constant collective pitch during which the increased rotor thrust (affected by any change in rotor speed) and aft disk tilt are used to decrease both the horizontal and vertical velocity components. The maximum pitch angle used by the pilot in the flare is often a matter of personal preference being affected by issues such as field of view. A maximum flare angle of 45° is used in the analysis given below as it gives a minimum theoretical result by ensuring that the sink rate is reduced by the greatest possible amount. At the end of this flare the aircraft should be close to the ground with its vertical velocity equal to zero (or at least below the design sink rate of the undercarriage) and with its horizontal velocity corresponding to autorotation at the angle of attack to which the rotor has been pitched. At this stage the pitch attitude will be typically much greater than that necessary to avoid a tail strike and considerable stored energy will remain in the rotor. The pilot will therefore rotate the helicopter nose-down towards a level attitude and use collective pitch to cushion the landing. Although in the idealized manoeuvre the pilot can use longitudinal cyclic and collective pitch in combination to level the aircraft whilst maintaining hovering thrust, at the expense of rotor speed, a different strategy is often adopted. The pilot will perform cyclic flare as previously described and will maintain it as long as possible. Eventually the flare effect will diminish to the point that the RoD reaches a minimum and starts to increase again. At this instant a rapid collective pull, or 'check', is made using some of the rotor energy to further reduce the horizontal and vertical velocity components. Shortly afterwards the helicopter is rotated nose-down and more collective pitch is used to minimize the sink rate as the aircraft is run on to the ground. This technique is summarized using the maxim '*flare-check-level*'.

2.14.2 Estimating the minimum touchdown speed

It is possible to estimate the theoretical minimum touchdown speed resulting from an idealized engine-off landing using some of the concepts introduced above [2.6 and 2.27]. The flare angle that can be employed will be affected by the maximum pitch rate that can be generated and the time the pilot has to level the aircraft. During the level manoeuvre, because the helicopter is still moving forward, it can be assumed that OGE hover power and thrust must be maintained even though the rotor speed is rapidly decaying. Since useable energy is stored in the rotor in the form of kinetic energy of rotation if the pilot could allow the rotor speed to decay to zero at touchdown then he could make use of the theoretical maximum, given by:

$$\text{maximum energy available} = \frac{1}{2} I\Omega_{\text{nom}}^2$$

Unfortunately this is not possible since as the rotor speed decays the angle of attack must be increased if the thrust produced is to remain constant. Consequently the thrust coefficient, or rotor blade lift coefficient, will rise, as N_R reduces, up to some maximum value beyond which rotor performance is not sustainable. Hence:

$$\text{useable energy} = \frac{1}{2} I(\Omega_{\text{nom}}^2 - \Omega_{\text{min}}^2) = \frac{1}{2} I\Omega_{\text{nom}}^2 \left(1 - \frac{\Omega_{\text{min}}^2}{\Omega_{\text{nom}}^2}\right)$$

Since:

$$\Omega_{min} = \Omega_{nom}\sqrt{\frac{\tilde{C}_{L\,nom}}{\tilde{C}_{L\,max}}}$$

then:

$$\text{useable energy} = \frac{1}{2}I\Omega_{nom}^2\left(1 - \frac{\tilde{C}_{L\,nom}}{\tilde{C}_{L\,max}}\right)$$

This energy is used to provide hover power and therefore the available time during which the pilot must level the helicopter will be given by:

$$\text{time available to level aircraft} = \Delta t = \frac{I\Omega_{nom}^2}{2P_{OGE}}\left(1 - \frac{\tilde{C}_{L\,nom}}{\tilde{C}_{L\,max}}\right)$$

Thus:

$$\theta_{max} = \alpha_{TPP_{max}} = q_{max}\Delta t = \frac{\gamma\Omega}{16}\Delta B_1 \Delta t$$

Therefore θ_{max} is the maximum fuselage pitch angle change that can be generated in the time available. If this calculated angle exceeds 45° then, as suggested above, 45° is used since this angle gives the theoretical maximum reduction in both horizontal and vertical velocities. (In many instances the pilot will be very reluctant to flare by more than 25% being concerned about either a tail strike or loss of forward vision.)

Having determined the flare angle, the forward speed achieved before the pilot begins to level the aircraft is then related to an equivalent stable autorotative condition. This simple approximation assumes that the flare reduces the sink rate to approximately zero. From Fig. 2.34, which shows the glide geometry, it can be seen that just before the helicopter is rotated nose-down the forward advance will equal the value of V_g, or $V_f \cos\theta_{max}$, obtained from steady autorotative data.

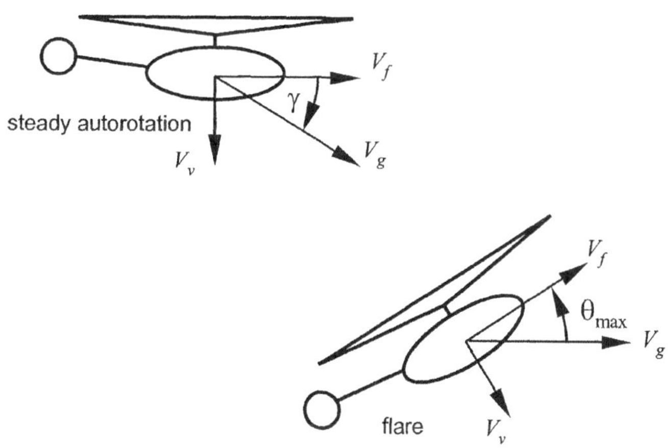

Fig. 2.34 EOL glide geometry.

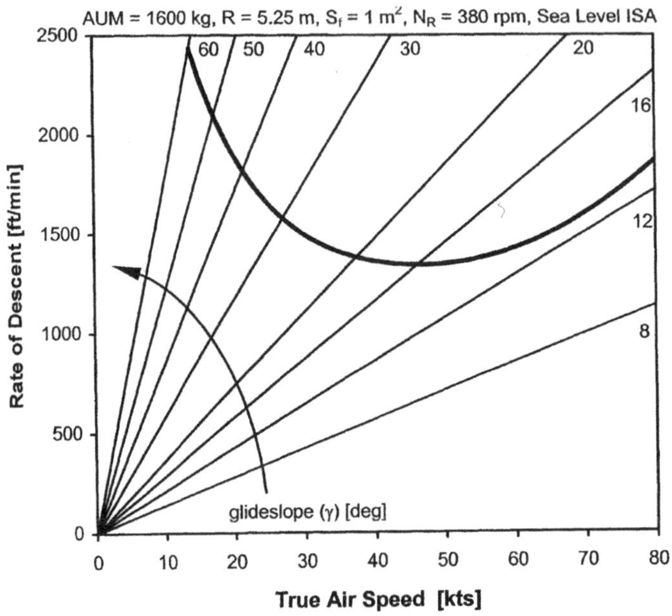

Fig. 2.35 Autorotative performance.

Figure 2.35 shows the steady autorotative performance for an example helicopter. At an AUM of 1600 kg this helicopter will be operating at a mean lift coefficient of 0.133, given by:

$$\tilde{C}_{\mathrm{L\,nom}} = \frac{2mg}{\rho\Omega^2_{\mathrm{nom}} bcR^3}$$

Under sea level, ISA conditions the rotor speed at 75% rotor radius equates to 0.45M and the hover power is 196.6 kW. If the rotor has an inertia of 1000 kgm^2 and is fitted with a NACA 0012 aerofoil then $C_{\mathrm{L\,max}}$ is approximately 1.0 and Δt equals 3.5 s. Such a large value of Δt gives the pilot plenty of time to level the aircraft. Therefore the only concern for the pilot when selecting a flare angle is ensuring that a reasonably low run-on speed is achieved without undue risk to the tail boom and empennage.

Just before the pilot begins to level the aircraft vertical forces are assumed to be in equilibrium. As the aircraft is levelled a steadily diminishing rearwards component of thrust continues to provide a decelerative force. The average deceleration is therefore give by:

$$a = \frac{F}{m} = \frac{T}{2m}\sin\theta_{\mathrm{max}}$$

But:

$$T = \frac{mg}{\cos\theta_{\mathrm{max}}}$$

Therefore, assuming a constant deceleration, the velocity reduction during the level manoeuvre is determined using:

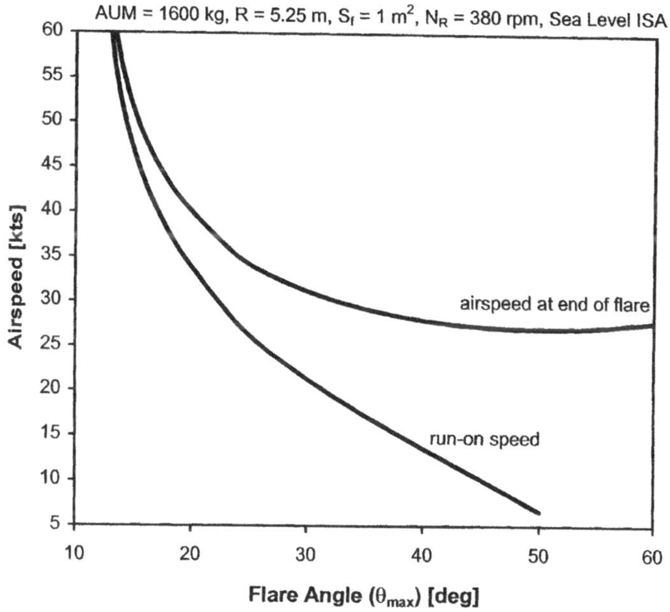

Fig. 2.36 Effect of flare angle on run-on speed.

$$\Delta v = a\Delta t = \frac{g}{2}\tan\theta_{max}\Delta t$$

This relationship indicates that if the pilot is able to select a high flare angle he can significantly reduce the run on speed. Figure 2.36 shows the reduction in speed that can be achieved at various flare angles.

In practice the method described here gives reasonable approximations to the minimum possible run-on speed although it ignores important actions the pilot will typically make in a real EOL. First it is common practice to allow N_R to increase during the flare, this capitalizes on the natural tendency of the rotor to accelerate and, at the expense of a slightly higher closure rate, puts energy into the rotor for use during the 'check'. During the check, a manoeuvre not included in the previous analysis, the collective pitch is rapidly increased causing a temporary rise in thrust and further reducing the glide velocity below that determined by the pitch attitude. Note also that some helicopters will accelerate if levelled completely.

Even in the most advantageous circumstances with tight control of airspeed and rotor RPM maintained during a period of stable autorotative flight, the pilot must still assess the optimum time to commence the flare manoeuvre with a high degree of accuracy. If the flare is commenced too high or too low the subsequent landing may be more hazardous than necessary! Equally if the airspeed on entry to the flare is excessive the flare will be less effective in reducing the horizontal velocity component resulting in a high run-on speed. Alternatively if the airspeed on initiation of the flare is too low insufficient rotor energy may be available to cushion the landing. Consequently within the qualification process for new single-engined helicopters there will be an EOL testing phase during which flares at various combinations of airspeed and height will be performed. Such tests will be conducted in order to determine which combination minimizes the difficulty of the complete engine-off landing.

2.15 THE AVOID CURVE

The foregoing paragraphs have discussed the factors affecting the rate of rotor speed decay following an engine failure and noted how timely action by the pilot can establish the helicopter in a stable autorotation. The flare manoeuvre has been described and using a relatively simplistic analysis of the trajectory of the helicopter it has been possible to determine the optimum height/speed combination at which to execute this manoeuvre. It is now necessary to consider the effect of airspeed on the risk associated with operating a single-engined helicopter close to the ground.

Given an appropriate combination of airspeed and height (AGL) the pilot will be able to manage the balance of kinetic energy (stored in both the rotor and the fuselage) and the potential energy to arrive at a *gate* condition from which a safe EOL is assured. However at low speed the pilot may have insufficient height, or potential energy, available to accelerate the aircraft to the gate speed. Alternatively when operating at low speed very close to the ground, say in a high hover, there may be insufficient kinetic energy available in the rotor to reduce the rate of descent to a survivable value. Equally a transit at high speed and low height may not give the pilot sufficient time to react to the engine failure cues and initiate a *zoom* climb to the EOL gate condition.

Consequently for most helicopters there exists a set of height–airspeed combinations which should be 'avoided' to prevent hazarding the aircraft in the event of total power loss. These critical parameters are typically presented graphically in an *avoid curve* or *height–velocity diagram*. The similar shape of most avoid curves makes it possible to produce a generalized non-dimensional curve, see Fig. 2.37, that can then be applied to a variety of rotorcraft and also scaled to take account of density altitude and gross weight [2.28].

2.15.1 Lower hover height

First estimation is made of the maximum height from which a vertical engine-off landing can be performed without damage to the undercarriage. This lower hover height, h_{LO}, is calculated by assuming that the whole manoeuvre is conducted at a RoD equal to the limiting sink rate for the landing gear, V_{LG}, and lasts as long as the energy stored in the rotor can be used to provide hover power. Since the rotor RPM will decay, as energy is bled away, the blade loading coefficient will rise even though the thrust produced will remain approximately constant. Thus the time available for the engine-off (which is analogous to the rotor decay time mentioned earlier) can be related to the ratio of the blade loading at the instant of engine failure to some nominal maximum value (Pegg [2.28] uses 0.2). Hence:

$$h_{LO} = \frac{V_{LG} I \Omega_{nom}^2 \left[1 - \sqrt{\dfrac{1}{0.2} \left(\dfrac{mg}{\rho A_b (R\Omega_{nom})^2} \right)} \right]}{2P_{IGE}}$$

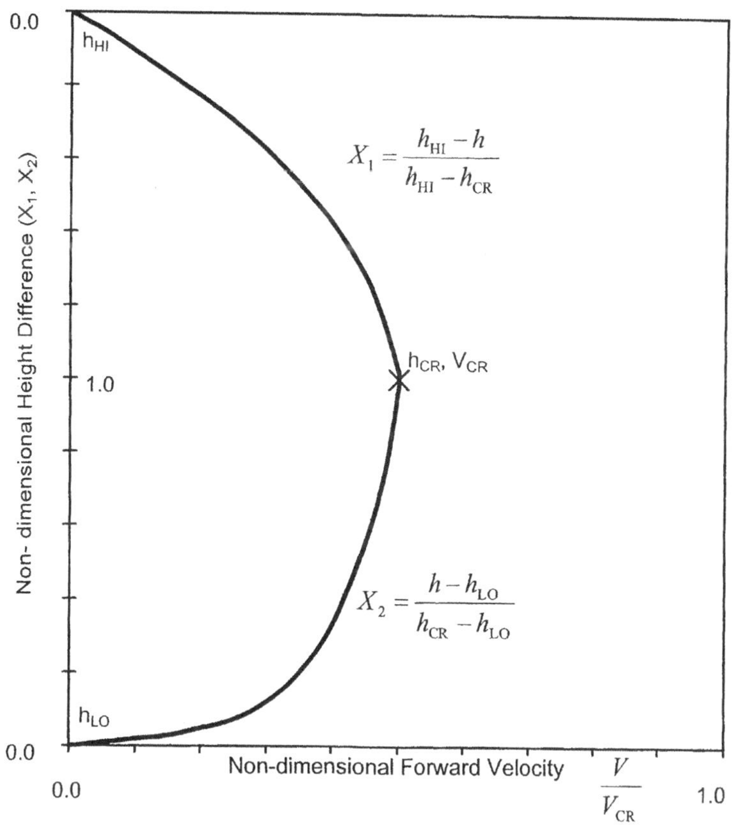

Fig. 2.37 Generalized avoid curve (adapted from [2.29]).

Consider a light helicopter fitted with an undercarriage stressed to accept a sink rate of 2 m/s and operating in SL-ISA conditions. Thus if $I = 1000$ kgm^2, $R = 5.25$ m, $A_b = 4.4$ m^2, $m = 1800$ kg, $\Omega_{nom} = 39.8$ rad/s, and $P_{IGE} = 191.7$ kW, then $h_{LO} = 9.6$ ft. It is of interest to compare this crude estimate with the results from a more sophisticated model. Using the equations developed above it is possible to predict the thrust developed and the torque required by the main rotor as a function of rotor speed and collective pitch. This data can then be used to predict the rate of descent, and subsequent height loss, following a power failure in the hover. Figure 2.38 shows the situation if the pilot elects to maintain the collective pitch fixed at its hover value.

Approximately 1.6 s after total power loss the helicopter is descending vertically at 2 m/s and has fallen 5 ft. Thus if the engine(s) were to fail at some higher hover height and the pilot failed to react, damage to the undercarriage would occur. In reality a higher value of h_{LO} is allowable since the pilot can be expected to utilize the energy stored in the rotor to cushion the landing. This situation is shown in Fig. 2.39. Here the pilot allows the RoD to build up to the onset of incipient vortex ring ($0.3v_{ih}$) before rapidly raising the collective lever to reduce the vertical velocity. The height lost before the RoD reduces back to 2 m/s is around 12 ft. Thus if the pilot increases collective pitch rapidly at just the right time he can suffer a total power failure whilst hovering at about 10 ft without damage to the aircraft.

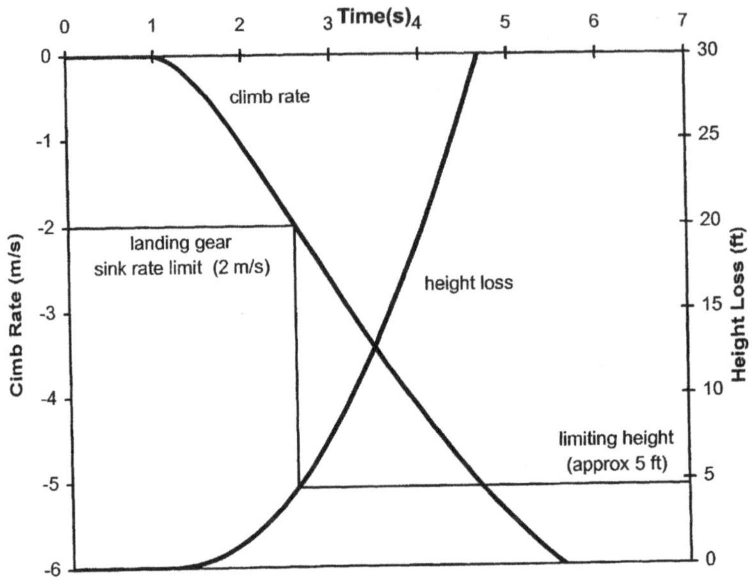

Fig. 2.38 Low hover EOL – no pilot action.

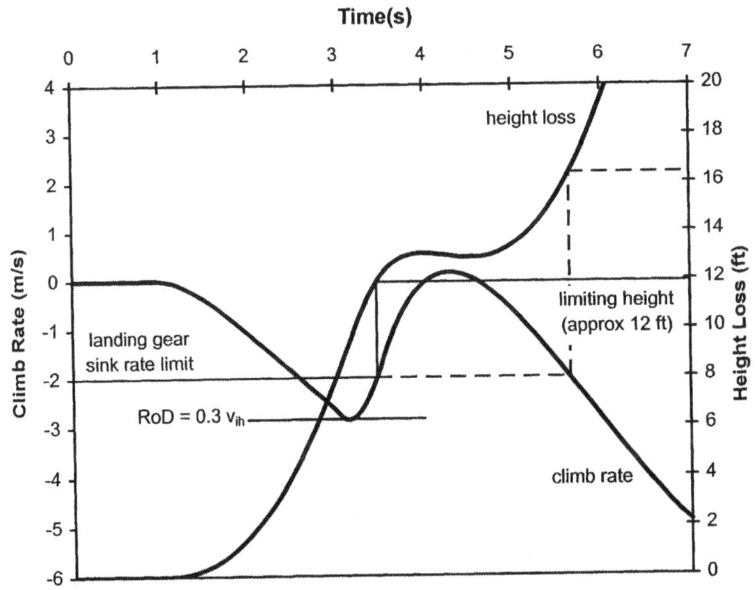

Fig. 2.39 Low hover EOL – pilot action.

2.15.2 The knee point

The knee point (h_{CR}, V_{CR}) is determined next by calculating the minimum power speed for the particular all-up-mass, rotor speed and atmospheric condition of interest. This point represents the minimum speed below which there exists a range of heights (AGL) that should be avoided. The minimum, or *critical* height, is not scaled and is taken as a fixed value regardless of aircraft configuration and only alters if the pilot delay time is changed. In FARs a one-second delay is assumed for height–velocity points above the knee and h_{CR} is taken as 95 feet. US Military Standards on the other hand assume a two-second delay and therefore h_{CR} is raised to 120 feet. Pegg [2.28] relates empirically the critical speed (V_{CR}) to the minimum power speed and the mean aerofoil lift coefficient (C_L/s) at the same speed. Although derived from a small data set this relationship does include the key variables. In most cases following an engine failure the pilot will attempt to accelerate to an EOL speed of around V_{MP} and the prevailing value of C_L/s will indicate the margin, below rotor blade stall, in which the pilot can operate. Note that C_L/s is given by:

$$\frac{C_L}{s} = \frac{2}{\mu^2}\frac{C_T}{s} = 2\left(\frac{V_T}{V_{MP}}\right)^2 \frac{mg}{\rho A_b V_T^2} = \frac{2mg}{\rho A_b V_{MP}^2}$$

For the example light helicopter ($A_b = 4.4\ m^2$, $m = 1800$ kg) the minimum power speed is 55 KTAS and thus $C_L/s = 8.2$. Consequently V_{CR} equals approximately 30 KTAS.

2.15.3 High hover height

If the pilot suffers a total power failure when hovering some distance from the ground his subsequent actions are somewhat different than those described above. He cannot simply bleed energy from the rotor in an attempt to arrest the rate of descent since the hover height is too high and he would achieve minimum sustainable rotor speed long before reaching the surface. Instead the pilot will initially dive the aircraft, increasing the kinetic energy of the vehicle at the expense of its potential energy, whilst attempting to maintain rotor speed within the power-off limits. At some altitude much lower than the original hover height he will conduct a cyclic flare EOL at a speed close to V_{MP}. Since V_{CR} is dependent on both aircraft and rotor performance (V_{MP} and C_L/s) it is possible to relate the high hover height (h_{HI}) to this critical speed [2.28 and 2.29]. For the example light helicopter, $h_{HI} = 276$ ft.

2.15.4 Scaled avoid curve

Using the four key data values: h_{LO}, h_{CR}, V_{CR} and h_{HI}, a scaled avoid curve can be constructed. Figure 2.40 shows such a curve for the light helicopter example and compares it with published data for an actual helicopter of broadly equivalent size and mass. Note that the avoid area expands somewhat if the AUM is raised from 1800 kg to 1900 kg (the limit of applicability of the published data) and a minimum power speed of 65 KTAS is used. Alternatively if the original V_{MP} is used with the higher AUM and a longer delay time is assumed then although modified charts are required to determine V_{CR} and h_{HI} [2.6] a closer approximation can be achieved, see also Fig. 2.40.

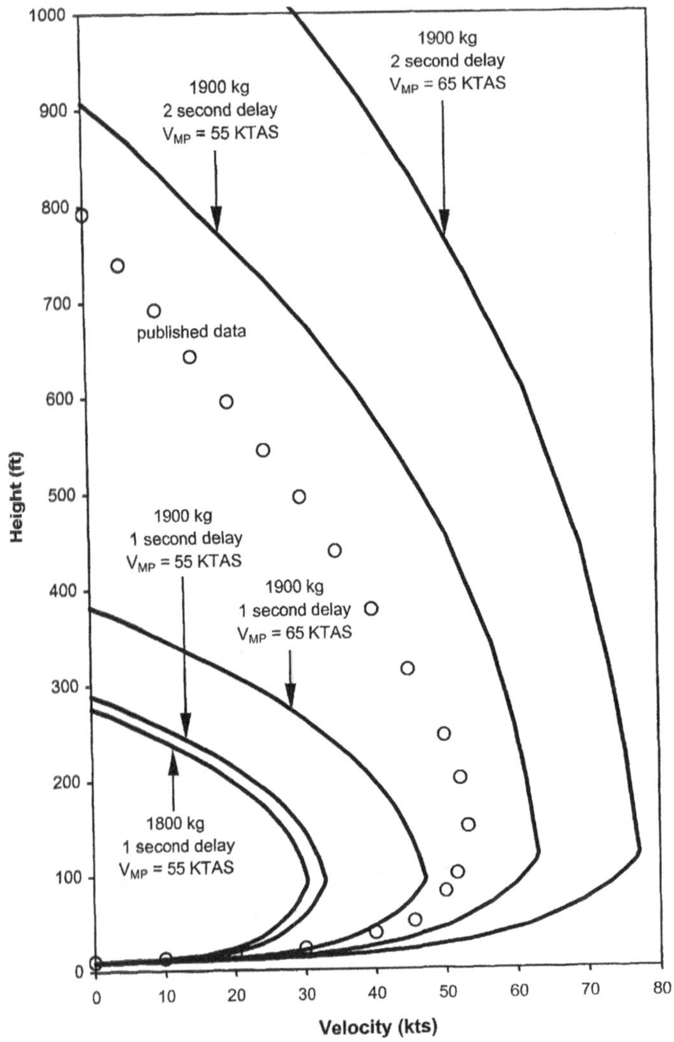

Fig. 2.40 Scaled avoid curves.

Chapter 3
Performance Testing

3.1 INTRODUCTION

Methods for measuring the steady-state performance of gas turbine-engined helicopters commonly use non-dimensional parameters [3.1]. These parameters consist of groups of relevant dimensional quantities arranged by means of dimensional analysis. Performance flight test then involves determining the relationship between pairs of non-dimensional parameters whilst the others are held constant. We shall see that this *experimental method* of testing reduces any limitations in the applicability of forward speed, rate of climb, power and fuel flow test data. This is because data converted into a non-dimensional form can be used to produce information relevant to atmospheric conditions and aircraft masses different from those actually tested. Consequently, with few exceptions, a relatively small number of tests at carefully chosen test sites can produce information relevant to much of the helicopter's flight envelope. The experimental method does however have some disadvantages in terms of planning and in the choice of non-dimensional grouping:

- Although the test method can yield large quantities of relevant data from a few test points [3.2] it requires detailed and careful pre-flight planning if the full utility of the method is to be achieved;
- The method can appear vague with alternative groupings possible;
- It is possible to require flight conditions, in terms of the non-dimensional groupings, that are outside the limitations of the aircraft;
- Performance limiting factors that depend on actual conditions may not be fully replicated although matching non-dimensional values have been targeted successfully.

3.2 NON-DIMENSIONAL PARAMETERS

The useful performance of any helicopter depends on the amount by which the power available exceeds that required. Helicopter performance is therefore measured in terms of the power required to maintain steady flight for various atmospheric conditions, over a range of weights and external configurations. Engine fuel flow data is also gathered to determine the helicopter's range performance. It has been shown that the factors affecting the performance of a family of geometrically similar helicopters with similar rotor blade profiles are:

- engine power;
- helicopter weight;

- rotor speed;
- forward speed;
- rate of climb;
- ambient atmospheric conditions.

Although dimensional analysis is used to show how these quantities can be related the basic non-dimensional groupings are rarely used. Since usually the performance of a single model of helicopter is considered at any given time, the linear dimensions of rotor radius, chord and disk area are omitted. For a similar reason ambient pressure, temperature and density are expressed as ratios of the standard sea-level conditions (δ, θ, σ). Likewise, rotor speed (Ω) is expressed as a percentage of some reference or standard value (Ω_0). This, of course, means that the groups have become *dimensional* although they still contain the required information. These modified groups are often termed '*normalized*', '*referred*' or '*reduced*'.

3.2.1 Derivation of the referred parameter relationships

Consider now the most general case of a helicopter in climbing flight at low level. As indicated earlier the power required to maintain a steady flight condition will depend on:

• weight (W)	[MLT^{-2}]
• forward speed (V)	[LT^{-1}]
• rate of climb (V_c)	[LT^{-1}]
• height (above ground level) (Z)	[L]
• local speed of sound (a)	[LT^{-1}]
• ambient density (ρ)	[ML^{-3}]
• rotor speed (Ω)	[T^{-1}].

Note that local speed of sound has been included as a means of accounting for compressibility effects on the lift and drag characteristics of the rotor blade. So:

$$P = f(W, V, V_c, Z, a, \rho, \Omega)$$

Using dimensional analysis yields:

$$P = f\left(\frac{W\Omega^2}{\rho a^4}, \frac{V}{a}, \frac{V_c}{a}, \frac{Z\Omega}{a}, \frac{\rho a^5}{\Omega^2}\right)$$

This can be rewritten as:

$$\frac{P\omega^2}{\sigma a^5} = f\left(\frac{W\omega^2}{\sigma a^4}, \frac{V}{a}, \frac{V_c}{a}, \frac{Z\omega}{a}\right)$$

where $\omega = \Omega/\Omega_0$ and $\sigma = \rho/\rho_0$.

Noting that a is a function of ambient temperature:

$$\frac{P\omega^2}{\sigma\theta^2\sqrt{\theta}} = f\left(\frac{W\omega^2}{\sigma\theta^2}, \frac{V}{\sqrt{\theta}}, \frac{V_c}{\sqrt{\theta}}, \frac{Z\omega}{\sqrt{\theta}}\right)$$

where $\theta = T/T_0$.

Reorganizing and collecting like terms produces the 'W/σ' *referred power relationship*:

$$\frac{P}{\sigma\omega^3} = f\left(\frac{W}{\sigma\omega^2}, \frac{V}{\omega}, \frac{V_c}{\omega}, Z, \frac{\omega}{\sqrt{\theta}}\right)$$

Since measuring air density is difficult, an alternative grouping can be obtained by replacing σ with δ/θ:

$$\frac{P}{\delta\sqrt{\theta}} = f\left(\frac{W}{\delta}, \frac{V}{\omega}, \frac{V_c}{\omega}, Z, \frac{\omega}{\sqrt{\theta}}\right)$$

This is called the 'W/δ' *referred power relationship* which, although easier to use since it lacks air density, cannot be used for rotorcraft with fixed rotor speed. Note that in both groupings the forward speed and rate of climb have been expressed as advance ratios.

3.2.2 Control of the referred groups

The use of the experimental method of data reduction depends on the ability to maintain at least one of the groups constant while varying the others, for example, if the relationship between $P/\sigma\omega^3$ and V/ω is to be investigated whilst maintaining $W/\sigma\omega^2$ and $\omega/\sqrt{\theta}$ constant. In any of the referred groups there is usually at least one parameter which is under our control and which can therefore be used to control the whole group. Table 3.1 shows each group in its *referred* or *normalized* form and the appropriate controlling parameter(s).

Table 3.1 Control of the referred groups.

Group	Referred form	Controlling parameter(s)
Forward speed	$\frac{V}{\omega}$	V or Ω
Rotor speed	$\frac{\omega}{\sqrt{\theta}}$	Ω or H_P
Weight	$\frac{W}{\delta}$	W or H_P
Weight	$\frac{W}{\sigma\omega^2}$	W, H_D or Ω
Vertical speed	$\frac{V_c}{\omega}$	V_c or Ω
Hover height	Z	Z

3.3 PERFORMANCE TEST PLANNING

A typical requirement for a test team is to determine the performance of a rotorcraft under a range of environmental conditions. These are usually quoted as an altitude (X ft Hp) with a given standard atmosphere pertaining (ISA $\pm Y$°C). The basic function of any performance test plan is therefore to establish sets of test points at the same referred weight (W/δ or $W/\sigma\omega^2$) and/or the same referred rotor speed ($\omega/\sqrt{\theta}$). Assuming that the take-off weight is known and that the helicopter is fitted with accurate fuel-gone indications it is possible to draw up a series of tables or charts to enable in-flight adjustment of altitude to maintain the referred weight constant as fuel is burnt. In addition, if the W/δ method is being used and an accurate rotor speed indicator is fitted, it is also possible to tabulate the rotor speed adjustments required to maintain $\omega/\sqrt{\theta}$ constant as the outside air temperature changes.

3.3.1 The W/δ method

Suppose that a test flight is to be conducted using an aircraft with a take-off weight of W_{TO} and that data is required at a referred weight (W/δ) of W_{REF}. At any point in the sortie if the fuel burn is W_{FUEL} then the relative pressure required for the desired W/δ is given by:

$$\delta = \frac{W_{TO} - W_{FUEL}}{W_{REF}}$$

The pressure altitude (in feet) associated with this relative pressure can be found using an inverse form of the ISA atmospheric equation [3.3]:

$$H_P = \frac{1 - \delta^{1/5.2559}}{6.8756 \times 10^{-6}}$$

Figure 3.1 shows an example of a planning chart based on these equations. Determining the effect of outside air temperature on the rotor speed required for a given referred rotor speed can be addressed in a similar manner. Suppose the standard rotor speed is Ω_0, the referred rotor speed required ($\omega/\sqrt{\theta}$) is ω_{REF} and the outside air temperature at the test altitude is T °C, then the relative rotor speed (ω) required is given by:

$$\omega = \omega_{REF}\sqrt{\frac{T + 273.15}{288.15}}$$

The actual rotor speed associated with this relative RRPM is obtained from:

$$\Omega = \omega \times \Omega_0$$

Figure 3.2 shows an example of a planning chart based on these equations. Note that the maximum and minimum power-on rotor speed limits might prevent the establishment of referred rotorspeeds significantly different from that depicted.

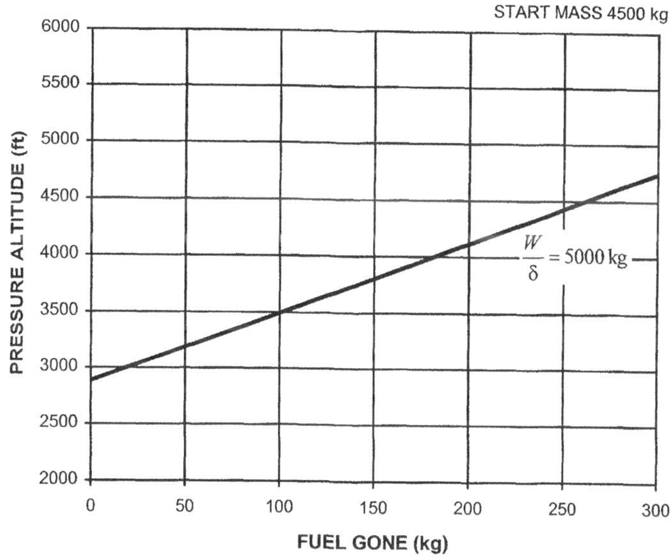

Fig. 3.1 Planning chart – W/δ method.

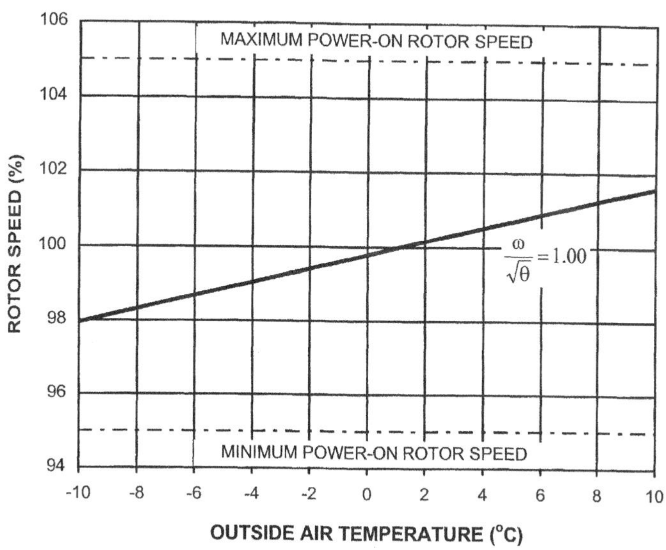

Fig. 3.2 Planning chart – $\omega/\sqrt{\theta}$ method.

3.3.2 The $W/\sigma\omega^2$ method

The task of deciding the pressure altitude required to establish a desired referred weight is complicated in this case by the additional requirement to account for the prevailing air temperature profile. Since this profile can only be estimated prior to flight, any planning tool can only be used to predict the required altitude. A final check on the actual referred weight at the instant the test conditions are established will be required. Suppose that the test aircraft has a standard rotor speed of Ω_0 and that data is required at some other rotor speed (Ω). The relative rotor speed (ω) is therefore obtained from:

$$\omega = \frac{\Omega}{\Omega_0}$$

Now suppose that a test flight is to be conducted using a take-off weight of W_{TO} and that data is required at a referred weight ($W/\sigma\omega^2$) of W_{REF}. At any point in the sortie, if the fuel burn is W_{FUEL} then the relative density required for the desired $W/\sigma\omega^2$ is given by:

$$\sigma = \frac{(W_{TO} - W_{FUEL})}{W_{REF}\omega^2}$$

Given an estimate of the temperature profile it is possible to determine the pressure altitude at which the required relative density is likely to occur. Test planning using this concept is more readily achieved by graphical means. The relative density at any pressure altitude is obtained from standard atmospheric equations and compared with that required for the test point(s). The atmospheric equations are:

$$\delta = (1 - 6.8756 \times 10^{-6} H_p)^{5.2559}$$

$$\theta = \frac{T + 273.15}{288.15}$$

$$\sigma = \frac{\delta}{\theta}$$

Figure 3.3 shows how test requirements can be plotted for ease of pre-flight planning. The example shows the pressure altitude range for the desired referred weight $\pm 0.5\%$.

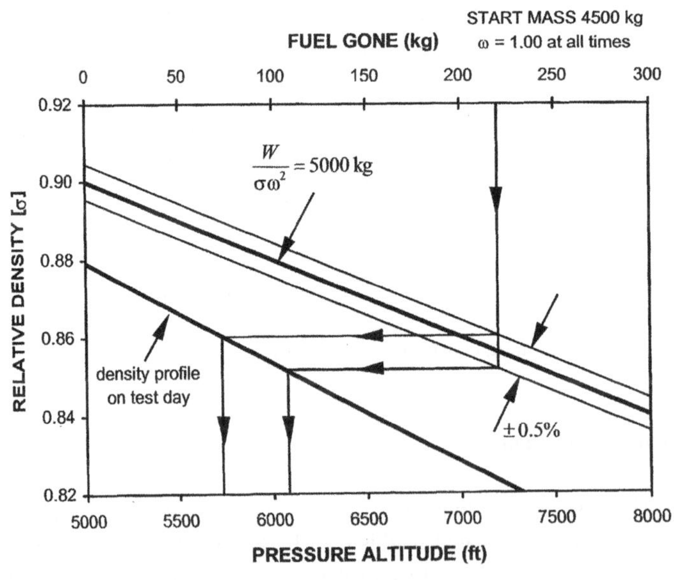

Fig. 3.3 Planning chart – $W/\sigma\omega^2$ method.

The style of planning chart shown in Fig. 3.3 has been successfully used to test helicopters with fixed rotor speed ($\Omega = \Omega_0$ at all times) as well as those that feature a ground adjustable datum rotor speed. It cannot be used, however, if the test requires maintenance of $\omega/\sqrt{\theta}$ as well as $W/\sigma\omega^2$ since in adjusting the rotor speed to set $\omega/\sqrt{\theta}$ the referred weight will be altered. The requirement to obtain performance information at constant $\omega/\sqrt{\theta}$ and $W/\sigma\omega^2$ occurs quite often, especially during the evaluation of *tip effects* (TE). In these situations an alternative approach is required. As mentioned earlier, if the test team can adjust the rotor speed in flight then the W/δ method may be used. Noting the relationship between W/δ and $W/\sigma\omega^2$ gives the alternative method:

$$\frac{W}{\sigma\omega^2} \times \left(\frac{\omega}{\sqrt{\theta}}\right)^2 = \frac{W}{\sigma\theta} = \frac{W}{\delta}$$

Therefore, the desired value of $W/\sigma\omega^2$ can be converted into a target value of W/δ. Provided this value can be obtained and the rotor speed can be adjusted to generate the desired $\omega/\sqrt{\theta}$ at the test altitude then the required value of $W/\sigma\omega^2$ can be tested indirectly.

3.3.3 Choice of suitable test site(s)

It is important to note that the range of available values of the referred parameters may be severely limited by the geographical position of the test site. When planning a test, therefore, the test team needs to determine the full range of all the relevant quantities (H_p, OAT, AUM, IAS, ROC/ROD, N_R) that may be experienced in operation. From this information, the required range of the referred parameters is derived. Having omitted any combinations of parameters that do not correspond with practical flight conditions, a test site must be chosen that provides the most suitable range of atmospheric conditions. Note that certain combinations of referred parameters need not be tested, such as a maximum value of W/δ and a minimum value of $\omega/\sqrt{\theta}$ as one represents the highest and the other the lowest operational altitude. It may be necessary to choose, or at least recommend, two or more sites so that the full range of test conditions is obtained.

3.4 LEVEL FLIGHT PERFORMANCE TESTING

Helicopters, along with most other air vehicles, spend a significant amount of time in a cruise configuration. Depending on the role of the rotorcraft, it may be required to cover the greatest distance or remain aloft for the maximum time on a given quantity of fuel. In order to determine the optimum airspeed for either requirement it is necessary to gather *level flight performance* test data. There are standard methods for measuring and presenting the performance of turbine-engined helicopters in terms of *referred parameters* [3.4].

Although it is common practice in a limited test programme to document simply the torque required, two major advantages accrue if a full set of engine performance data is gathered in the initial phase of a large test programme. Assumptions based on constant SFC can be relaxed; and complex fuel massflow measuring equipment need not be fitted on subsequent trials as the variation of fuel burn with power and engine life will have been determined.

It is important that the performance of a helicopter in level forward flight throughout its speed range is well documented. Power required and fuel usage information is needed to determine the helicopter's suitability for a given role, to check the performance against specification requirements, and to provide data for inclusion in the Operating Data Manual (ODM). Generally, a test establishment will carry out spot checks of the helicopter manufacturer's data but occasionally a full performance evaluation will be required covering a wide range of conditions of AUM, altitude, outside air temperature (OAT) and rotor RPM.

The helicopter operator will use the ODM performance data to determine optimum altitudes, airspeeds and rotor speeds to fly for maximum range and endurance under all likely conditions of AUM and OAT. In addition, the maximum and minimum level flight airspeeds available under given conditions and the achievable range or radius of action will be obtained from this published data.

3.4.1 Range performance

As will be seen, the ability of a helicopter to cover distance efficiently depends on the forward airspeed flown. This dependence is overcome by expressing the range performance of an aircraft in terms of its *specific air range* (SAR). Specific air range is defined as the distance covered (nautical miles) per unit of fuel and typical units are nm/kg or nm/lb. Now the distance covered at constant TAS (V) is given by Vt, where t is the time spent in the cruise condition. Likewise, the total fuel burn will equal Gt, where G is the fuel flow rate. So the definition of specific air range becomes:

$$\mathrm{SAR} = \frac{\textit{distance covered}}{\textit{mass of fuel used}} = \frac{Vt}{Gt} = \frac{V}{G}$$

Introducing the *specific fuel consumption* (SFC), s, which is the fuel flow rate per unit power (kg/kW h), gives:

$$\mathrm{SAR} = \frac{V}{G} = \frac{V}{sP}$$

The power required for level flight can be divided into four components: the induced power (P_i); the parasite power (P_{par}); the rotor profile power (P_{pr}), and the power required to drive the tail rotor, accessories and overcome transmission losses (P_{losses}). If the lost power is assumed to always equal a small and fixed percentage of the total power required, then:

$$\mathrm{SAR} \approx \frac{V}{s(P_i + P_{par} + P_{pr})}$$

An important factor in the analysis of range performance is the variation in SFC. However, if it is assumed to be constant over the range of interest then SAR can be related directly to the variation of power required with TAS. Specific air range will be a maximum when V/P is a maximum or when P/V is a minimum. Consequently, in the absence of any variation in SFC the speed for maximum range may be obtained from the power curve. It has already been shown that for a simple helicopter the power required by the main rotor can be expressed in the following form (Equation (2.17)):

$$C_P = C_{Pi} + C_{Ppar} + C_{Ppr} + C_{Pm} + C_{Ps}$$

Using typical empirical data, assuming small disk tilts and neglecting the effects of reversed flow, compressibility and blade stall:

$$C_P = 1.2\lambda C_T + \frac{1}{2}\mu^3\frac{f}{A} + \frac{1}{8}sC_D(1 + 4.3\mu^2)$$

or:

$$P = 1.2W\sqrt{-\frac{V^2}{2} + \sqrt{\frac{V^4}{4} + \left(\frac{W}{2\rho A}\right)^2}} + \frac{1}{2}\rho V^3 f + \frac{1}{8}\rho bcRV_T^3\left[1 + 4.3\left(\frac{V}{V_T}\right)^2\right]C_D$$

Thus, the range performance will depend in some manner on the weight of the helicopter, the rotor speed, the external configuration and the true airspeed.

3.4.1.1 *Factors affecting range performance*

Consider now the change in power required as weight is increased at constant true airspeed, V. The rotor thrust, T, must rise correspondingly and this will cause an increase in the inflow velocity, v_i. In addition, if the higher collective setting required at higher AUM leads to increases in drag due to Mach effects (reducing M_{crit}) then C_D will rise above its nominal value. Therefore the power required must increase with weight and consequently the SAR will reduce. So:

$$\text{SAR} \propto \frac{1}{W}$$

Alternatively, consider the variation of power required with true airspeed for a given weight. At a fixed weight the thrust required will be approximately constant assuming a modest parasite drag coefficient. Therefore the induced velocity will reduce as airspeed increases whereas the parasite power increases rapidly and the profile power increases but at a more gentle rate. Consequently the overall trend in the total power required is the familiar U-shaped curve. This curve can be used to determine the speed for best range (V_{MR}) since a line drawn from the origin to a point on the curve is proportional to P/V. Therefore the lowest P/V (highest SAR) will occur at the point where the line is tangential to the curve, see Fig. 3.4.

It is instructive to determine the factors that affect V_{MR}. Typically the maximum range speed will be in the moderate to high forward speed bracket so the high-speed approximation may be used:

$$v_i = \frac{v_{i_h}^2}{V} = \frac{W}{2\rho AV}$$

Thus the total power required can be written as:

$$P = 1.2\frac{W^2}{2\rho AV} + \frac{1}{2}\rho V^3 f + \frac{1}{8}\rho bcRV_T^3(1 + 4.3\mu^2)C_D$$

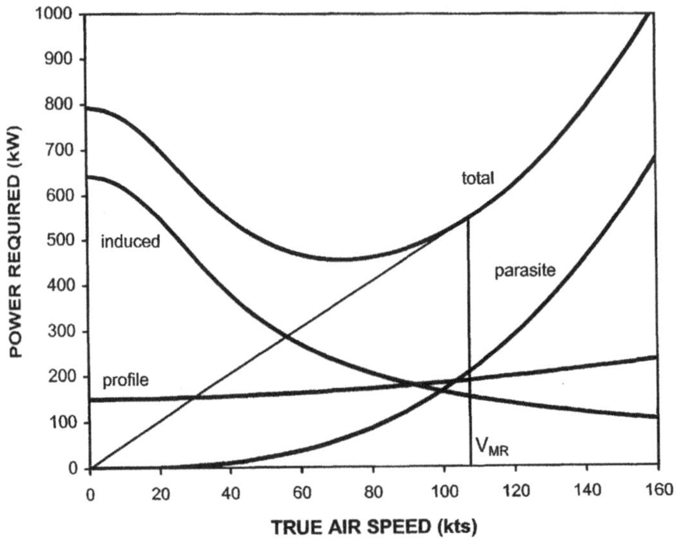

Fig. 3.4 Estimating V_{MR} from variations in power for level flight.

Now maximum SAR will occur at the speed at which P/V is a minimum, that is when:

$$\frac{d}{dV}\left(\frac{P}{V}\right)=0$$

If it is assumed that the profile drag is approximately constant with speed then:

$$\frac{d}{dV}\left(\frac{P}{V}\right)=\frac{0.6W^2}{\rho AV^2}+\frac{1}{2}\rho V^2 f=0$$

Then:

$$V_{MR}^4=\frac{1.2W^2}{\rho^2 fA}$$

$$V_{MR}\propto\sqrt{W}$$

So the speed for best range, assuming SFC and profile drag is approximately constant, will be proportional to $\sqrt{W}$.

3.4.1.2 *Flying at best range speed*

Figure 3.5 shows typical curves of referred power ($P/\sigma\omega^3$) against referred TAS (V/ω). Assuming constant specific fuel consumption and ignoring the effects of referred RRPM ($\omega/\sqrt{\theta}$), this figure can be used to determine the ranges of altitude, RRPM and TAS that will ensure maximum SAR at any weight. Since SFC is constant then SAR will be directly proportional to V/P.

If referred TAS is divided by referred power then a trend that indicates the variation of SAR with referred weight ($W/\sigma\omega^2$) can be produced, as shown in Fig. 3.6. From

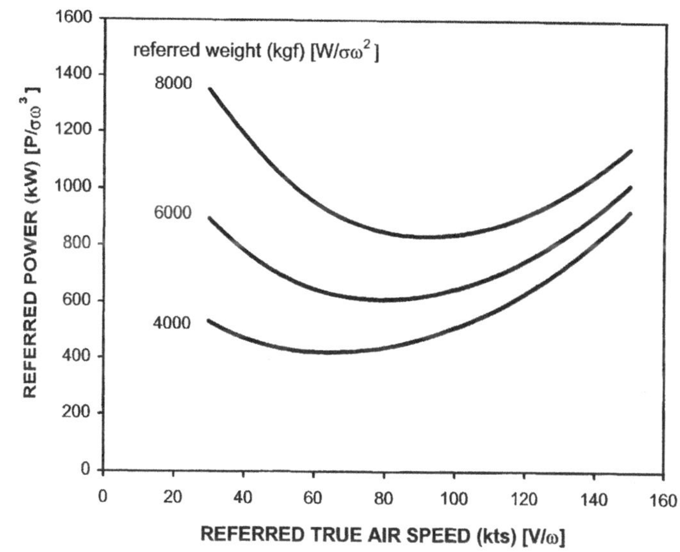

Fig. 3.5 Variation of referred power with referred weight and TAS.

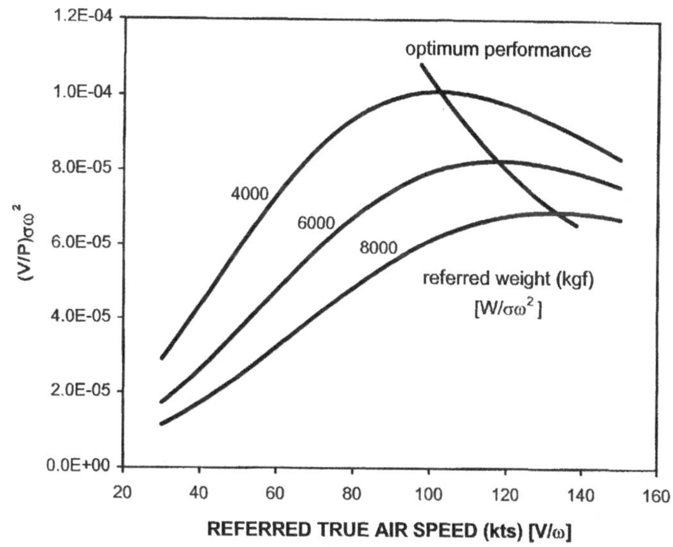

Fig. 3.6 Variation of referred SAR function with referred weight and TAS.

this figure it is possible to determine the optimum value of $V\sigma\omega^2/P$ and the referred TAS at which it occurs for each referred weight.

Suppose the flight conditions required to ensure maximum range at an actual aircraft mass of, say, 5500 kg are to be determined. The pilot can change airspeed, N_R and altitude (relative density) therefore the variation of V/P with $\sigma\omega^2$ must be evaluated. Using the optimum $V\sigma\omega^2/P$ and $W/\sigma\omega^2$ data obtained from Fig. 3.6 it is

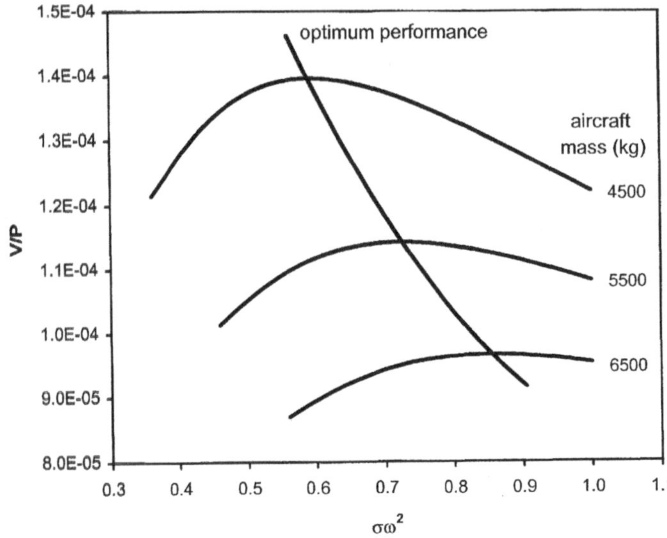

Fig. 3.7 Variation of SAR function with referred weight and TAS.

possible to plot the variation of V/P with $\sigma\omega^2$ for a particular mass, see Fig. 3.7. This figure shows that, for a given weight, there is a unique value of $\sigma\omega^2$ which will give the highest V/P and the best range performance. In other words, for each actual weight there will be a unique referred weight ($W/\sigma\omega^2$) or C_T which will give the best range performance. The practical outcome of this situation is that to ensure maximum SAR the pilot must change altitude as fuel is burnt and adjust rotor speed to suit the prevailing ambient conditions.

Consider the choice of range TAS. From Fig. 3.6 the variation of V/ω, for best range, with $W/\sigma\omega^2$ can be determined. This relationship can be used to select the TAS for each actual weight as density altitude is increased using practical rotor speeds. In this way even though the range performance may be sub-optimal due to power-on RRPM considerations the most appropriate range speed is being used. However, if airspeed limitations, perhaps due to airframe or rotorhead stresses, are introduced it can be seen that the range performance may be further reduced by the inability to fly at a high enough TAS, see Fig. 3.8.

The combined effect on V/P (or SAR) of imposing power-on RRPM and TAS limits is also shown in Fig. 3.9. It can be seen from this figure that the optimum range performance is only achieved at low weight and high density altitude, or at high weight and low density altitude. Nonetheless, even when flying optimally at high weight the actual SAR will be lower than that recorded at lower weights.

3.4.2 Endurance performance

The maximum endurance is obtained at a flight condition that results in the lowest fuel flow rate or minimum fuel consumption. The basic theory is simple: for constant SFC the fuel flow rate (G) is directly proportional to power required and therefore the

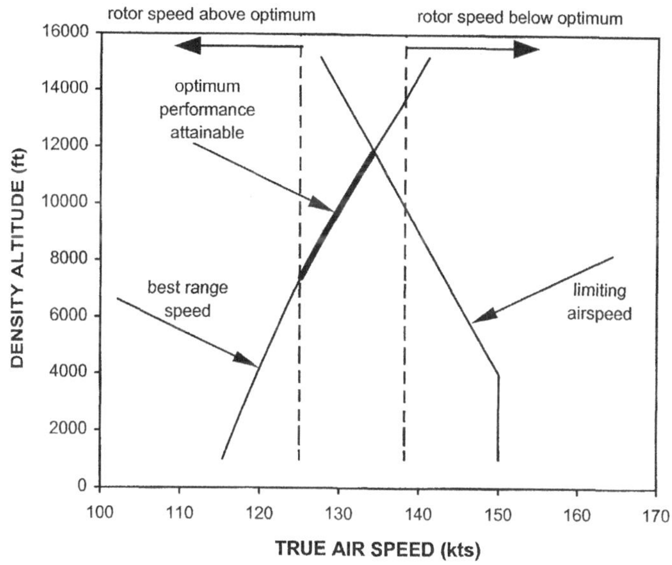

Fig. 3.8 Flight conditions required for optimum range performance (5500 kg).

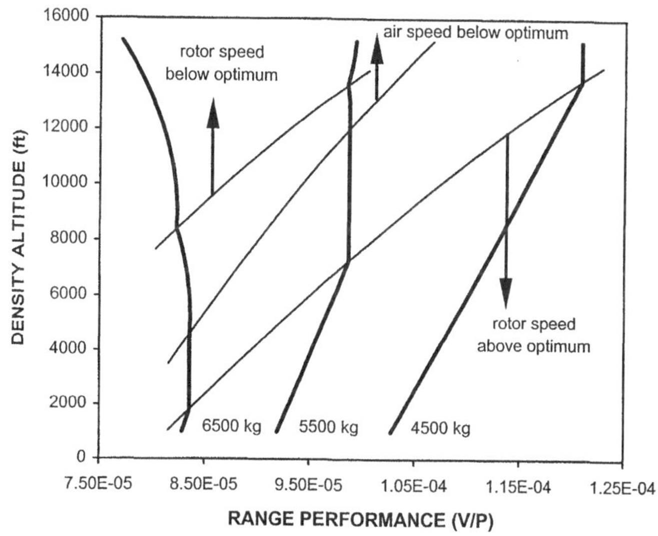

Fig. 3.9 Achievable range performance.

true airspeed for maximum endurance (V_{ME}) occurs at the speed for minimum power, see Fig. 3.10.

3.4.3 Effects of variable specific fuel consumption

The foregoing discussion has assumed constant SFC but a real engine will almost certainly not conform to this. Specific fuel consumption will almost certainly vary

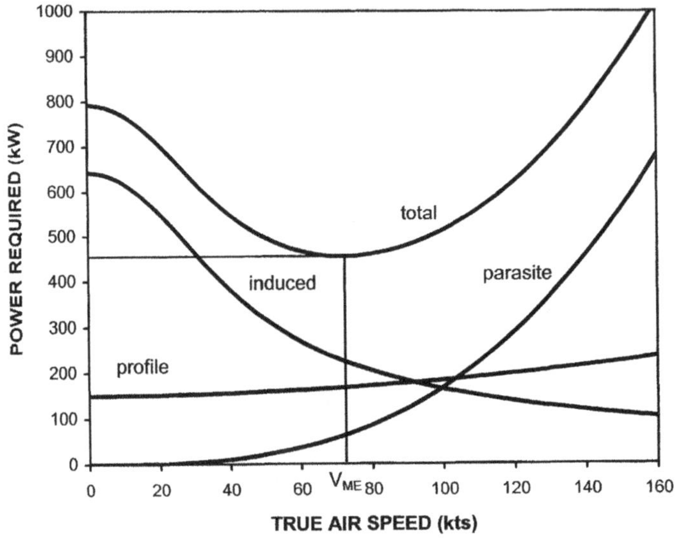

Fig. 3.10 Estimating V_{ME} from variations in power for level flight.

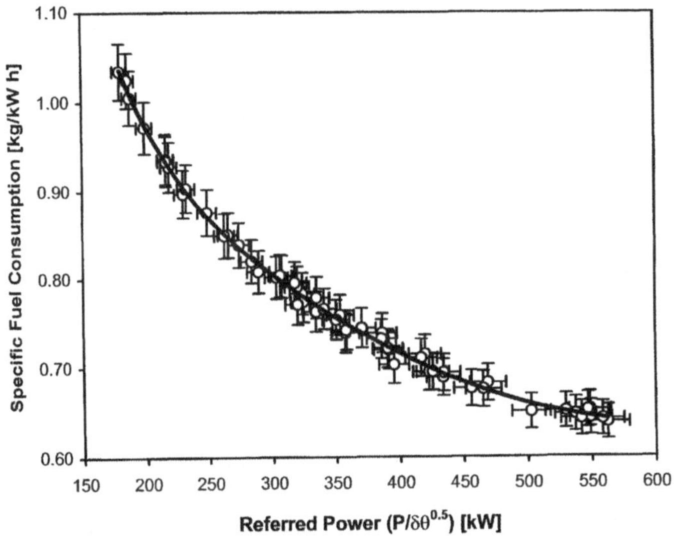

Fig. 3.11 Variation of SFC with referred power – test data.

with power, altitude and RRPM, usually falling as power and altitude increase and as RRPM decreases. Figure 3.11 shows a typical variation of SFC with referred power ($P/\delta\sqrt{\theta}$). As power increases, the engine operates more efficiently and SFC decreases. It is, therefore, often better to set a higher RRPM and fly at a higher true airspeed than that predicted by the simple theory if maximum SAR is to be achieved. Likewise, the endurance may be improved by operating at a slightly higher rotor speed. Based on research by Langdon [3.5] it can be shown that the extra power required to fly is offset by the improved SFC.

3.4.4 Reduction of level flight performance data

It has been already been shown that the power required by a helicopter can be written in terms of referred groups as:

$$\frac{P}{\sigma w^3} = f\left(\frac{W}{\sigma\omega^2}, \frac{V}{\omega}, \frac{V_c}{\omega}, Z, \frac{\omega}{\sqrt{\theta}}\right)$$

$$\frac{P}{\delta\sqrt{\theta}} = f\left(\frac{W}{\delta}, \frac{V}{\omega}, \frac{V_c}{\omega}, Z, \frac{\omega}{\sqrt{\theta}}\right)$$

In level flight, $V_c = 0$ and generally the aircraft will be outside ground effect. Thus:

$$\frac{P}{\sigma w^3} = f\left(\frac{W}{\sigma\omega^2}, \frac{V}{\omega}, \frac{\omega}{\sqrt{\theta}}\right)$$

$$\frac{P}{\delta\sqrt{\theta}} = f\left(\frac{W}{\delta}, \frac{V}{\omega}, \frac{\omega}{\sqrt{\theta}}\right)$$

It is interesting to compare the latter relationship with the power coefficient formula reviewed earlier:

$$C_P = 1.2\lambda C_T + \frac{1}{2}\mu^3\frac{f}{A} + \frac{1}{8}s(1 + 4.3\mu^2)C_D + C_{Pm} + C_{Ps}$$

It has already been shown that $\lambda = f(\mu)$. Additionally it can be surmised that the power increment due to retreating blade stall will be a function of rotor speed and forward speed (advance ratio) hence $C_{Ps} = f(\mu)$. Also the power loss due to compressibility will depend on rotor speed, forward speed *and* Mach number, so $C_{Pm} = f(\mu, M_B)$. Thus, the power coefficient formula can be written as:

$$C_P = f(C_T, \mu, M_B)$$

Noting that for a given helicopter:

$$C_P = \frac{P}{\rho A V_T^3} \propto \frac{P}{\rho\Omega^3} \propto \frac{P}{\sigma\omega^3}$$

$$C_T = \frac{T}{\rho A V_T^2} \propto \frac{W}{\rho\Omega^2} \propto \frac{W}{\sigma\omega^2}$$

$$\mu = \frac{V}{V_T} \propto \frac{V}{\Omega} \propto \frac{V}{\omega}$$

$$M_B = \frac{V_T}{a} \propto \frac{\Omega}{\sqrt{T}} \propto \frac{\omega}{\sqrt{\theta}}$$

The direct correlation between performance expressed in coefficient terms and performance expressed using referred parameters can be seen readily.

Since SFC is rarely constant with power it is necessary to document the variation of *referred fuel flow rate* with referred TAS, referred weight and referred rotor speed

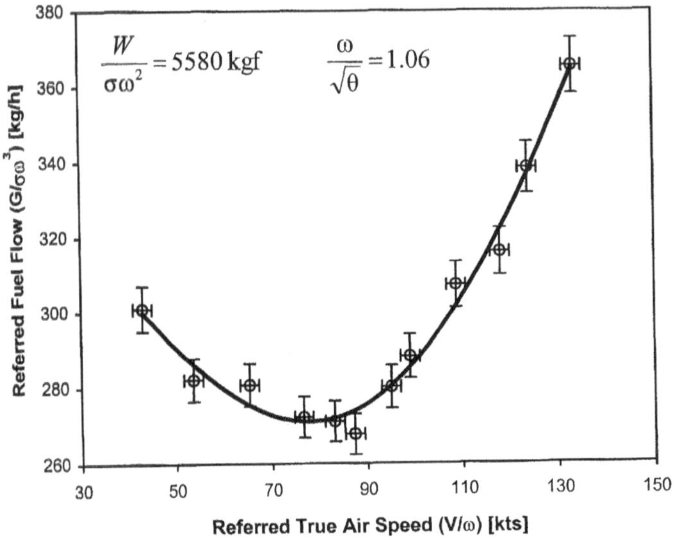

Fig. 3.12 Variation of referred fuel flow with referred TAS – test data.

to determine the endurance performance. Since the fuel flow rate of a gas turbine engine will be dependent on the power generated it can be assumed that $G = \mathrm{f}(P)$. Thus referred fuel flow is directly related to referred power:

$$\frac{G}{\sigma\omega^3} = \mathrm{f}\left(\frac{W}{\sigma\omega^2}, \frac{V}{\omega}, \frac{\omega}{\sqrt{\theta}}\right)$$

$$\frac{G}{\delta\sqrt{\theta}} = \mathrm{f}\left(\frac{W}{\delta}, \frac{V}{\omega}, \frac{\omega}{\sqrt{\theta}}\right)$$

Figure 3.12 shows a typical variation of referred fuel flow ($G/\sigma\omega^3$) with V/ω for a single value of $W/\sigma\omega^2$ and $\omega/\sqrt{\theta}$.

Likewise, it is necessary to document how *referred specific air range* varies with referred weight, referred true airspeed and referred rotor speed. Since SAR is equal to the ratio of true airspeed and fuel flow, their referred forms can be used in order to find the referred form of specific air range. Thus:

$$\mathrm{SAR} = \frac{V}{G} = \frac{\omega\left(\frac{V}{\omega}\right)}{\mathrm{f}\left[\left(\frac{P}{\sigma\omega^3}\right)\sigma\omega^3\right]} = \frac{\omega\left(\frac{V}{\omega}\right)}{\mathrm{f}\left(\frac{W}{\sigma\omega^2}, \frac{V}{\omega}, \frac{\omega}{\sqrt{\theta}}\right)\sigma\omega^3} = \mathrm{f}\left(\frac{W}{\sigma\omega^2}, \frac{V}{\omega}, \frac{\omega}{\sqrt{\theta}}, \frac{1}{\sigma\omega^2}\right)$$

$$= \mathrm{f}\left(\frac{W}{\sigma\omega^2}, \frac{V}{\omega}, \frac{\omega}{\sqrt{\theta}}, \frac{\theta}{\theta\sigma\omega^2}\right) = \mathrm{f}\left(\frac{W}{\sigma\omega^2}, \frac{V}{\omega}, \frac{\omega}{\sqrt{\theta}}, \frac{1}{\theta\sigma}\right) = \mathrm{f}\left(\frac{W}{\sigma\omega^2}, \frac{V}{\omega}, \frac{\omega}{\sqrt{\theta}}, \frac{1}{\delta}\right)$$

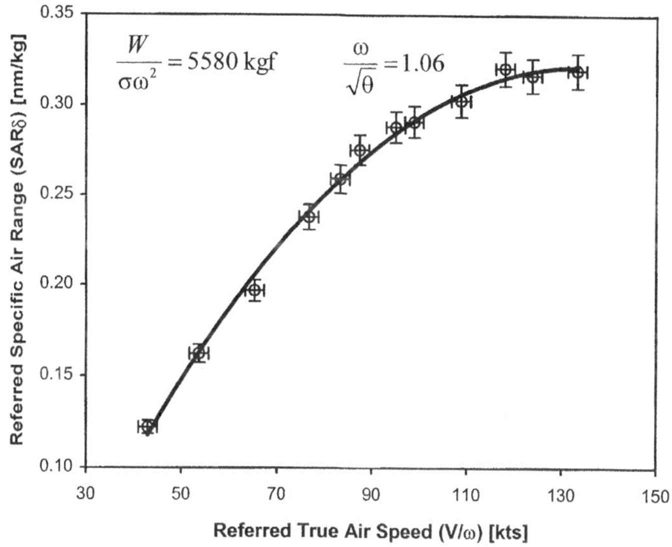

Fig. 3.13 Variation of referred SAR with referred TAS – test data.

and:

$$\text{SAR}\,\delta = f\left(\frac{W}{\sigma\omega^2}, \frac{V}{\omega}, \frac{\omega}{\sqrt{\theta}}\right)$$

Figure 3.13 shows a typical variation of referred SAR (SAR δ) with V/ω for a single value of $W/\sigma\omega^2$ and $\omega/\sqrt{\theta}$.

3.4.5 Choice of referred grouping

The choice of referred grouping is partly dependent on the test aircrew's ability to adjust rotor speed in flight. Another factor to consider is the test objectives. If there is a requirement to assess the magnitude of compressibility and retreating blade stall effects (tip effects) then it is prudent to use the $W/\sigma\omega^2$ method regardless. Consider the referred power required by a typical conventional helicopter as a function of referred true airspeed. If test data is plotted using the $P/\sigma\omega^3$ grouping the absence of any tip effects, implied by a constant drag coefficient, will be immediately apparent by the coincidence of the curves. This is because all the referred parameters in this grouping contain RRPM (ω) and so any increase in power in proportion to the increased rotor speed will be removed by the referring process. However, any increase in power not in proportion to RRPM (due in reality to an increased average drag coefficient) will be unaffected by the referring process and will be evident when plotted, see Fig. 3.14.

3.4.6 Engine characteristics

The manufacturer's predictions of engine performance will be based on test bed data. It is important, therefore, during a comprehensive performance trial that engine

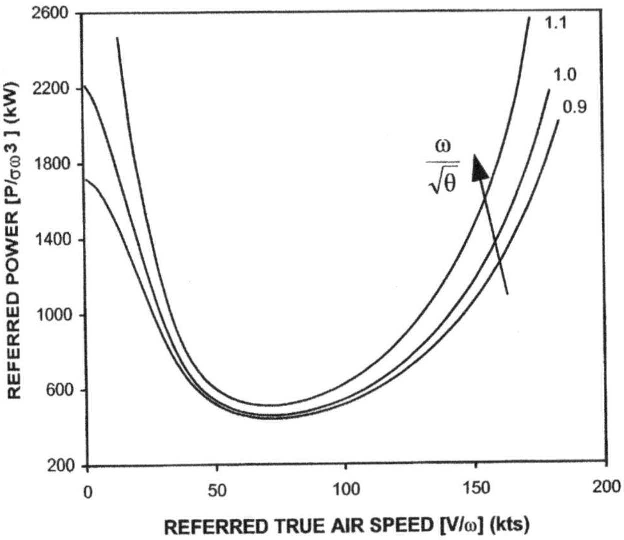

Fig. 3.14 Effect of referred rotor speed on power required.

parameters are recorded on every flight so that these predictions can be confirmed. Additionally, the effects of inlet losses and air bleeds and of any deterioration throughout a prolonged flying period can be ascertained. If manufacturer's engine performance data is not available then the effects of varying $\omega/\sqrt{\theta}$ and any other engine configuration parameter, such as the IGV or BCV position (Δ), on power available should be considered. Although experience has shown that these parameters do not generally alter the unique relationship between $P/\delta\sqrt{\theta}$, $G/\delta\sqrt{\theta}$, and $N/\sqrt{\theta}$ over the working range of the engine they cannot be ignored in the absence of reliable data. They should be dealt with as follows. If $\omega/\sqrt{\theta}$ is critical then curves of referred power against referred N will vary with $\omega/\sqrt{\theta}$ due to changes in the efficiency of the free power turbine (FPT). Tests can be made at high, low and intermediate values of $\omega/\sqrt{\theta}$; interpolation will then be possible between them. If rotor speed does not vary, sorties can be flown at extremes of OAT to establish the required spread of values of $1/\sqrt{\theta}$. If Δ is critical, generally because the IGVs are not fully open over the working range of the engine, differential corrections will have to be calculated to obtain the correct values of power and fuel flow.

When undertaking an evaluation of the level flight performance accurate calibration of the engine, the power instruments and the fuel flow measurement system is essential. Usually the engine will be bench tested at the manufacturer before and after the tests so that the results can be compared with the manufacturer's predictions based on accurate test bed data. Some deterioration of engine condition is inevitable during a prolonged trial and so the post-trial engine bench tests are as important as the pre-trial bench tests. Test bed data is also used for the correction of the effects of engine bleeds, although the bleed ports can be blanked off to ensure that they do not affect the results. Roots and Blake [3.6] give more details on the use of calibrated engines and the importance of gathering engine data whilst conducting evaluations of aircraft performance.

3.4.7 General test conditions

Tests are normally made at three or four values of referred weight, which are selected to give a good coverage of results from the maximum to the minimum permitted AUM. Normally only a central CG is used, unless the CG range of the helicopter is large as changes in CG may affect the results significantly. Datum testing is conducted with the helicopter in a clean configuration before it is evaluated in its normal role fit. Accurate testing can only be conducted in clear air, away from clouds, fronts and turbulence. The best results are often obtained late in the evening, or at dawn, when the air is generally more stable [3.6].

A limited performance trial, targeting only three values of referred weight for example, may be conducted economically by using the same mass and testing across three altitude bands. To get the best coverage of results, one height should be as low as practicable and one should be close to the ceiling of the aircraft, with one intermediate altitude. The rotor RPMs to be used will depend upon the test methods selected and the aircraft's engine and rotor governing system. If the rotor RPM cannot be varied in flight, tests are done at the normal governed RPM using the $W/\sigma\omega^2$ weight group. Relative rotor speed (ω) may also be ignored in the case of a constant-speeding rotor governing system, such as a FADEC equipped helicopter, as opposed to the system described above that does not allow the pilot to compensate for the effects of static droop in flight. Use of the W/σ group permits a more compact presentation of results but does not portray the effects of changes in rotor Mach number. It should therefore be borne in mind that significant errors could arise if the results are used to predict performance in temperatures markedly different from those in which the tests were conducted.

Data is normally gathered throughout the speed range from the lowest speed that gives a sensible indication up to V_H, in 10 kt increments, with 5 kt increments around V_{MP} and close to V_H. Engine air bleeds and other secondary controls (cabin heating, engine anti-icing, oil cooler shutters, blade de-icing and other heavy electrical loading) should be pre-set as specified by the test schedule or adjusted to maintain optimum engine conditions. Datum results are initially obtained with all secondary controls off, and fuel consumption penalties then determined with these systems switched on.

3.4.8 Flight test techniques

Prior to take-off the helicopter loading is checked and the correct graphs, tables or hand-held computer program made available showing height to fly versus AUM or fuel gone and OAT. Care needs to be taken in selecting the combination of altitude and AUM used to obtain a given target referred weight as they both may limit the maximum and minimum speed that can be tested. Equally, consideration must be given to the order in which the speeds are flown so that the maximum coverage is obtained. It may therefore be necessary to either reballast or refuel the aircraft between a series of runs to contain the test height band within reasonable limits. It is worth remembering that provided the relevant referred parameters are kept constant, data gathered from several flights may be plotted together. This is one of the main advantages of the experimental method.

Before a test run is commenced, 1013 mbar is set on the altimeter sub-scale to ensure that it reads pressure altitude, and the position of all relevant secondary controls checked. The likely AUM of the aircraft at the start of the data run is calculated and a climb made to the altitude required to target the desired referred weight based on that estimated AUM. At this altitude the correct rotor RPM is set, if applicable, for the observed OAT. AUM, altitude, OAT and rotor RPM should then be re-checked and adjusted as necessary. Note that the target referred weight should be achieved approximately halfway through the conduct of the test point. If altimeter PECs are significant then due allowance will have to be made when reading the altimeter. If the atmospheric conditions are stable and the $W/\sigma\omega^2$ method is being used time can be saved by first climbing to the top of the anticipated altitude band required for the complete sortie whilst documenting the outside air temperature. This provides the test aircrew with a complete air density profile thereby reducing the iteration required to obtain the target referred weights.

The following parameters are recorded when the helicopter is stabilized in level flight at the required airspeed:

- altitude;
- outside air temperature;
- airspeed;
- torque;
- rotor RPM;
- engine parameters (temperature and speed);
- fuel state;
- fuel flow;
- sideslip.

The test condition should be maintained for a minimum of one minute, ideally taking further readings of the above parameters at the middle and end of the period to determine a mean value of each parameter; the accuracy of the test condition itself and the fuel flow in the absence of appropriate instrumentation. In a given aircraft, the accuracy with which a test run can be flown will depend on the atmospheric conditions, the airspeed and the altitude. The following accuracies are typical:

- airspeed – no worse than ± 1 kt
- altitude – no worse than ± 20 ft of desired altitude with no perceptible change during the run
- rotor speed – essentially fixed
- slipball – central
- referred weight – within 0.5% of target.

In this context, the vertical speed indicator (VSI) rarely gives sufficiently accurate indications of zero climb or descent. On suitably equipped rotorcraft; maximum fuel flow measurement accuracy will be obtained by starting the stopwatch as a fuel counter *clicks over* and stopping timing, after approximately one minute, as *click over* of the final reading occurs, see Roots and Blake [3.6] for alternative methods.

3.4.9 Planning a level flight performance trial

3.4.9.1 Requirements

It is important that flight trials are carefully planned and that certain essential conditions are fulfilled if the maximum information is to be obtained from the minimum flight hours. The planning task may be summarized by the following questions:

- What test conditions are required?
- What characteristics (temperature and altitude) prevail at the proposed test site(s)?
- What test conditions can be obtained at the proposed test site(s)?
- What range of ballast and fuel states is required to complete the matrix of test conditions?
- How many flight hours are required?

3.4.9.2 *Example: planning flight tests at constant* $W/\sigma\omega^2$ and $\omega/\sqrt{\theta}$

Suppose a helicopter is to be tested with an AUM range of 4000 kg to 5000 kg that is cleared to operate between sea level and 10 000 ft pressure altitude The rotor speed can be varied in flight and has a power-on RRPM range of between 95% and 105% (standard RRPM is 100%). It is expected that in service the helicopter will experience temperature profiles between ISA – 30°C and ISA + 30°C. At the home establishment the trials will be conducted within the following constraints:

- Although the test aircraft can be ballasted to 5000 kg, a sensible minimum start mass is 4200 kg (2 crew and 45 minutes endurance).
- The helicopter has been authorized to operate at pressure altitudes up to 12 000 ft for short periods to complete data runs as required.
- Average temperature profiles experienced at the home establishment are between ISA – 20°C and ISA + 15°C.
- Data gathering will not start below pressure altitudes of 2000 ft or above pressure altitudes of 10 000 ft.
- Referred rotor speeds will be selected using an ISA atmosphere as this is most representative of summer conditions at the test establishment.

From the test specification, it is possible to determine the range of referred parameters required to address the test objectives, see Table 3.2 and Table 3.3. Likewise the range of referred parameters available at the test site can be determined, see Tables 3.4 and 3.5. Comparing the test requirements with the available referred parameters reminds us that since rotor speed affects referred weight it will be impossible to target the maximum possible referred weight and rotor speed simultaneously. Therefore, a lower referred weight will have to be targeted for *tip effects* testing.

The interdependency of the test parameters requires an integrated approach to the selection of test parameters. Start by selecting a maximum referred weight of 7000 kgf for tip effects evaluation. This weight is the approximate mean referred weight available at 10 000 ft. From the variation of referred rotor speed at 10 000 ft, 1.00, 1.04 and 1.08 can be selected as the target referred rotor speeds. By assuming a fuel burn per data run (say 450 kgf) and an ISA atmosphere, the altitudes required to set the desired $W/\sigma\omega^2$ can be identified, see Table 3.6. (Note that $W/\sigma\omega^2$ multiplied by $(\omega/\sqrt{\theta})^2$ equals W/δ, therefore a pressure altitude can be identified for any actual mass.) It can be seen that these targets are impractical since a pressure altitude above 12 000 ft is required to reduce the relative density as the rotor speed is increased to give the higher referred RRPM. Likewise as the altitude is raised to account for fuel burn the rotor

Table 3.2 Operational variation of referred weight (kgf).

	Atmosphere											
	ISA – 30°C				ISA				ISA + 30°C			
	sea level		10 000 ft		sea level		10 000 ft		sea level		10 000 ft	
Ω	95	105	95	105	95	105	95	105	95	105	95	105
δ	1.000	1.000	0.688	0.688	1.000	1.000	0.688	0.688	1.000	1.000	0.688	0.688
θ	0.896	0.896	0.827	0.827	1.000	1.000	0.931	0.931	1.104	1.104	1.035	1.035
σ	1.116	1.116	0.831	0.831	1.000	1.000	0.738	0.738	0.906	0.906	0.664	0.664
ω	0.95	1.05	0.95	1.05	0.95	1.05	0.95	1.05	0.95	1.05	0.95	1.05
4000	3971	3250	5331	4364	4432	3628	6002	4913	4894	4006	6673	5462
5000	4963	4063	6664	5455	5540	4535	7502	6141	6117	5007	8341	6828

Table 3.3 Operational variation of referred rotor speed (%).

Rotor speed (%)	Atmosphere ISA – 30°C sea level	ISA – 30°C 10 000 ft	ISA sea level	ISA 10 000 ft	ISA + 30°C sea level	ISA + 30°C 10 000 ft
95	1.004	1.045	0.950	0.984	0.904	0.934
100	1.057	1.100	1.000	1.036	0.952	0.983
105	1.109	1.155	1.050	1.088	0.999	1.032

speed must be reduced in order to keep the desired referred RRPM. A revised test matrix can be developed but it will inevitably be more conservative, see Table 3.7.

From Table 3.7 it can be seen that a referred weight of 6300 kgf can be tested at three referred rotor speeds (1.01, 1.03, 1.05). On the assumption that tip effects are not present, additional referred weights to test at a fixed mean referred RRPM, say 1.03, can be selected. Table 3.8 shows a suitable range of referred weights.

3.5 VERTICAL PERFORMANCE TESTING

From a practical viewpoint the difficulty with vertical performance testing is that of establishing and holding a steady zero-airspeed flight condition in which accurate measurements can be made. In order to establish such a condition in free flight it is necessary to use some means other than the ASI to indicate a true hover or to rely on a fixed, ground reference. The latter method is simplest but means that the tests must be carried out relatively close to the ground in as near still air conditions as possible (wind speeds less than 3 kts). Ballast or cable tension may be required to obtain a sufficient spread of hover weights.

The climb performance of a helicopter can be divided into two separate areas: vertical performance and forward flight climb and descent performance. The latter is used to determine the best rate of climb, optimum climb schedule and service ceiling. The former, together with hover performance data, is used to give the pilot the fullest possible information on the aircraft's limits in axial flight by showing how vertical rate of climb varies with altitude, OAT, weight, wind speed and RRPM at a specified engine rating. The techniques employed here are reduced power verticals, maximum power verticals or low airspeed verticals. The likelihood that the conditions will be such that the tests can be carried out in still air is remote so it is normal practice to either climb vertically relative to the ground in low windspeed conditions and correct the results to zero windspeed or to climb vertically from a free-air hover.

3.5.1 Free flight hover testing

As derived above, the general power required relationships are:

$$\frac{P}{\delta\sqrt{\theta}} = f\left(\frac{W}{\delta}, \frac{V}{\omega}, \frac{V_c}{\omega}, Z, \frac{\omega}{\sqrt{\theta}}\right) \quad \text{or} \quad \frac{P}{\sigma\omega^3} = f\left(\frac{W}{\sigma\omega^2}, \frac{V}{\omega}, \frac{V_c}{\omega}, Z, \frac{\omega}{\sqrt{\theta}}\right)$$

Table 3.4 Variation of referred weight (kgf) available at test site.

	Atmosphere											
	ISA – 20°C				ISA				ISA + 15°C			
	2000 ft		10 000 ft		2000 ft		10 000 ft		2000 ft		10 000 ft	
Ω	95	105	95	105	95	105	95	105	95	105	95	105
δ	0.917	0.917	0.862	0.862	0.986	0.986	0.931	0.931	1.038	1.038	0.983	0.983
θ	0.930	0.930	0.688	0.688	0.930	0.930	0.688	0.688	0.930	0.930	0.688	0.688
σ	1.014	1.014	0.798	0.798	0.943	0.943	0.738	0.738	0.895	0.895	0.699	0.699
ω	0.95	1.05	0.95	1.05	0.95	1.05	0.95	1.05	0.95	1.05	0.95	1.05
4200	4589	3756	5832	4774	4936	4041	6302	5159	5197	4254	6654	5447
5000	5463	4472	6943	5684	5877	4810	7502	6141	6187	5064	7922	6485

Table 3.5 Variation of referred rotor speed (%) available at test site.

Rotor speed (%)	Atmosphere ISA – 20°C sea level	ISA – 20°C 10 000 ft	ISA sea level	ISA 10 000 ft	ISA + 15°C sea level	ISA + 15°C 10 000 ft
95	0.992	1.023	0.957	0.984	0.932	0.958
100	1.044	1.077	1.007	1.036	0.981	1.008
105	1.097	1.131	1.057	1.088	1.030	1.059

Table 3.6 Test conditions required to match original target referred parameters.

Target $W/\sigma\omega^2$	7000	7000	7000
Target $\omega/\sqrt{\theta}$	1.00	1.04	1.08
W/δ	7000	7571	8165
Start AUM (kg)	4900	4900	4900
Test altitude (ft)	9543	11 556	13465
Expected OAT (°C)	−3.9	−7.9	−11.7
Test RRPM (%)	96.7	99.8	102.9
End AUM (kg)	4450	4450	4450
Test altitude (ft)	12011	13987	15862
Expected OAT (°C)	−8.8	−12.7	−16.4
Test RRPM (%)	95.8	98.9	101.9

Table 3.7 Test conditions required to match modified target referred parameters.

Target $W/\sigma\omega^2$	6300	6300	6300
Target $\omega/\sqrt{\theta}$	1.01	1.03	1.05
W/δ	6427	6684	6946
Start AUM (kg)	4900	4900	4900
Test altitude (ft)	7315	8342	9341
Expected OAT (°C)	0.5	−1.5	−3.5
Test RRPM (%)	98.4	100.0	101.6
End AUM (kg)	4450	4450	4450
Test altitude (ft)	9823	10831	11813
Expected OAT (°C)	−4.5	−6.4	−8.4
Test RRPM (%)	97.5	99.1	100.7

For the hover, $V = V_c = 0$, and in the free flight case ground effect is not an issue, therefore:

$$\frac{P}{\delta\sqrt{\theta}} = f\left(\frac{W}{\delta}, \frac{\omega}{\sqrt{\theta}}\right) \quad \text{or} \quad \frac{P}{\sigma\omega^3} = f\left(\frac{W}{\sigma\omega^2}, \frac{\omega}{\sqrt{\theta}}\right)$$

Rotor speed effects are evaluated by obtaining data at the same referred weight but over a range of $\omega/\sqrt{\theta}$. If the variation in ground effect with height is required, separate curves can be obtained for each value of hover height, Z, usually defined as the wheel

Table 3.8 Test conditions required to match additional referred parameters.

Target $W/\sigma\omega^2$	4400	5500	6600
Target $\omega/\sqrt{\theta}$	1.03	1.03	1.03
W/δ	4668	5835	7002
Start AUM (kg)	4200	4900	4900
Test altitude (ft)	2894	4753	9550
Expected OAT (°C)	9.3	5.6	−3.9
Test RRPM (%)	102.0	101.3	99.6
End AUM (kg)	3750	4450	4450
Test altitude (ft)	5935	7308	12018
Expected OAT (°C)	3.2	0.5	−8.8
Test RRPM (%)	100.9	100.4	98.7

or skid clearance above ground. If detailed results are not necessary it is usually sufficient to obtain results IGE and/or OGE. If hover height is not considered and $\omega/\sqrt{\theta}$ effects are insignificant then the results will be presented as a single curve of $P/\sigma\omega^3$ against $W/\sigma\omega^2$. This method of presentation is used *regardless* of the type of helicopter since rotor RPM changes should affect both parameters in the same proportion and a single curve should result. Had $P/\delta\sqrt{\theta}$ versus W/δ been used then, even with negligible tip effects, separate curves would still result for each $\omega/\sqrt{\theta}$ (see the discussion on tip effects given below).

At each of several values of $\omega/\sqrt{\theta}$, torque and rotor speed are measured at sufficient values of weight to cover the desired range of $W/\sigma\omega^2$. At each test point OAT, pressure altitude and engine data should also be recorded. Prior planning is required to ensure that the desired range of referred weights will be achieved. Simple theory predicts that if rotor RPM effects can be ignored, that is the profile drag coefficient is constant, then:

$$\frac{P}{\sigma\omega^3} = \frac{Q}{\sigma\omega^2} + K_1\left(\frac{W}{\sigma\omega^2}\right)^{3/2} + K_2$$

where K_1 and K_2 are constants.

If therefore $P/\sigma\omega^3$ is plotted against $(W/\sigma\omega^2)^{3/2}$ the result will be a straight line provided K_2 actually remains constant. Thus, in the absence of tip effects, the results obtained from the free-air hovering tests can be smoothed by the following method:

- Plot $P/\sigma\omega^3$ against $(W/\sigma\omega^2)^{3/2}$.
- Draw the best straight line through the points.
- For desired values of $W/\sigma\omega^2$, calculate $(W/\sigma\omega^2)^{3/2}$ and use the smoothed line to read off corresponding values of $P/\sigma\omega^3$.
- Plot $P/\sigma\omega^3$ against $W/\sigma\omega^2$.

In Fig. 3.15 a typical set of test data is presented. A straight line fit has been applied along with error bars based on a 3% tolerance. Assuming that this is the accepted level of accuracy for the test it can be seen that the trend line falls within the error bars. Thus a single straight line fit is appropriate and RRPM effects can be ignored. This fit has been used to transfer the trend information to Fig. 3.16. (Note that the line presented in Fig. 3.16 is not straight since the independent variable is now $W/\sigma\omega^2$.)

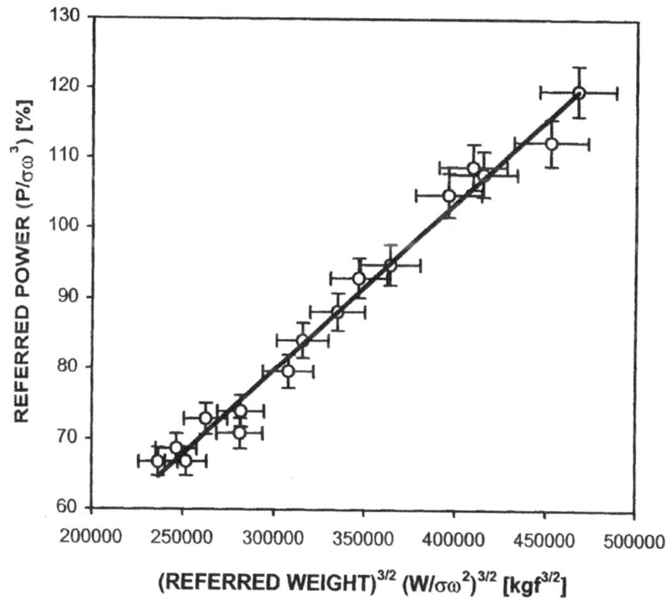

Fig. 3.15 Typical hover performance test data.

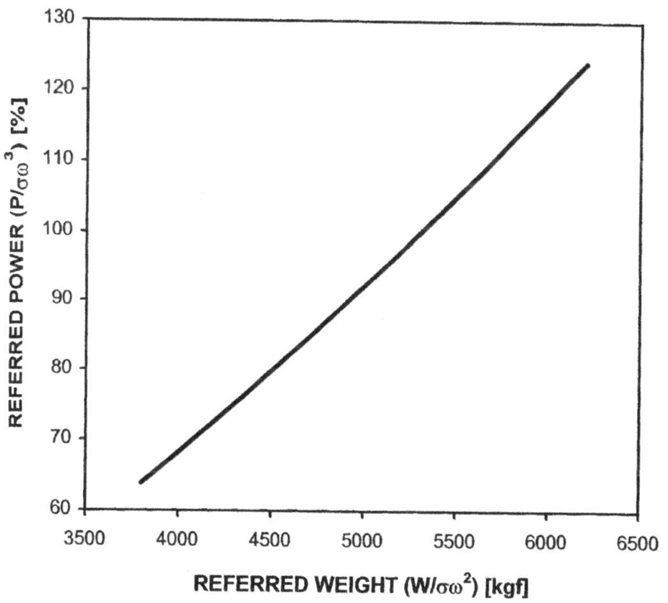

Fig. 3.16 Smoothed hover performance data.

3.5.2 Tethered hover testing

The same referred groups are used for tethered hovering tests as for free-air hover trials. In this case, however, hover height is an important parameter. The variation of

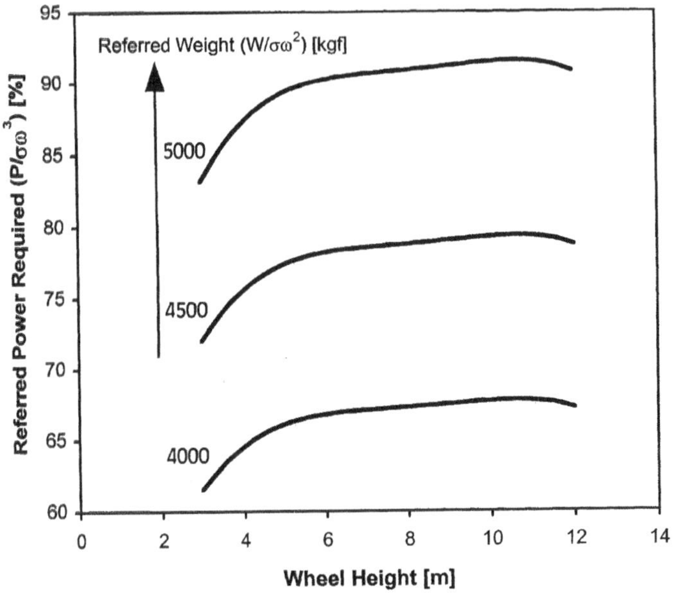

Fig. 3.17 Typical tethered hover data.

ground effect with hover height can be measured accurately using tethering cables of different lengths. The relationship used is either:

$$\frac{P}{\delta\sqrt{\theta}} = f\left(\frac{W}{\delta}, Z, \frac{\omega}{\sqrt{\theta}}\right) \quad \text{or} \quad \frac{P}{\sigma\omega^3} = f\left(\frac{W}{\sigma\omega^2}, Z, \frac{\omega}{\sqrt{\theta}}\right)$$

As discussed above the latter equation is most often used. The helicopter is established in a tethered hover with the cable taut and power is increased until the maximum permitted tension is achieved. Having taken all the relevant data, power is reduced in stages (data being recorded at each stage) until cable tension is almost lost. The cable tension simulates varying aircraft weights. Hover height is controlled by the cable length; various cable lengths providing a good range of heights is desirable. Results are usually presented as a carpet plot of $P/\sigma\omega^3$ against $W/\sigma\omega^2$ and Z. The 'smoothing' process described for free air hovering tests may be used for tethered hovering trials. Figure 3.17 illustrates the form in which the results finally appear. It is important to remember that tethered hovering trials *must* be carried out in as near still air conditions as possible, less than 3 kts is usually taken as acceptable [3.7].

3.5.3 Tip effects evaluation

Tip effects are a complex subject and determining their presence can be difficult and time-consuming. There are, however, advantages in proving that they are absent. If the effect of referred RRPM ($\omega/\sqrt{\theta}$) on referred power ($P/\sigma\omega^3$) can be ignored then there exists the possibility of generating a greater range of referred weight ($W/\sigma\omega^2$) through changes in RRPM. There is also the added benefit that when determining mission suitability the effect of OAT on referred RRPM can be ignored. Although for simplicity this discussion is restricted to the free-air hover, tip effects should be investigated at high forward speed.

3.5.3.1 *Source of tip effects*

Before the process of investigating tip effects can be described in detail it is important to consider what is meant by the term. A tip effect is an actual increase in power arising from a change in rotor RPM that cannot be eliminated by referring the parameters using the $W/\sigma\omega^2$ method. Consider the effect of a change in rotor speed on the power required to hover at a given weight. Assuming that power can be given by:

$$P = Tv_{i_h} + \frac{1}{8}\rho bc\Omega^3 R^4 C_D$$

and that it is referred by dividing by $\rho\Omega^3$, which is directly proportional to $\sigma\omega^3$, then:

$$\frac{P}{\sigma\omega^3} = K\left(\frac{P}{\rho\Omega^3}\right) = K\left(\frac{Tv_{i_h}}{\rho\Omega^3} + \frac{1}{8}bcRC_D\right)$$

Thus as the rotor speed is varied the contribution to referred power arising from profile power will be constant provided the average profile drag coefficient (C_D) remains unchanged. If, on the other hand as a result of changes in RRPM, the blades 'see' different AOAs and Mach numbers which cause variations in C_D then alterations in referred power will arise. Before ascribing changes in referred power, at constant weight, solely to tip effects it is necessary to consider the effect of RRPM on induced power. Recalling that the induced velocity in the hover is given by:

$$v_{i_h} = \sqrt{\frac{T}{2\rho A}} = \left(\frac{W}{2\rho A}\right)^{1/2}$$

Thus the contribution to referred power from induced power is:

$$\frac{Tv_{i_h}}{\rho\Omega^3} = \frac{T}{\rho\Omega^3}\left(\frac{W}{2\rho A}\right)^{1/2} = \sqrt{\frac{1}{2A}}\left(\frac{W}{\rho\Omega^2}\right)^{3/2}$$

Therefore the referred power relationship can be written as:

$$\frac{P}{\sigma\omega^3} = K\left[\sqrt{\frac{1}{2A}}\left(\frac{W}{\rho\Omega^2}\right)^{3/2} + \frac{1}{8}bcRC_D\right] = K\left[K_1\left(\frac{W}{\rho\Omega^2}\right)^{3/2} + K_2C_D\right]$$

Therefore:

$$\frac{P}{\sigma\omega^3} = K_3\left(\frac{W}{\sigma\omega^2}\right)^{3/2} + K_4C_D$$

Thus in order to assess the size of any tip effect it is only necessary to document the variation of referred power ($P/\sigma\omega^3$) with rotor speed at *constant* referred weight ($W/\sigma\omega^2$). Since several test points may be established at differing altitudes it is necessary to use *referred RRPM* ($\omega/\sqrt{\theta}$) instead of rotor speed.

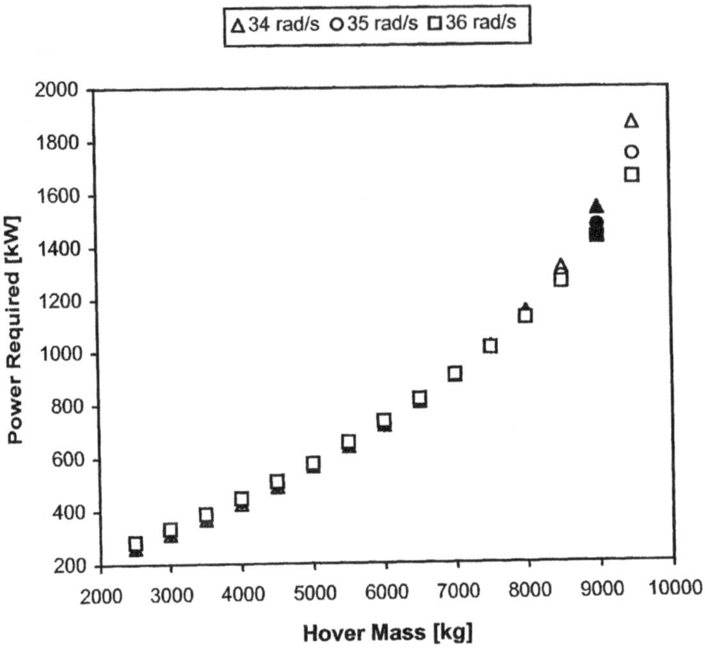

Fig. 3.18 Hover performance data – raw.

3.5.3.2 *Developing a test philosophy using the free-air method*

As already mentioned tip effect evaluations consist of establishing a series of hovers so that the variation of power (referred) with RRPM (referred) at constant weight (referred) can be documented. It is important, therefore, that the appropriate test philosophy is used to ensure that an adequate range of test data is obtained. Consider the situation that arises when a test team elects to establish a series of hovers and during each hover they vary the RRPM by ± 1 rad/s from nominal. (They conduct a *rotor sweep.*) Assume for simplicity that all the hovers are conducted at the same height (but at different weights) and that the fuel burn during each rotor sweep can be ignored. Figure 3.18 shows a typical set of raw data. The team then processes the data to yield the required referred parameters, see Fig. 3.19. Any assessment of the importance of tip effects has been complicated since the actual weight at which each rotor sweep was conducted has yielded three referred weights as shown by the three highlighted points in each figure. In practice, due to the uncertainties associated with actual flight test, the team will usually be unsure if the differences in referred power are due solely to tip effects. The correct test philosophy, therefore, is to plan a series of hovers at differing weights and altitudes that will generate a consistent set of referred weights.

As an example of the correct approach to tip effect evaluation consider the planning required for a single test point based on a nominal condition of 4000 kg at 5000 ft (ISA) and a rotor sweep of 95% to 105%. At 5000 ft the relative density is 0.8617 so the referred weight resulting from the nominal conditions (100% RRPM) is 4000/ 0.8617 = 4642 kg. Suppose sometime later the weight has reduced by 50 kg to 3950 kg and the team wish to establish the 95% RRPM test point. Knowing that the

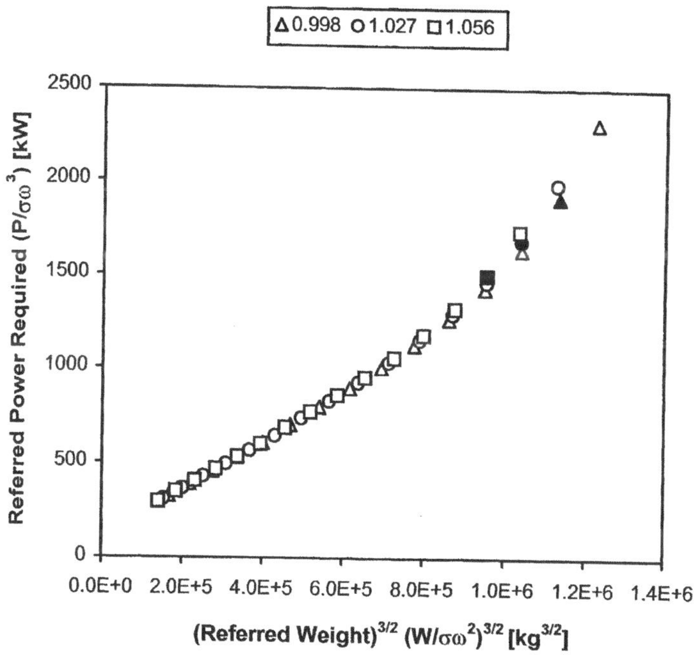

Fig. 3.19 Hover performance data – referred.

referred weight required is 4642 kg and that ω will be 0.95 ($\omega^2 = 0.9025$) it is possible to determine the relative density required to give the desired referred weight when the actual weight is 3950 kg:

$$\frac{W}{\sigma\omega^2} = 4642 = \frac{3950}{\sigma(0.95)^2}$$

$$\sigma = \frac{3950}{4642(0.95)^2} = 0.9429$$

This relative density will be found at around 2000 ft. Likewise, if later in the sortie when the helicopter weight reaches 3900 kg, the team wish to establish the 105% RRPM a test point at an altitude of approximately 9000 ft would be required to generate the desired referred weight. It should be clear, therefore, that an accurate estimate of air density (OAT at regular pressure altitudes) is an important factor in successful planning.

So far in this discussion actual RRPM has been considered as the controlling parameter. In fact the test objective is to assess the effect on referred power of *referred RRPM* ($\omega/\sqrt{\theta}$) at constant referred weight. To allow for data interpolation, and possible extrapolation, it will be necessary for the test team to gather data from a series of test points over a range of referred weights with at least two values of referred RRPM established at each referred weight. Since OAT and therefore $\sqrt{\theta}$ will vary with altitude the variation of the actual RRPM required to meet the test objectives will need to be determined and compared with any published limitations. The example introduced above will be used to illustrate this point. Assume that the required referred

RRPM range is 0.95 to 1.05 (equating to 95% RRPM to 105% RRPM at sea level ISA) and that the nominal rotor RPM is 100%. Since in this example the temperature profile is known it is possible to use:

$$\frac{W}{\sigma\omega^2} = \frac{W}{\delta} \times \frac{\theta}{\omega^2} = \frac{W}{\delta}\left(\frac{\sqrt{\theta}}{\omega}\right)^2 = 4642$$

Thus for the three referred RRPMs required (0.95, 1.00 and 1.05) a value of W/δ can be found which will dictate a particular pressure altitude for a given weight. For example, if $\omega/\sqrt{\theta}$ equals 1.00 and the AUM is 4000 kg, then:

$$\frac{W}{\delta} = \frac{4000}{\delta} = 4642$$

and

$$\delta = \frac{4000}{4642} = 0.8617$$

This relative pressure can be found at approximately 4100 ft H_p. At this altitude the OAT will be 280 K and therefore the required RRPM for $\omega/\sqrt{\theta} = 1.00$ is given by:

$$\omega = \left(\frac{\omega}{\sqrt{\theta}}\right) \times \sqrt{\theta} = 1.00 \times \sqrt{\frac{280.00}{288.15}} = 0.9858$$

Thus an actual RRPM of 98.6% will be required at 4100 ft to generate the desired referred RRPM and weight. Similarly if $\omega/\sqrt{\theta} = 0.95$ is set when the AUM is 3950 kg a pressure altitude of 1600 ft and RRPM of 94.5% will be required. Finally if $\omega/\sqrt{\theta} = 1.05$ is desired at 3900 kg a height of 7300 ft and RRPM of 102.3% will be needed.

In order to meet all the test objectives it is therefore evident that the test team must plan each sortie in detail. It is usual to use altitude to generate *changes* in referred weight as well as changing altitude to maintain the desired referred weight for different referred RRPMs. If a sensible choice of referred weight(s) is made it *may* be possible to maintain hover height and change rotor speed thereby generating a test point at a different referred weight which will be completed at some other altitude. A sample set of test points is shown in Table 3.9 and an example plot of the relevant referred data

Table 3.9 Typical vertical performance test point matrix.

Start mass (kg)	Fuel gone (kg)	Hp (ft)	OAT (°C)	RRPM (%)	$W/\sigma\omega^2$ (kgf)	$\omega/\sqrt{\theta}$
4750	50	8000	−0.8	102.1	5750	1.05
4750	75	5100	4.9	97.2	5750	0.99
4750	100	6900	1.3	99.6	5760	1.02
4750	125	4500	6.1	100.4	5240	1.02
4750	150	7400	0.3	102.3	5500	1.05
4750	175	3300	8.5	97.9	5260	0.99
4750	200	6300	2.5	99.8	5510	1.02
4750	225	6700	1.7	102.6	5260	1.05
4750	250	4900	5.3	97.3	5490	0.99

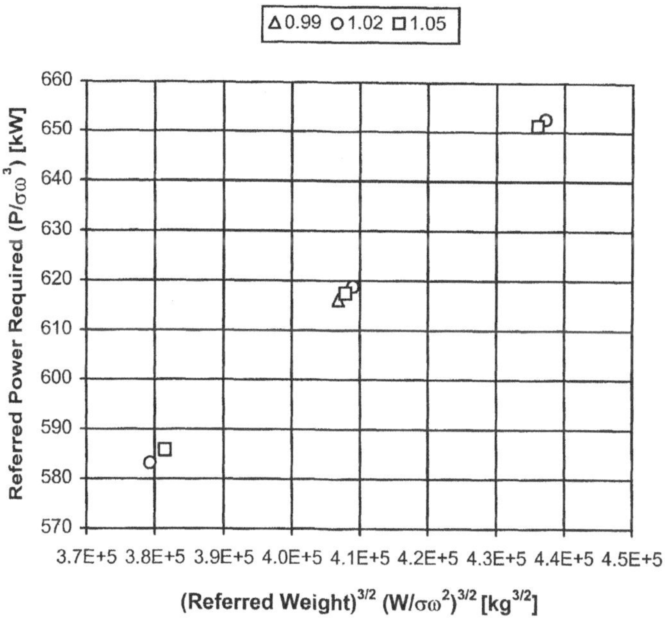

Fig. 3.20 Referred hover data – correct methodology.

is at Fig. 3.20. Note that it is now possible to make a judgement on the magnitude of any tip effects.

3.5.4 Vertical climb testing

For vertical climb performance testing, $V = 0$ and Z is not fixed, so for climbs outside of ground effect the general power required relationship can be reduced to:

$$\frac{P}{\sigma\omega^3} = f\left(\frac{W}{\sigma\omega^2}, \frac{V_c}{\omega}, \frac{\omega}{\sqrt{\theta}}\right)$$

which can be rearranged more sensibly as:

$$\frac{V_c}{\omega} = f\left(\frac{P}{\sigma\omega^3}, \frac{W_c}{\sigma\omega^2}, \frac{\omega}{\sqrt{\theta}}\right)$$

Vertical climb testing is achieved by establishing the aircraft in an OGE hover, at a altitude about 500 ft below that corresponding to the selected value of $W/\sigma\omega^2$ and with the rotor speed set for the desired $\omega/\sqrt{\theta}$. Power is increased incrementally up to the maximum available and the vertical rate of climb generated at each increment is measured. The climb begins below the required test altitude so that data can be logged at the required $W/\sigma\omega^2$. As fuel is used, the target altitude is increased to maintain $W/\sigma\omega^2$ constant. Ideally a 'smoothed' rate of climb is obtained from altitude/time data recorded at 100 ft intervals during the climb. This so-called *reduced power vertical*

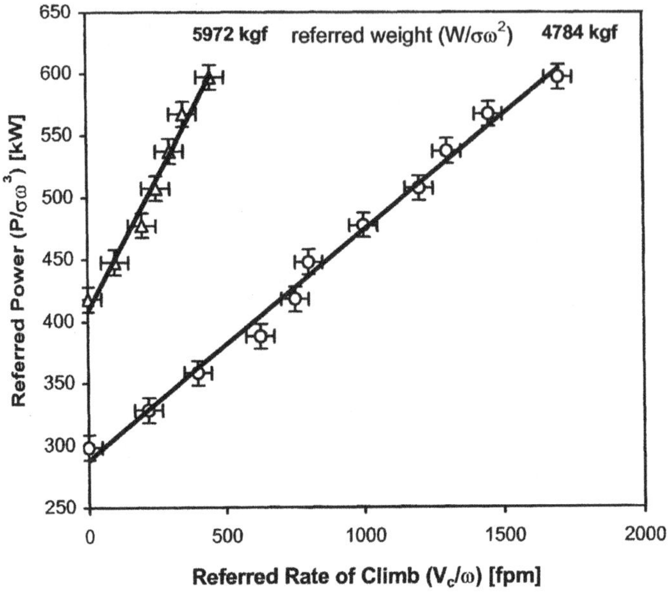

Fig. 3.21 Referred vertical climb test data – RPV method.

(RPV) technique can be replaced by a quicker, but less accurate, alternative based on data obtained at maximum torque only (the *maximum power vertical* – MPV method).

The smoothed rate of climb is used to plot $P/\sigma\omega^3$, or $Q/\sigma\omega^2$, against V_c/ω, as shown in Fig. 3.21, for a range of constant values of $W/\sigma\omega^2$. Ideally these tests should all be performed at the same value of $\omega/\sqrt{\theta}$ to eliminate the possibility of tip effects affecting the data. This will typically require the flight crew to adjust the rotor speed as the climb is commenced to account for static droop and the reduction of temperature with altitude. Ease of test planning (selection of convenient hover altitudes) and test conduct (use of the MPV method) will generally lead to the collection of vertical climb data at a range of referred weights over which there has been imprecise control, see Fig. 3.22. Since mission weights may not have been targeted, it is necessary to develop a more usable form for this data. For selected values of V_c/ω a chart such as Fig. 3.23 can be produced showing corresponding values of $P/\sigma\omega^3$ at various values of $W/\sigma\omega^2$.

3.5.5 General test conditions and methodology

Tests are normally planned to allow the widest possible range of referred AUM and generally start at maximum altitude and high AUM (maximum fuel and ballast) and progress to a 'light and low' condition via a descending series of hovers. It may be advantageous to have the aircraft cleared to operate at an overload weight for the purposes of these tests. The presence of external stores and/or the operation of movable aerodynamic surfaces may influence vertical performance. Where this is the case, a range of configurations should be tested at a number of nominal atmospheric conditions. Centre of gravity (CG) position is not usually critical to vertical perfor-

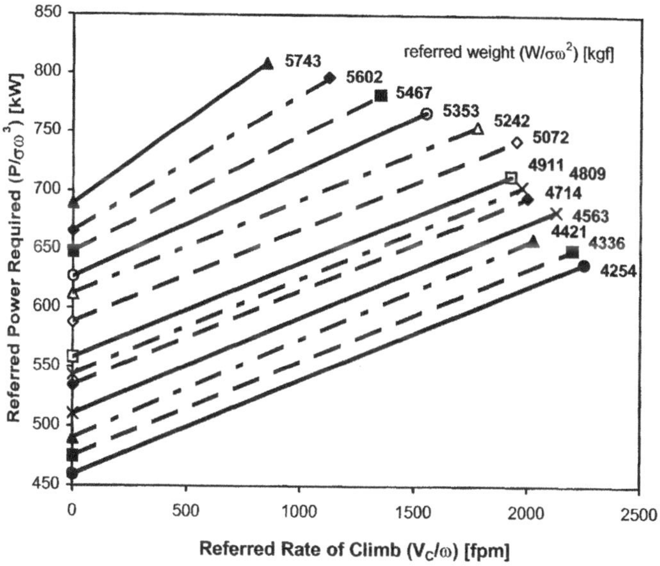

Fig. 3.22 Referred vertical climb test data – MPV method.

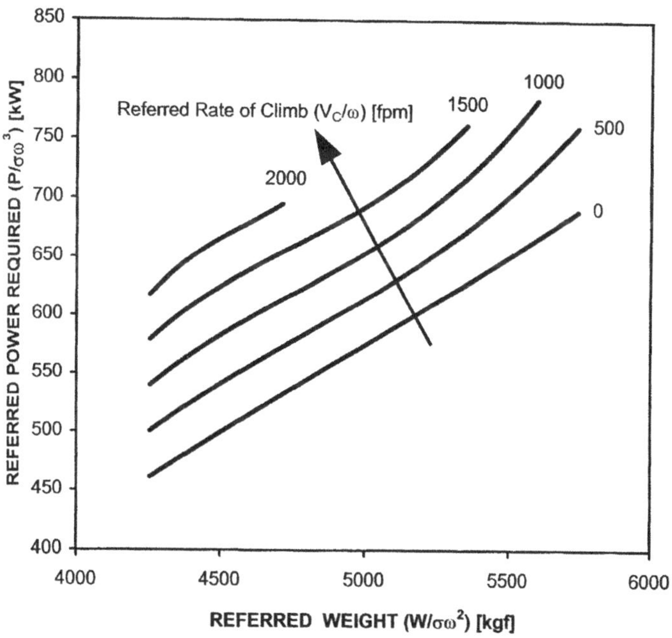

Fig. 3.23 Smoothed vertical climb data.

mance and so a nominal mid-position is normally chosen. The use of airbleeds will reduce the available engine power and consequently the performance of the aircraft when engine-limited. Most testing is conducted with airbleeds OFF but comparative tests with various combinations of heaters, anti-ice and other significant electrical loading may be required. Inside ground effect (IGE) hover points should be flown at a variety of wheel/skid heights to determine the magnitude of ground effect and the IGE/OGE boundary. Rotor speed may be adjusted within the continuous power-on range in order to maintain $\omega/\sqrt{\theta}$ at the desired value. For vertical performance testing, the effect of relative humidity (RH) is usually ignored. However, if the aircraft is required to operate in hot conditions it may be necessary to account for it. For example, at 30°C and sea level, 100% RH makes the true DA about 1000 ft higher than that calculated from H_p and OAT alone.

If only one test site is available, the range of referred weights attainable will be restricted by the local ambient conditions, although seasonal variations may be exploited on a long test programme. A comprehensive performance trial will normally be conducted in temperate conditions and then supplemented by further testing in hot and high conditions. Additionally data may be recorded during ice and snow trials. IGE tests are further restricted by the actual elevation of the test site(s). The necessity to conduct ground-referenced IGE tests at high elevation and in a wide range of temperatures often entails testing at a remote site with a commensurate increase in the flight time budget. Careful planning may be required to achieve the maximum possible referred weight range within the constraints of limited engine performance. It may be necessary to fly 'very heavy and low' rather than 'heavy and high' for a particular condition due to the engine's inability to produce sufficient power at high altitude.

Since the advance ratio is zero in the hover, it is unlikely that there will be significant compressibility effects in most hover conditions unless the ambient temperature is very low and/or the altitude is high. At high thrust coefficients, however, the drag rise Mach number at the blade tip may be sufficiently low to cause a significant increase in drag and thus of power required. Equally, there may be local blade stalling which will contribute to an overall increase in power over and above that predicted by simple theory. Tip effects (TE) are rarely significant except in the case of a dramatic increase in the operational weight or significant change to the theatre of operations. It may be possible, therefore, to reduce the total amount of testing by conducting specific tests designed to expose these effects. Tip effects are most evident when high $\omega/\sqrt{\theta}$ (high N_R and low OAT) is combined with a high AUM. The widest possible OAT range is therefore required for TE testing and this is most easily achieved by hovering at different altitudes. For rotorcraft with fixed N_R, test conduct becomes problematic since, for a given atmospheric condition, there will be a unique relation between θ and σ. Thus, for any given value of $W/\sigma\omega^2$ the only parameter that can be varied independently is the all-up-mass (W). Consequently, a considerable number of changes in mass and a succession of climbs and descents will be required to obtain data for a range of referred rotor speeds at the desired referred weights. If specific testing shows that TE are insignificant, subsequent performance testing may be accelerated by changing rotor speed to increase the available range of $W/\sigma\omega^2$ for a given ambient condition and ignoring the effect of ambient temperature on blade tip Mach number.

3.5.6 Flight test techniques

3.5.6.1 Free-air hovering

Clearly the free-air hover technique necessitates some form of omnidirectional low airspeed measurement system so that stabilized vertical flight, relative to the local air mass, can be achieved in the absence of outside references. A variety of systems have been developed; some are restricted to the research environment while others have been produced commercially, primarily for use within fire control computation in weapon systems. Most systems drive a low velocity indicator (LVI) display in the cockpit that comprises a 'cross-hair' presentation of longitudinal and lateral airspeed. A selection of methods is described below.

The *Hovermeter* was a joint CEV/A&AEE development aimed specifically at trials work. It relies on the fact that the rotor downwash in the hover is effectively vertical, with only small swirl and convergence components. A pair of vanes, one pivoted about the *X*-axis and the other about the *Y*-axis, are mounted in the downwash clear of fuselage turbulence effects. When a small airspeed causes the downwash angles to change, the vanes deflect from their neutral position and cause a commensurate change in cockpit indications. The primary shortcoming of this system is that it cannot compensate for variations in downwash velocity caused by rotor thrust changes.

The *Omnidirectional Air Data System* (Pacer/OADS) is fitted to the AH 64A Apache and the HH-65 Dolphin. It comprises a pair of total pressure ports mounted on each end of a rotating boom that is concentric with the rotor mast and mounted above the rotor disk. A comparison of the two pressure measurements, via a central diaphragm device, determines the magnitude of any horizontal airspeed. A synchro unit allows the resolution of the direction by comparing the phase of the pressure differential signal with a reference.

The *Helicopter Air Data System* (HADS) is fitted to some marks of the AH 1 Cobra and the AH 64D Apache. It comprises a boom-mounted swivelling pitot-static head, which is designed to align itself with the local airflow. Once calibrated, the angles of the swivel and the magnitude of the pitot pressure can be resolved to give omni-directional airspeed indications.

The *Vitesse Indiquee Mesure Installation d'essais* (VIMI) system was developed by the CEV for research but a sophisticated application of the device has been suggested for use in the Tiger attack helicopter programme. The VIMI relies on the principle that the rotor, and therefore the swashplate, will assume a unique orientation relative to absolute vertical when the helicopter is in the hover at a particular AUM. In its simplest form the VIMI compares pitch and roll attitudes to cyclic control positions and resolves a flight director display which tells the pilot what control inputs are required to achieve a hover. However, airspeed is not the only influence on attitude and control position in the hover and so a simple VIMI system can only be optimized for one AUM and CG condition. A more sophisticated system would employ a microprocessor to compute the effects of AUM, CG and tail rotor roll (pedal position).

With the possible exception of the VIMI, the big drawback to all of these systems is that they require independent calibration for a particular installation. This is normally achieved by conducting a series of low airspeed points using a pace vehicle. A range of AUM and IGE and OGE points may have to be flown. In the case of the HADS, this process is known as the 'characterization' of a particular installation. The effect of air density will need to be taken into account to ensure that a system that has been characterized at low altitudes gives accurate indication at higher altitudes.

The installation and calibration of a low airspeed sensing system is clearly impractical for all but the most prolonged test programmes. The formation method, which entails flying the test aircraft close to a suitably equipped 'pace' aircraft and using it as an external hover reference, provides a practical alternative. The optimum separation is twice the rotor diameter of the larger aircraft. Any closer is not only dangerous but can also produce distortion of the results due to downwash interference. The greater the separation the more difficult it becomes to hold a steady relative position. One minor complication is that if the changes in hover power requirement with fuel burn of the two aircraft are significantly different, relative vertical motion can develop over the duration of a test point despite the maintenance of constant power settings. The formation method may be extended to cover vertical climb testing provided the helicopters have compatible vertical agility. In this case the test aircraft initiates the desired ROC and the reference aircraft matches the ensuing climb. It is best to have two crew members in the reference aircraft so that the pilot can concentrate on maintaining vertical flight whilst his colleague observes the test aircraft and advises on power settings. Thus the test aircraft can maintain a constant power setting while the reference aircraft remains responsible for maintaining plan position.

When entering a free-air hover using internal references extreme caution must be exercised to avoid vortex ring, loss of control due to inadvertent rearwards flight and overtorqueing. Our experience suggests that the best technique is to approach the hover from slightly below the desired altitude. The initial deceleration to, say, ten knots less than V_{imp} may be conducted quite rapidly but thereafter it is crucial to anticipate the requirement for additional power as speed is reduced further. A constant decelerative attitude (say about 5 degrees nose-up) is maintained until the longitudinal airspeed is virtually zero at which point a hover attitude is positively selected whilst making due allowance for tail rotor roll and with sufficient power applied to prevent a ROD. If insufficient power is available, the test point should be rejected by selecting a positive accelerative attitude and flying away. Once a coarse hover has been established, small attitude/trim adjustments may be made to refine the zero airspeed condition. Once the hover is established, small collective adjustments in either direction may be made to refine the altitude. To guard against an inadvertent entry into the vortex-ring condition a ROD of greater than about 300 ft/min should not be allowed to develop. Equally the minimum hover height (AGL) should be chosen with due consideration of the anticipated height loss involved in recovering from inadvertent entry into the vortex-ring condition.

The actual method of test for free-air hovering is self-evident. The test aircraft is simply stabilized at the highest desired referred weight, using whatever reference is available, and data is recorded. A rotor speed sweep may be conducted at each hover either to increase the range of tested $W/\sigma\omega^2$ or to evaluate TE. Remember that a change in N_R will always affect $W/\sigma\omega^2$ and will require a collective adjustment to maintain zero vertical speed. This procedure is then repeated as required at decreasing altitudes until the desired range of referred weight has been achieved. The normal aircraft VSI is used to determine the desired ROC. There may well be inaccuracy in the VSI readings, especially near the hover, and so frequent cross checking to the altimeter is required. Once experience is gained with a particular installation, it becomes simple to set up, for example, a 50 ft/min ROD (indicated) to actually achieve a hover.

3.5.6.2 Ground-referenced hovering

Ground-reference vertical performance test techniques are primarily used to determine IGE performance although a limited amount of OGE data may also be gathered. The range of referred weights will be restricted by the elevations of the available test sites.

The method is often very time-consuming since frequent ballast adjustments are required to both establish and maintain a range of referred weights.

The method obviously requires extremely light wind conditions, typically less than 3 kts. A mast-mounted anemometer located at the hover height is often used to ensure that this condition prevails. A pace vehicle may be used in light to moderate winds to achieve zero airspeed provided the wind strength and direction are steady and an into-wind track can be found. Caution must be exercised to avoid downwash interference as the test point is established. It is often best to give the pace vehicle driver the opportunity to determine what ground speed gives the best average still-air condition with the helicopter well clear. This ground speed can then be maintained as the helicopter approaches to establish a good relative position reference. The most usual height reference is the aircraft's radar altimeter. Zero-errors are eliminated by subtracting the residual reading obtained when the aircraft is on the ground from any height datum. An alternative method is the use of a weighted rope whereby a crew member voice-marshals the aircraft in the vertical axis to keep the end of the rope just brushing the ground. External references may also be used but these will be subject to parallax errors.

A series of ground-referenced hovers are flown at different heights to document ground effect and to determine the IGE/OGE boundary. Once this has been accomplished, testing will concentrate on role-relatable hover heights (particularly in the case of rotorcraft destined to conduct roles requiring a load lifting or winching capability). A test session will normally be commenced at maximum AUM. If desired, a rotor speed sweep can be conducted at each hover. The AUM should be maintained within about 1% of the nominal value and so frequent reballasting (perhaps once every 10 minutes) will be required to compensate for fuel burn. Once the range of heights has been completed at a particular weight, ballast is reduced and the process repeated. Accurate data requires the minimum of pilot interference during the test point itself. Consequently the collective is not moved during data gathering and the aircraft is allowed to drift up and down gently as necessary. If height excursions exceed about ±10% of the datum the weather will generally be unsuitable for accurate testing.

3.5.6.3 *Tethered hovering*

Tethered hovering is a versatile technique and is widely used. The helicopter is attached to the ground by a cable of known length fitted with a tensiometer. A wide range of effective weights can be evaluated in a short time as the rotor thrust can be varied from a low to a high value without the need for reballasting. Although the minimum and maximum weights that can be studied are usually the same as can be flown using the ground referenced method, the big advantage of this method is that height control ceases to be a problem. Therefore given good weather conditions and access to a range of cable lengths, accurate IGE and OGE performance figures can be obtained in a short time. A further possible advantage is that of safety: a multi-engine helicopter may be flown throughout the test at an actual AUM for which it has OEI hover performance. However, unless a high altitude test facility is available, data at the higher referred weights will be lacking. As with the ground referenced method, tethered hovering tests demand very light winds. Turbulence or wind fluctuations tend to play havoc with hovering accuracy and the cable tension.

Disadvantages of this method are those associated with the items of additional equipment and extra personnel required to conduct a potentially hazardous test technique in reasonable safety. A tensiometer, usually a strain gauge device, is placed in series with the aircraft's cargo hook. It is obviously undesirable to allow the tensiometer to fall to the ground from any significant height and so test procedures will have to be conducted accordingly. Some installations employ a modified cargo hook and the tensiometer remains attached to the aircraft at all times. The tensiometer typically drives a cable tension indicator that is temporarily mounted in the cockpit where it can be seen by both the observer and the pilot. More sophisticated test installations will be capable of providing the pilot with a 'cross-hairs' type indication of cable angle. A single sensitivity setting may be inappropriate for all cable lengths, in particular they are often too sensitive for use with short cable lengths. The cockpit indicator is placed so that the pilot can easily scan to it whilst retaining a good view of external hover references. Ground marshallers may be required in the absence of a cable angle indicator and, possibly, as a back-up to such a device, to assist the pilot in maintaining the cable vertical. A range of cable lengths designed to give the required wheel/skid height will be required. The cables need to be proof loaded and certificated to the maximum likely working tension. The tethering point should be in a clear area well away from structures that could cause recirculation. The surface should be level and smooth and, preferably, hardened to facilitate handling the cables. As with the cables the actual tie-down ring should be proof-loaded and certificated. Well-briefed and practised ground handlers will be required to facilitate cable changes. Ground observers are required to observe the cable angle and indicate corrections to the pilot accordingly. The employment of a site controller, in radio contact with the pilot, is an effective risk reduction measure as is the carriage of a rear crew member to monitor the cable angle and advise the pilot on positioning.

The cable should be as close to vertical as possible when data is taken, the maximum error which can be tolerated is typically $\pm 10°$. Aircraft plan position must thus be held very accurately, this is particularly important when short cable lengths are used. The optimum method of maintaining plan position varies according to cable length. For short cable lengths (about 20 ft or less) the best hover cues are the normal external references. The exclusive use of a cable angle indicator may well lead to overcontrolling. For very short cables the necessary accuracy may be achieved by observing the position of the aircraft's shadow relative to a feature on the ground. A cable angle indicator becomes more 'user-friendly' with cable lengths greater than about 20 ft. The best technique may be to identify the correct position using the indicator and then to maintain that position using normal hover techniques.

The cable tension, and therefore the effective AUM of the aircraft, can be varied very quickly and so the most expedient test method is to explore a range of tensions for a given cable length, N_R and loading and then to change N_R, cable or ballast before repeating the process. For a particular condition the cable tension will vary from a maximum that is limited by the effective mass, hook limits or power, to a minimum limited by the requirement to maintain a small nominal tension to aid precise height control.

Since the force applied to the helicopter by the cable tension is below the CG, an excessive cable angle is destabilizing therefore the normal and emergency cargo release mechanisms must be checked and be functional before commencing these tests. After the cable is attached, the cargo master is usually selected to ON to facilitate a rapid emergency release. Usually the after take-off checks are performed in an IGE hover over the tether point before the aircraft is climbed vertically until the cable goes taut.

It is important to avoid snatching the cable since this could cause either structural damage or a premature failure of the cable. Once minimum cable tension has been established, power is progressively applied to achieve the desired value. Longitudinal compensation is often required as the effective CG changes with increases in hook load. Once the test condition has been stabilized it is unwise to chase the cable tension, it being better to leave the collective fixed and take an average tension reading. Data recording will facilitate this process. It may be advantageous to allow the co-pilot/observer to control the collective while the pilot concentrates on maintaining an accurate plan position. If the aircraft drifts well out of position it is important to allow the cable to go slack immediately before resuming the correct plan position and reapplying tension. Helicopters have been rotated into the ground by a combination of excessive cable angle, high cable tension and low control power. On completion of the tests, it is vital that the pilot receives a verbal assurance from the groundcrew that the cable has been disconnected from the aircraft before being marshalled from the test site.

3.5.6.4 Vertical climbs

Both reduced power verticals and maximum power verticals can be conveniently combined with OGE hover tests. Once the hover data has been taken, the desired torque increment can be added and the aircraft timed through an altitude band of around 400 ft, or for 20 seconds. Split times are recorded at intermediate altitudes so that a mean, smoothed ROC can be determined. A more accurate result can be obtained by adjusting the mean altitude to ensure that the referred weight remains essentially constant for each hover/RPV/MPV combination. If static droop is present, N_R will change as power is added and so an adjustment will have to be made to regain the desired referred rotor speed once the climb has been stabilized. It is easy to underestimate the altitude that will be required to achieve this. A further complication arises if the test team attempts to target a constant value of referred power available as this will involve adjustment of the collective as altitude increases and air temperature and density fall. In practice such adjustments are both difficult to make accurately and are very time-consuming. Consequently, an average torque value is recorded at a fixed collective position. If desired, TE can be evaluated using similar techniques to those applicable to the hover case.

A plot of referred ROC versus referred power for a given referred weight reveals that the relationship between power increment and ROC is approximately linear and that the lines are broadly parallel. A considerable amount of time and effort can therefore be saved by conducting only hover and MPV points at a given referred weight. With careful planning a very efficient use of time can be achieved as follows:

(1) Conduct a hover (H_1) at the altitude (H_{p1}) giving the highest desired referred weight (W_1) and establish the hover power.
(2) Descend and establish H_2 at H_{p2} ($=H_{p1} - 500$ ft) and W_2.
(3) Set maximum power and conduct MPV_1, timing the ROC in the altitude band $H_{p1} - 100$ ft to $H_{p1} + 300$ ft. The extra 100 ft allows for a small increase in H_p to compensate for the fuel burned between H_1 and MPV_1 in order to maintain W_1.
(4) Descend and establish H_3 at H_{p3} ($=H_{p2} - 500$ ft) and W_3. Conduct MPV_2 as above.
(5) Repeat the process until the desired data is acquired.

This method relies on the expeditious establishment of successive test conditions. Only about 5 minutes can elapse between the hover and MPV points for a given referred weight otherwise the fuel burn will be excessive and more than 100 ft will be required to compensate for it. It is important to ensure that $\omega/\sqrt{\theta}$ is kept constant. In particular the effects of static droop must be accounted for when conducting MPVs by setting a slightly higher N_R in the hover before power is applied. During the MPV, adjustments are made to the N_R to achieve the required value of $\omega/\sqrt{\theta}$ at the point where data is taken.

3.5.6.5 *One-engine inoperative testing*

All vertical performance test methods remain valid in an OEI situation but particular attention must be paid to the engine rating used. It may be necessary to get dispensation to operate at, for example, a 5-minute power rating for relatively prolonged periods in order to accomplish adequate testing. Under normal circumstances the power required will not change OEI and the engine's performance will still be obtainable from test bed data. However, some aircraft require a different N_R to be used OEI and so extra testing will be required in this case. As a first estimate it may be permissible to use data gathered during 'all engines operative' testing and simply apply the OEI limits to get the desired performance information. However, actual OEI testing should be conducted later to confirm the estimates. OEI operation may in fact consume more power because of asymmetric loading of the transmission and adverse engine inlet conditions.

3.6 CLIMB PERFORMANCE TESTING

Despite the majority of helicopter roles requiring flight only at low level there is sometimes a need to climb rapidly and efficiently to altitude. Although the main object of *partial* climb performance testing is the determination of the optimum climb speed and associated rate of climb, of equal importance is the variation of these parameters with density altitude, rotor speed, power available and aircraft mass. Associated tests are the documentation of pressure errors and the assessment of engine and rotor governing characteristics. The pressure errors may differ markedly from those recorded during level flight tests because of changes in main rotor wake strength and direction.

3.6.1 Analysis of a climbing helicopter using momentum theory

Before describing how the experimental method can be adapted for use in climb performance testing it is worthwhile to use simple momentum theory to analyse the factors affecting a helicopter in climbing flight. Figure 3.24 shows the change in velocity that occurs as flow passes through a climbing actuator disk.

As before, analysis begins by balancing the forces acting on the centre of gravity:

$$T\cos\gamma_d = mg + d_v = mg + \frac{1}{2}\rho V_v^2 S_v$$

$$T\sin\gamma_d = d_f = \frac{1}{2}\rho V_f^2 S_f$$

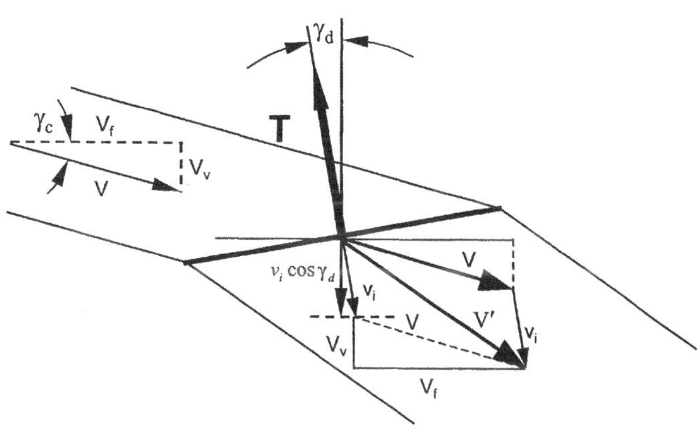

Fig. 3.24 Momentum disk theory applied to a climbing rotor.

Therefore:

$$\tan\gamma_d = \frac{\rho V_f^2 S_f}{2mg + \rho V_v^2 S_v}$$

Note how the airspeed of the helicopter has been broken down into horizontal and vertical components, each responsible for drag generation (calculated using the concept of drag area). Applying the elliptic wing analogy:

$$T = 2\rho A v_i V'$$

Now:

$$V' = \sqrt{V_f + v_i \sin\gamma_d)^2 + (V_v + v_i \cos\gamma_d)^2}$$

$$= \sqrt{V_f^2 + 2V_f v_i \sin\gamma_d + v_i^2 \sin^2\gamma_d + V_v^2 + 2V_v v_i \cos\gamma_d + v_i^2 \cos^2\gamma_d}$$

$$V' = \sqrt{V^2 + 2v_i(V_f \sin\gamma_d + V_v \cos\gamma_d) + v_i^2}$$

and:

$$v_i = \frac{T}{2\rho A V'} = \frac{T}{2\rho A\sqrt{V^2 + 2v_i(V_f \sin\gamma_d + V_v \cos\gamma_d) + v_i^2}} \qquad (3.1)$$

Equation (3.1) can be solved iteratively to yield the variation of induced velocity with forward speed and vertical velocity, see Fig. 3.25. From the figure it can be seen that for typical forward velocities ($\approx 3\,v_{ih}$) the variation of induced velocity with ROC is negligible. This is a very convenient result since it means that the extra power required to climb can be simply added to that required for level flight. Thus the climb power is given by:

$$P_{climb} = 1.2Tv_i + \frac{1}{2}\rho V_f^3 S_f + \frac{1}{8}\rho bcRC_D V_T^2(1 + 4.3\mu^2) + TV_v + \frac{1}{2}\rho V_v^3 S_v \qquad (3.2)$$

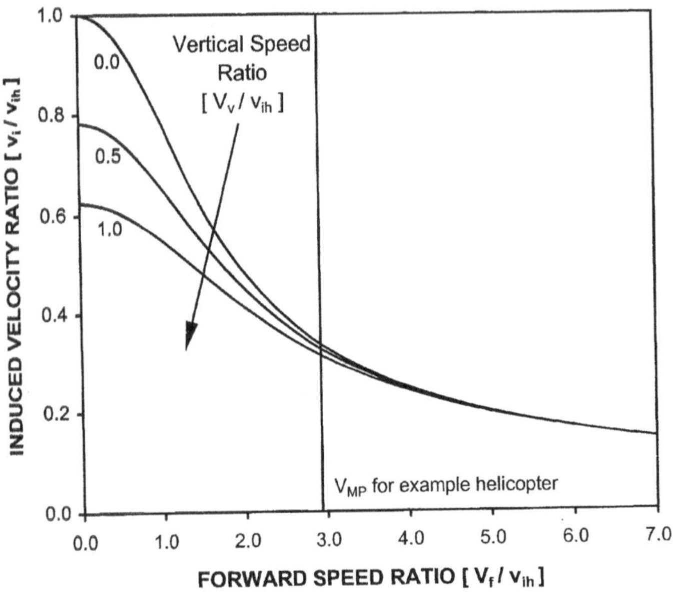

Fig. 3.25 Variation of induced velocity with RoC and TAS.

Alternatively if the power available, in excess of the power required for level flight (PFLF), is known then:

$$P_{\text{excess}} = TV_v + \frac{1}{2}\rho V_v^3 S_v$$

and if the parasitic drag and download are small compared with all-up-weight then:

$$V_v = \frac{P_{\text{excess}}}{mg} \tag{3.3}$$

Equation (3.3) suggests that the climb performance is directly proportional to the excess power available and therefore the forward speed for maximum ROC (V_Y) will equate to the minimum power speed (V_{MP}).

3.6.1.1 *Estimating the climb performance*

The accuracy of this approximation can be verified by an example. Consider a helicopter operating at SL-ISA with a 6.5 m radius 4-bladed rotor of 0.4 m chord. If the horizontal drag area is 2 m^2 and the main rotor profile drag coefficient is 0.01 the variation of PFLF with TAS and AUM will be as shown in Fig. 3.26. Suppose the maximum power available under SL-ISA conditions is 1 MW. Then from Equation (3.3) the variation of RoC with TAS and AUM can be determined with relative ease, see Fig. 3.27. Comparing this approximate method with the results from a fuller approach (Fig. 3.28) shows that the two are in close agreement for airspeeds in excess of V_{MP}.

3.6.1.2 *Problems with wasted power*

Using data from flight tests (see Fig. 3.29) it is clear that estimating climb performance from level flight data is not straightforward in practice. The difficulty lies in determining

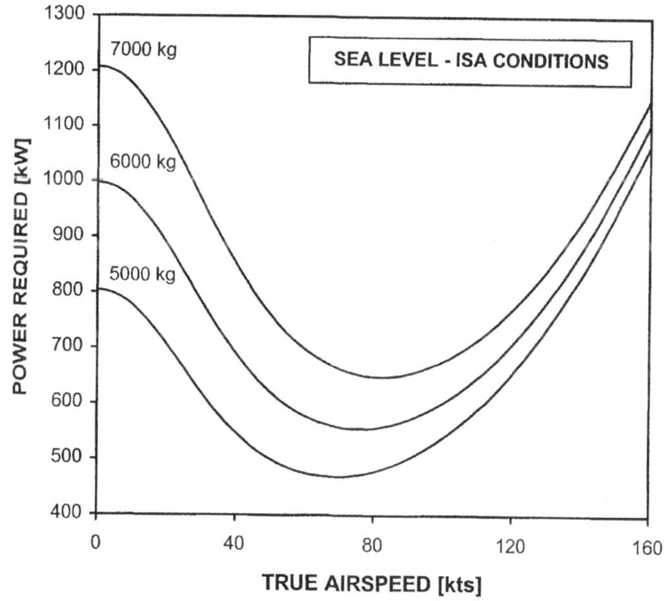

Fig. 3.26 Variation of power for level flight with AUM.

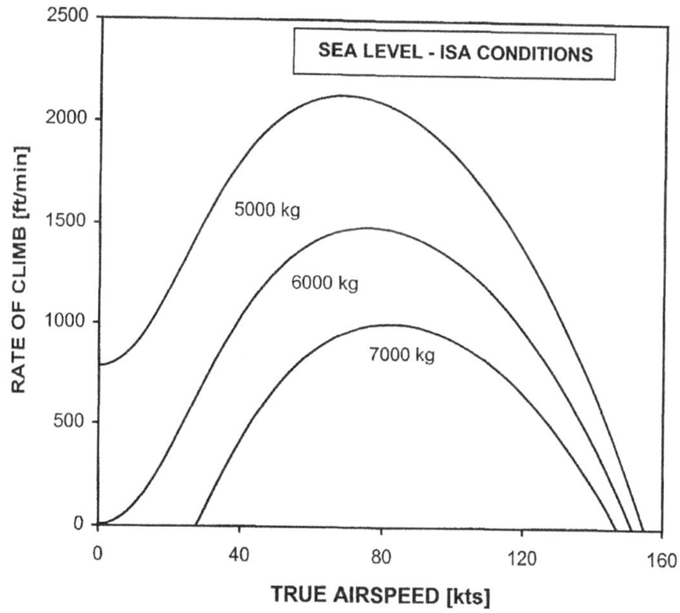

Fig. 3.27 Variation of RoC with TAS and AUM.

the percentage of excess power that is available as climb power. For an accurate estimate it is necessary to determine the precise amount of power wasted through transmission losses and in generating the extra anti-torque force needed as a consequence of the greater collective pitch applied to the main rotor. The simplest

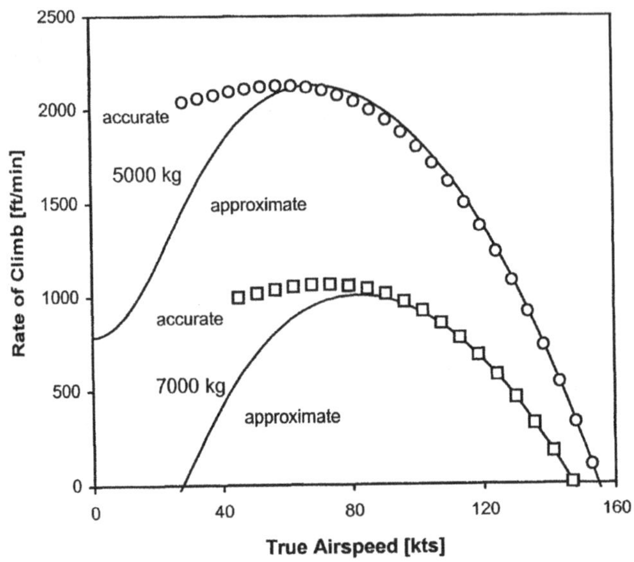

Fig. 3.28 Estimates of climb performance.

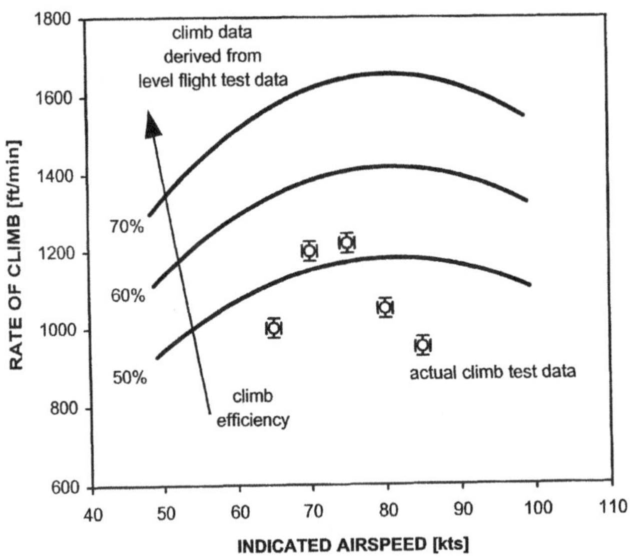

Fig. 3.29 Estimates of climb performance based on level flight data.

approach of assuming that all the excess power available is taken up as climb power will typically lead to an overestimate of the actual rate of climb, although the speed for best rate of climb will usually be predicted quite accurately. More accurate estimates of climb performance can be made by incorporating the concept of a climb efficiency factor (η_c) [3.8]:

$$V_v = \frac{\eta_c P_{excess}}{mg}$$

3.6.1.3 *Measuring climb performance*

In the case of partial climbs ground effect is not considered and the parameter $\omega/\sqrt{\theta}$ cannot be controlled since the outside air temperature changes as the rotorcraft climbs. Therefore, the familiar power required relationship can be written as:

$$\frac{V_c}{\omega} = f\left(\frac{P}{\delta\sqrt{\theta}}, \frac{W}{\delta}, \frac{V}{\omega}\right)$$

$$\frac{V_c}{\omega} = f\left(\frac{P}{\sigma\omega^3}, \frac{W}{\sigma\omega^2}, \frac{V}{\omega}\right)$$

The equations can be re-written in an alternative form introducing a power-to-weight ratio:

$$\frac{V_c}{\omega} = f\left(\frac{P}{W\sqrt{\theta}}, \frac{W}{\delta}, \frac{V}{\omega}\right)$$

$$\frac{V_c}{\omega} = f\left(\frac{P}{W\omega}, \frac{W}{\sigma\omega^2}, \frac{V}{\omega}\right)$$

3.6.2 Flight test techniques

The main problem with using an 'experimental' approach to reducing climb performance data is the difficulty associated with maintaining the relevant referred parameters constant. It is therefore common practice to use test day conditions and refer the data afterwards, if required, rather than target specific combinations of referred parameters during the testing itself. There are two techniques that may be adopted to minimize the effect of reducing aircraft mass on the measured climb performance:

- Mean-all-up-mass method;
- Constant mean altitude and power/weight ratio method.

The *mean-all-up-mass method* involves documenting the rate of climb through the same altitude band (typically 1000 ft) with the test airspeeds being flown in a particular order. Starting with the lowest, the test airspeeds are progressively increased up to the maximum *then* progressively reducing intermediate speeds are evaluated back down to the minimum, see Fig. 3.30. A mean line drawn between two curves, corresponding to the test points from minimum to maximum and vice-versa represents the climb performance for a mass equal to the average for a complete run.

The *constant power/weight ratio* method is slightly different. Although the altitude band remains the same, as with the previous method, the power (or torque) is reduced progressively as fuel is burnt to maintain the power/weight ratio constant. The test speed is changed progressively from minimum to maximum *or* vice-versa, see Fig. 3.31. Figure 3.32 compares the results obtained using both methods. Note the similarity in the curve fit obtained.

As in other tests that involve measurement of the vertical speed, it is common practice to determine the rate of climb by documenting incremental altitude and

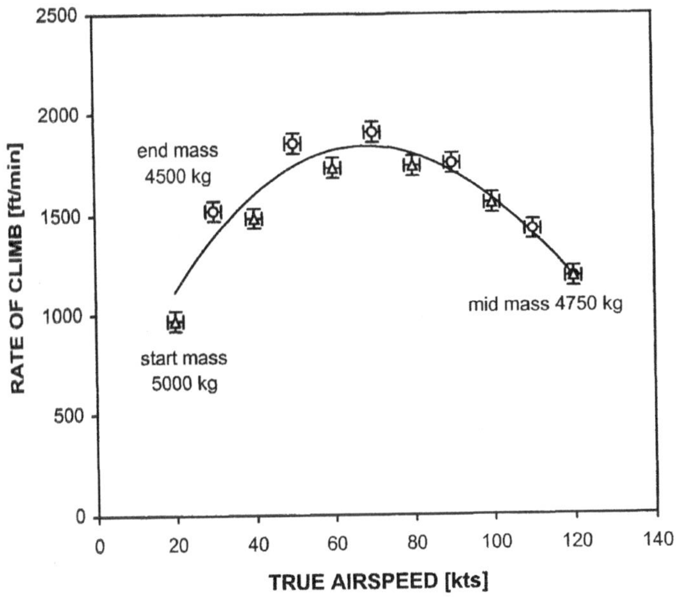

Fig. 3.30 Climb performance – mean-all-up-weight method.

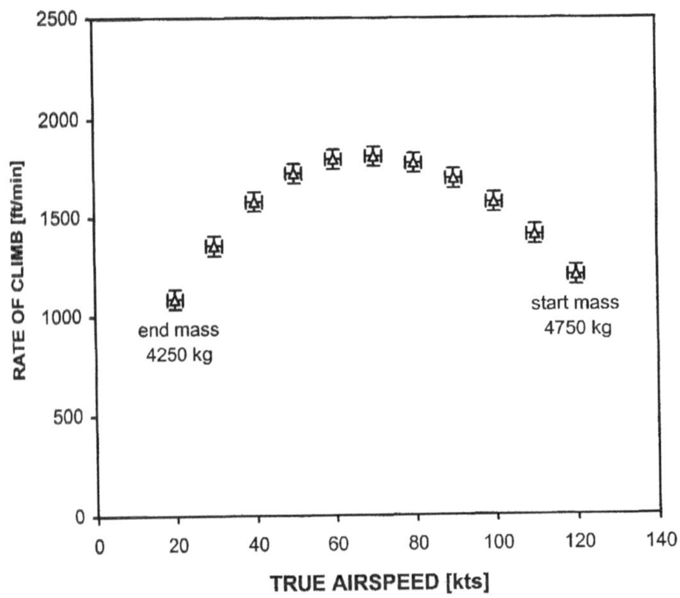

Fig. 3.31 Climb performance – constant power/weight ratio method.

elapsed time. Charts of altitude versus time are plotted for each speed tested and used to obtain a *smoothed* rate of climb by means of a straight-line fit. This process compensates for data scatter and highlights any non-linearities associated with the helicopter drifting away from trim. An analogous process can also be used to determine

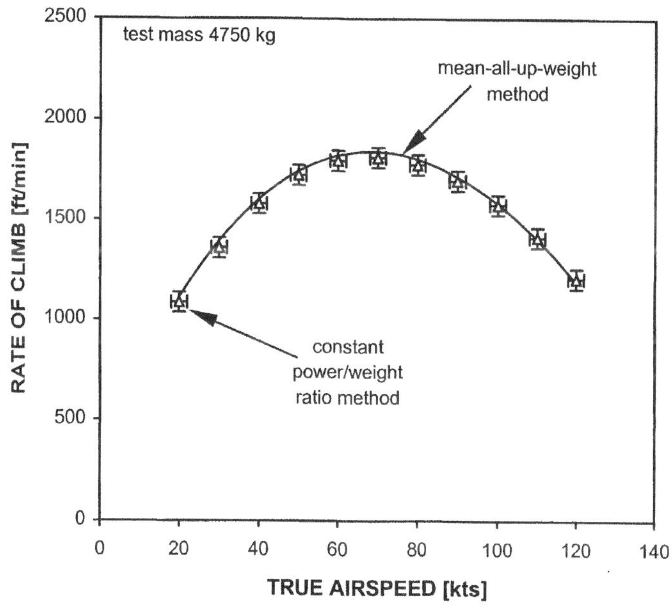

Fig. 3.32 Climb performance – comparison of methods.

the rate of descent in a flight idle glide or an autorotation. Data may be gathered economically if climbs and descents are combined using a *sawtooth* flight profile.

Consolidated plots can be compiled once data has been gathered for a range of conditions. Typically data is presented for a range of altitudes at the same all-up-mass or for a range of all-up-masses at the same altitude, see Figs 3.33 and 3.34 which show typical variations of climb performance with mass and altitude. These consolidated plots are then used to draw-up climb schedules for assessment during *ceiling climb* tests.

3.6.3 Ceiling climb tests

The variation in rate of climb with altitude and/or AUM in forward flight can be used to generate a range of climb schedules: for best rate of climb, for best angle of climb or minimum time to height. Consequently, it is often necessary to validate these schedules during a series of ceiling climbs. In addition to determining if the expected performance is realized, these tests are also used to establish whether any modification to the schedule is required for handling reasons.

Using a climb schedule, such as that shown in Fig. 3.35, a sustained climb is made to the maximum permitted altitude or the service ceiling (defined as ROC = 100 ft/min), whichever is the lower. Data is recorded at periodic intervals based on height achieved and includes:

- Pressure altitude;
- Elapsed time;
- Indicated airspeed;
- Fuel state;

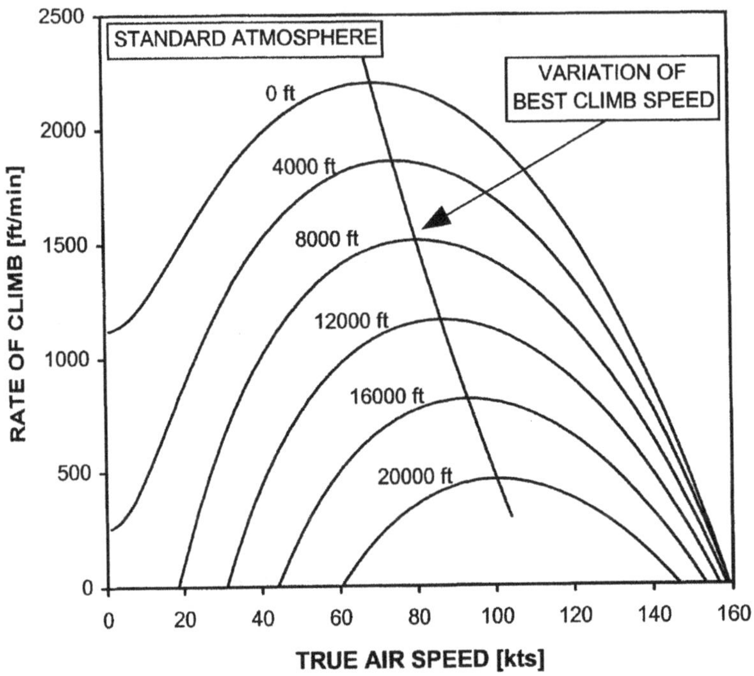

Fig. 3.33 Variation of climb performance with altitude.

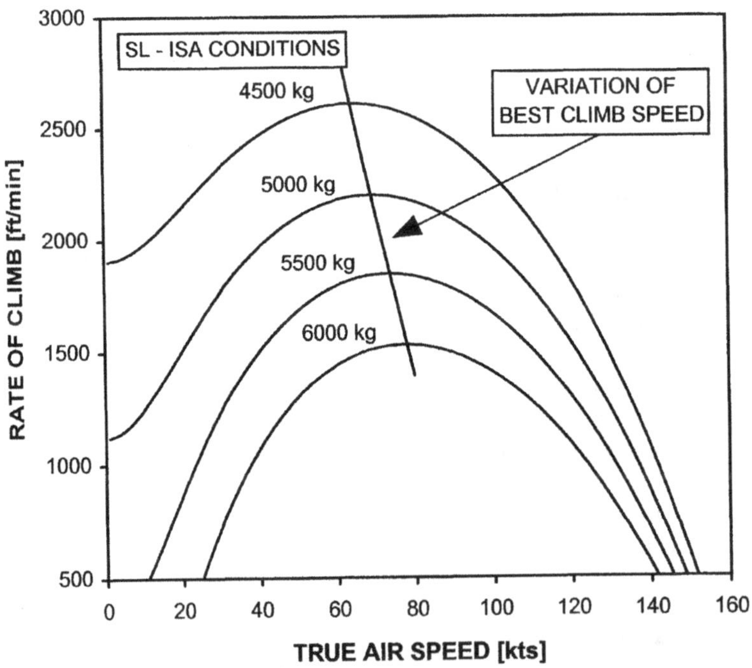

Fig. 3.34 Variation of climb performance with mass.

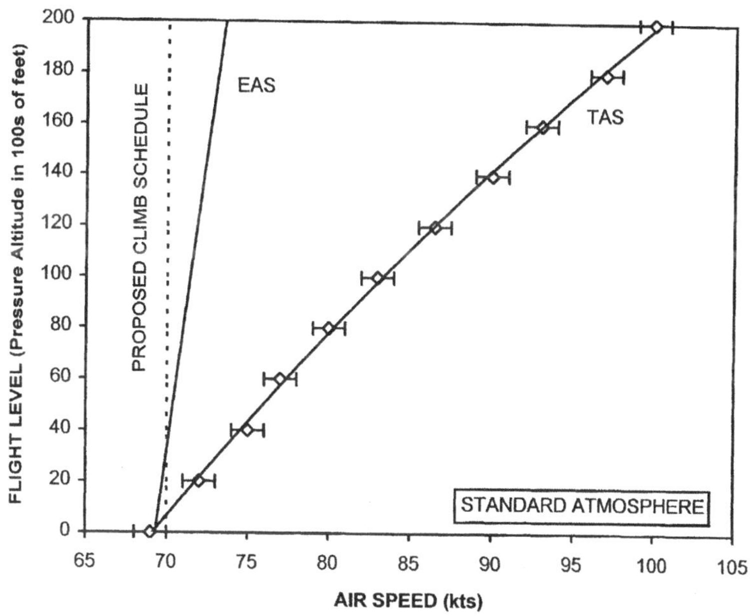

Fig. 3.35 Derived climb schedule.

- Indicated rate of climb;
- Power parameter (torque or collective pitch);
- Engine speed and temperature;
- Rotor speed;
- Control positions;
- Vibration.

At low altitudes height intervals of 1000 ft are commonly used. The interval reduces to every 500 ft and then every 200 ft or even every 100 ft as the excess power diminishes.

3.7 DETERMINATION OF PERFORMANCE LIMITED PARAMETERS

Quite often one of the objectives of a flight test programme is to determine the parameter that will limit the performance under the atmospheric conditions likely to be experienced in the role. Under certain atmospheric conditions, usually hot and high, the engines, rather than the transmission will limit the performance. It is therefore necessary to determine the precise limiting factor for the conditions specified. The method is relatively simple but requires access to engine performance data such as that shown in Figs 3.36 and 3.37. The analysis proceeds by obtaining the actual limiting values from the aircraft documentation; these values will be called Q_{LIMIT}, T_{LIMIT} and N_{LIMIT}. Using the pressure altitude and air temperature specified the *transmission limited referred power* (TLRP) $= Q_{LIMIT}\omega/\delta\sqrt{\theta}$ is calculated. The *engine temperature limited referred power* (ETLRP) is obtained using Fig. 3.21 and the

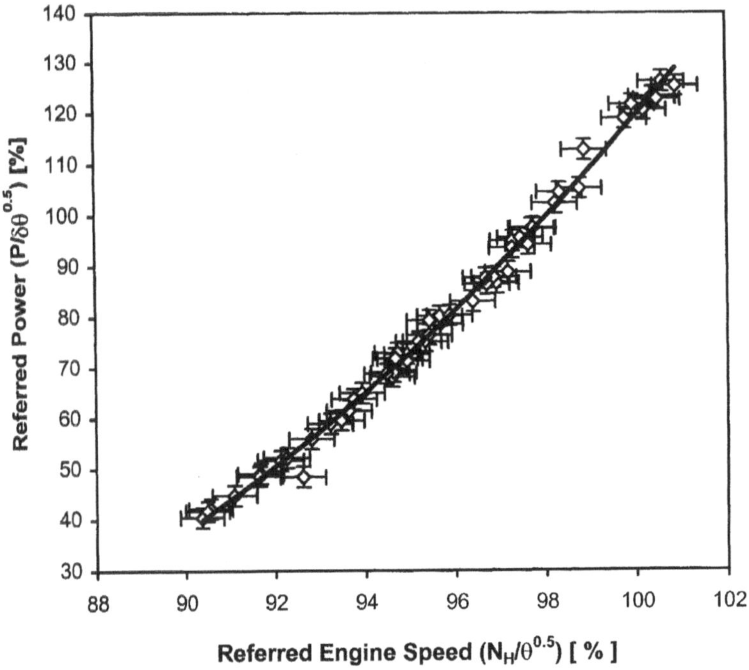

Fig. 3.36 Engine test data – power versus speed.

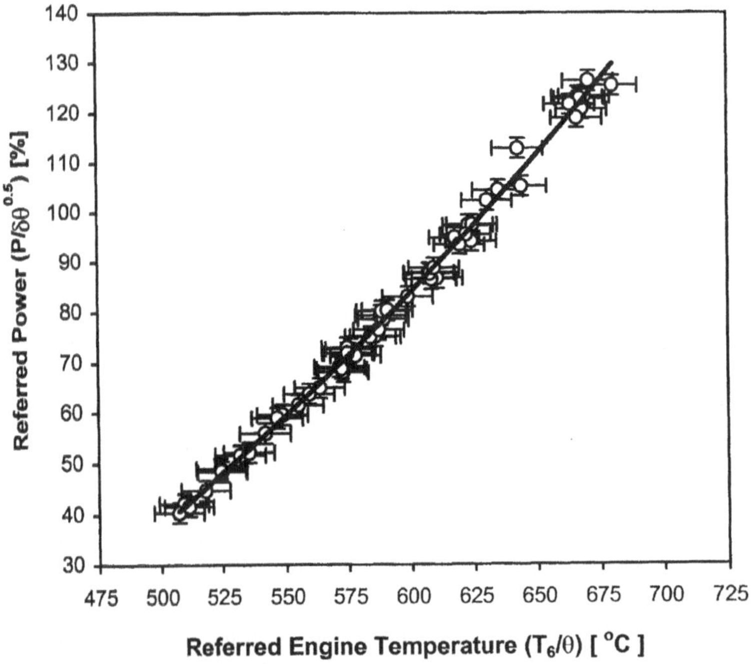

Fig. 3.37 Engine test data – power versus temperature.

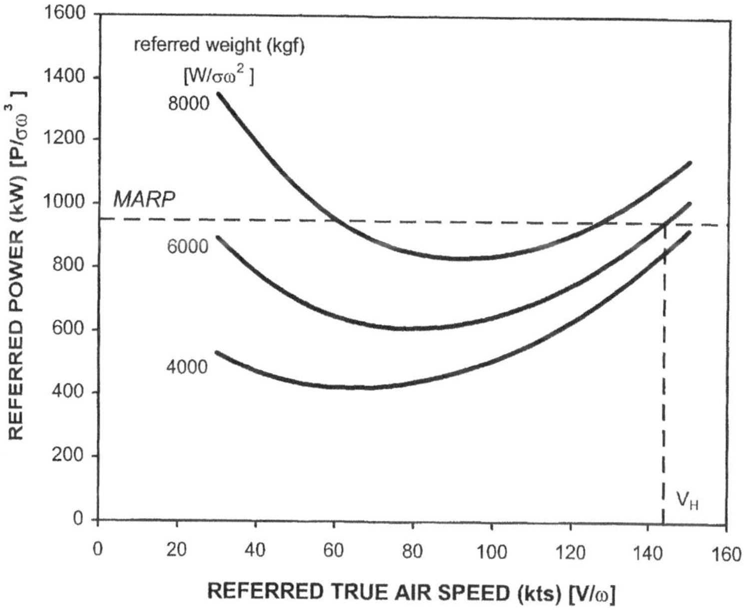

Fig. 3.38 Limiting performance – level flight speed.

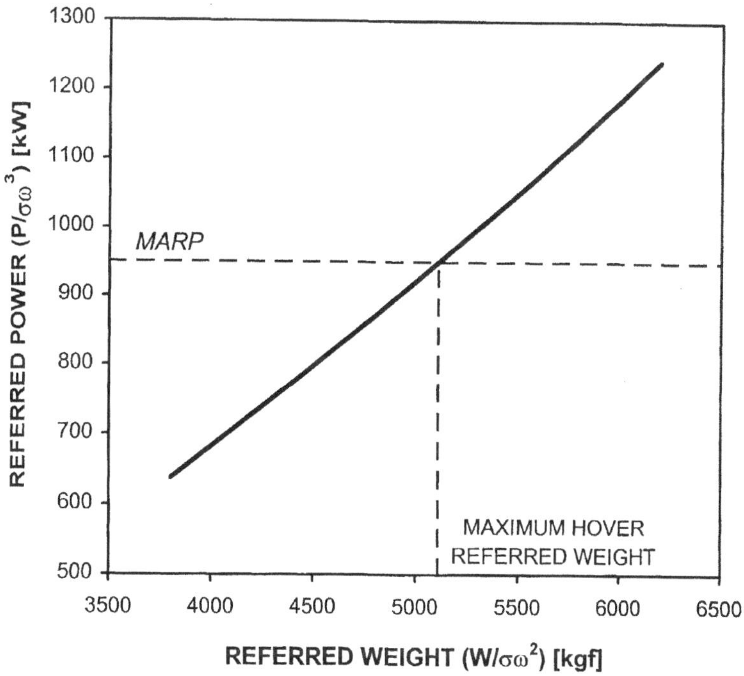

Fig. 3.39 Limiting performance – hover mass.

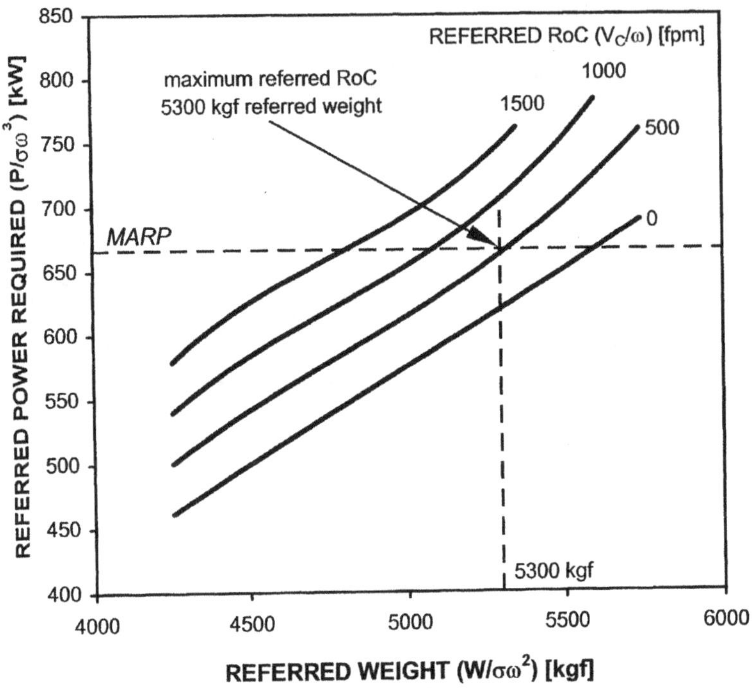

Fig. 3.40 Limiting performance – vertical rate of climb.

appropriate value of T_{LIMIT}/θ. Likewise the *engine speed limited referred power* (ESLRP) is found using Fig. 3.22 and the appropriate value of $N_{LIMIT}\sqrt{\theta}$. If either ETLRP or ESLRP is less than TLRP then the performance will be engine-limited under the conditions specified.

It is now possible to determine the maximum performance available by choosing the lowest limited referred power, now called LRP_{MIN}, and for a given rotor speed calculating the maximum available referred power (MARP) from:

$$\mathrm{MARP} = \frac{P_{MAX}}{\sigma\omega^3} = \frac{\mathrm{LRP_{MIN}}}{(\omega/\sqrt{\theta})^3}$$

This value can then be used to determine a variety of performance limited parameters, such as: the maximum level flight speed, V_H, (see Fig. 3.38), maximum hover mass (Fig. 3.39) or maximum vertical rate of climb at a given hover mass (Fig. 3.40).

Chapter 4
Stability and Control Theory

4.1 INTRODUCTION

The detailed study of helicopter stability and control is a complex matter that is beyond the scope of this book. Therefore, a simpler approach will be taken making use of several common assumptions: the rotor speed remains constant; in disturbed flight the rotor behaves as if the motion were a sequence of steady conditions; and the lateral/directional and longitudinal motions are decoupled. The rotor is, therefore, regarded as responding instantaneously to speed and angular rate changes. Stability and control theory aims primarily at finding the factors involved in designing an aircraft with satisfactory flying qualities and in order to make an accurate assessment of an aircraft's handling qualities it is important that these factors are understood. The main rotor provides the largest contribution to the total stability and the way in which it flaps is most important. Consideration of rotor flapping in the hover and forward flight, before an investigation of helicopter stability, will be instructive.

Practical helicopter and autogyro flight was not possible until rotor hinges were fitted to relieve the large bending stresses and rolling moments that arise in forward flight. The most important of these hinges is the *flapping hinge* that allows the blade to flap, that is to move out of the plane of rotation. However a blade that is free to flap experiences large Coriolis moments [4.1] in the plane of rotation and therefore a further hinge called the *drag* or *lag hinge* is required to relieve these moments. Finally, another hinge, the *feathering hinge*, is provided to allow adjustment of the blade pitch, or feathering, angle. A rotor incorporating these hinges is referred to as fully articulated. The sequence of hinges is not always the same. The flapping hinge is usually the most inboard hinge but some helicopters have intersecting flapping and drag hinges and some have the drag hinge outboard of the feathering hinge. Also, as described below, hinge axes are not always mutually perpendicular.

It is advantageous to provide aerodynamic damping to the flapping motion of a rotor blade. A blade that is free to flap will move such that the combination of the lift, blade weight and centrifugal force are in equilibrium. However a further *spring effect* can arise if reduced lift occurs with increased flap. Such a positive pitch-flap coupling can be engineered by arranging for the blade pitch to be reduced with increased flap angle by means of a skew of the flap hinge line so that it is no longer perpendicular to the radial axis of the blade. The angle of skew is referred to by the symbol δ_3, hence the name *the delta three hinge*. The blades of a two-bladed rotor are usually mounted as a single unit on a teetering hinge. There are no drag hinges as underslinging the rotor greatly reduces the lagging moments generated by the Coriolis effect. When the underslung rotor flaps the radial velocities of points above the hinge line are negative whereas those below are positive. The corresponding Coriolis forces are of opposite sign and if the hinge height is chosen carefully the moment at the blade root can be reduced to a second-order effect. Preconing also helps in this respect.

Several problems are associated with articulated rotor heads. The bearings of the hinges and dampers operate under very high centrifugal loads, requiring frequent servicing and maintenance. When the number of blades is large the hub can become extremely complex and bulky, especially if automatic blade folding is incorporated; thereby contributing a large proportion to the total drag of the helicopter. Improvements in blade design and construction have led to the development of the *hingeless rotor*. The flapping and lagging hinges have been dispensed with, the motion in these two senses being allowed by flexible elements within the hub and blade root.

4.2 AERO-DERIVATIVES FOR ROTORCRAFT

During the different phases of a helicopter's flight it is subject to a number of forces and moments that must be balanced if the aircraft is to remain in trim. Alternatively should the pilot wish to control the helicopter by modifying its flight path he must generate an out-of-balance condition by moving the flight controls from their trim position. On the other hand atmospheric disturbances may upset the force and moment balance and cause the flight path to change unintentionally. The nature of the out-of-balance condition following such a gust affects the stability characteristics of the rotorcraft by dictating whether the helicopter will return to trim without pilot intervention. The amount by which flight controls and atmospheric disturbances modify the forces and moments acting on a rotorcraft are the key to determining its stability and control characteristics. Typically, these forces and moments can be changed by any of the following, alone or in combination:

(1) Disturbances in linear speeds or angular rates;
(2) Changes in main rotor and tail rotor blade pitch;
(3) Movement of the centre of gravity position.

Measuring or predicting how a given force or moment will change as a result of variations in the above parameters is fundamental to determining the handling qualities of a particular air vehicle. A *shorthand* based on the *derivative* has therefore been developed to simplify discussion of these effects.

4.2.1 The derivative

The concept of the derivative is best understood by means of a simple example. Consider an aircraft in trimmed straight and level flight at an angle of attack (AOA) α_0. If the variation of the lift and drag can be simply portrayed, as in Fig. 4.1, it is possible to determine their trimmed values (L_0 and D_0).

Suppose as a consequence of a gust the AOA is increased by $\delta\alpha$ to α_1. From Fig. 4.1 it is a simple matter to determine L_1 and D_1. If, however, only small changes in AOA occurred it would be possible to calculate changes in lift and drag using the local slope around the trim point. Thus:

$$L_1 - L_0 = \delta L = \frac{dL}{d\alpha}\delta\alpha \quad \text{and} \quad D_1 - D_0 = \delta D = \frac{dD}{d\alpha}\delta\alpha$$

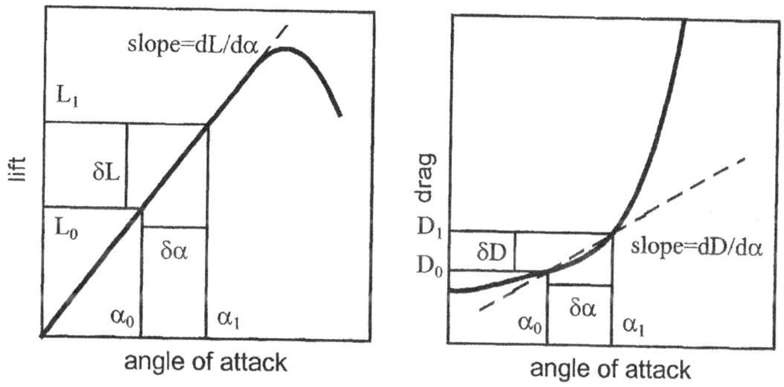

Fig. 4.1 Variation of lift and drag with angle of attack (AOA).

Since these slopes represent the rate of change of lift and drag with AOA around the trim point, they can be used directly to calculate the change in these forces. As shall be seen later, all equations of motion assume the aircraft is in equilibrium prior to being disturbed by a gust or pilot action. Thus in our example, α, L and D should represent the *change* in AOA, lift and drag from their equilibrium values. Therefore, written more formally:

$$L=\frac{\mathrm{d}L}{\mathrm{d}\alpha}\alpha \quad \text{and} \quad D=\frac{\mathrm{d}D}{\mathrm{d}\alpha}\alpha$$

Alternatively, taking note of the symbols used for vertical and horizontal forces and standard *aero-derivative* notation:

$$L=-Z=L_{\alpha}.\alpha=-Z_{\alpha}.\alpha \quad \text{and} \quad D=-X=D_{\alpha}.\alpha=-X_{\alpha}.\alpha$$

In fact, modifications to the AOA can result from changes in forward speed (u) and from changes in vertical speed (w). Thus more generally:

$$X=X_u.u+X_w.w \quad \text{and} \quad Z=Z_u.u+Z_w.w$$

4.2.2 Derivation of aerodynamic derivatives

So far, the discussion has been restricted to force changes caused by modifications to the angle of attack. Also small changes (*small perturbations*) and linear rates of change have been assumed. A completely general approach would be to include other variables and to allow for non-linear changes. A series of Taylor expansions provide us with a convenient method of expressing these more complex relationships. Therefore, with reference to the *body axes* [4.2]:

$$X_{\mathrm{a}}=X_{\mathrm{ae}}+\frac{\partial X_{\mathrm{a}}}{\partial u}.u+\frac{\partial^2 X_{\mathrm{a}}}{\partial u^2}.\frac{u^2}{2!}\ldots\text{etc.}+\frac{\partial X_{\mathrm{a}}}{\partial v}.v+\frac{\partial^2 X_{\mathrm{a}}}{\partial v^2}.\frac{v^2}{2!}\ldots\text{etc.}+\frac{\partial X_{\mathrm{a}}}{\partial w}.w+\frac{\partial^2 X_{\mathrm{a}}}{\partial w^2}.\frac{w^2}{2!}\ldots\text{etc.}$$

$$+\frac{\partial X_{\mathrm{a}}}{\partial p}.w+\frac{\partial^2 X_{\mathrm{a}}}{\partial p^2}.\frac{p^2}{2!}\ldots\text{etc.}+\frac{\partial X_{\mathrm{a}}}{\partial q}.q+\frac{\partial^2 X_{\mathrm{a}}}{\partial q^2}.\frac{q^2}{2!}\ldots\text{etc.}+\frac{\partial X_{\mathrm{a}}}{\partial r}.r+\frac{\partial^2 X_{\mathrm{a}}}{\partial r^2}.\frac{r^2}{2!}\ldots\text{etc.}$$

$$+\frac{\partial X_{\mathrm{a}}}{\partial \dot{u}}.\dot{u}+\frac{\partial^2 X_{\mathrm{a}}}{\partial \dot{u}^2}.\frac{\dot{u}^2}{2!}\ldots\text{etc.}+\frac{\partial X_{\mathrm{a}}}{\partial \dot{v}}.\dot{v}+\frac{\partial^2 X_{\mathrm{a}}}{\partial \dot{v}^2}.\frac{\dot{v}^2}{2!}\ldots\text{etc.}+\frac{\partial X_{\mathrm{a}}}{\partial \dot{w}}.\dot{w}+\frac{\partial^2 X_{\mathrm{a}}}{\partial \dot{w}^2}.\frac{\dot{w}^2}{2!}\ldots\text{etc.}$$

$$+\frac{\partial X_a}{\partial \dot{p}}.\dot{p}+\frac{\partial^2 X_a}{\partial \dot{p}^2}.\frac{\dot{p}^2}{2!}\ldots\text{etc.}+\frac{\partial X_a}{\partial \dot{q}}.\dot{q}+\frac{\partial^2 X_a}{\partial \dot{q}^2}.\frac{\dot{q}^2}{2!}\ldots\text{etc.}+\frac{\partial X_a}{\partial \dot{r}}.\dot{r}+\frac{\partial^2 X_a}{\partial \dot{r}^2}.\frac{\dot{r}^2}{2!}\ldots\text{etc.}$$

$$+\text{higher-order terms}$$

where

- X_a = aerodynamic X-force
- X_{ae} = equilibrium X-force generated when the aircraft is in trim
- u = change in longitudinal speed from trim
- v = change in lateral speed from trim
- w = change in vertical speed from trim
- p = change in roll rate from trim
- q = change in pitch rate from trim
- r = change in yaw rate from trim
- $\dot{u}$ = longitudinal acceleration
- $\dot{v}$ = lateral acceleration
- $\dot{w}$ = normal acceleration
- $\dot{p}$ = roll acceleration
- $\dot{q}$ = pitch acceleration
- $\dot{r}$ = yaw acceleration.

If the small perturbation assumption is retained it is again possible to linearize factors in this more complex relationship. Thus if second and subsequent terms in each Taylor series are assumed to be negligible then:

$$X_a = X_{ae}+\frac{\partial X_a}{\partial u}.u+\frac{\partial X_a}{\partial v}.v+\frac{\partial X_a}{\partial w}.w+\frac{\partial X_a}{\partial p}.p+\frac{\partial X_a}{\partial q}.q+\frac{\partial X_a}{\partial r}.r$$
$$+\frac{\partial X_a}{\partial \dot{u}}.\dot{u}+\frac{\partial X_a}{\partial \dot{v}}.\dot{v}+\frac{\partial X_a}{\partial \dot{w}}.\dot{w}+\frac{\partial X_a}{\partial \dot{p}}.\dot{p}+\frac{\partial X_a}{\partial \dot{q}}.\dot{q}+\frac{\partial X_a}{\partial \dot{r}}.\dot{r}$$

Now applying standard *aero-derivative* notation:

$$X_u=\frac{\partial X_a}{\partial u}\quad X_v=\frac{\partial X_a}{\partial v}\quad X_w=\frac{\partial X_a}{\partial w}\ldots\text{etc.}$$

Which leads to:

$$X_a = X_{ae}+X_u.u+X_v.v+X_w.w+X_p.p+X_q.q+X_r.r+X_{\dot{u}}.\dot{u}+X_{\dot{v}}.\dot{v}$$
$$+X_{\dot{w}}.\dot{w}+X_{\dot{p}}.\dot{p}+X_{\dot{q}}.\dot{q}+X_{\dot{r}}.\dot{r}$$

And applying the same analysis to the other forces and moments acting on a helicopter yields the following set of aero-derivatives:

$$\begin{array}{cccccccccccc}
X_u & X_v & X_w & X_p & X_q & X_r & X_{\dot{u}} & X_{\dot{v}} & X_{\dot{w}} & X_{\dot{p}} & X_{\dot{q}} & X_{\dot{r}}\\
Y_u & Y_v & Y_w & Y_p & Y_q & Y_r & Y_{\dot{u}} & Y_{\dot{v}} & Y_{\dot{w}} & Y_{\dot{p}} & Y_{\dot{q}} & Y_{\dot{r}}\\
Z_u & Z_v & Z_w & Z_p & Z_q & Z_r & Z_{\dot{u}} & Z_{\dot{v}} & Z_{\dot{w}} & Z_{\dot{p}} & Z_{\dot{q}} & Z_{\dot{r}}\\
L_u & L_v & L_w & L_p & L_q & L_r & L_{\dot{u}} & L_{\dot{v}} & L_{\dot{w}} & L_{\dot{p}} & L_{\dot{q}} & L_{\dot{r}}\\
M_u & M_v & M_w & M_p & M_q & M_r & M_{\dot{u}} & M_{\dot{v}} & M_{\dot{w}} & M_{\dot{p}} & M_{\dot{q}} & M_{\dot{r}}\\
N_u & N_v & N_w & N_p & N_q & N_r & N_{\dot{u}} & N_{\dot{v}} & N_{\dot{w}} & N_{\dot{p}} & N_{\dot{q}} & N_{\dot{r}}
\end{array}$$

Table 4.1 Force derivatives.

X_u Y_u Z_u	Forward, side and vertical force due to forward velocity
X_v Y_v Z_v	Forward, side and vertical force due to lateral velocity
X_w Y_w Z_w	Forward, side and vertical force due to vertical velocity
$X_{\dot{u}}$ $Y_{\dot{u}}$ $Z_{\dot{u}}$	Forward, side and vertical force due to forward acceleration
$X_{\dot{v}}$ $Y_{\dot{v}}$ $Z_{\dot{v}}$	Forward, side and vertical force due to lateral acceleration
$X_{\dot{w}}$ $Y_{\dot{w}}$ $Z_{\dot{w}}$	Forward, side and vertical force due to forward acceleration
X_p Y_p Z_p	Forward, side and vertical force due to roll rate
X_q Y_q Z_q	Forward, side and vertical force due to pitch rate
X_r Y_r Z_r	Forward, side and vertical force due to yaw rate
$X_{\dot{p}}$ $Y_{\dot{p}}$ $Z_{\dot{p}}$	Forward, side and vertical force due to roll acceleration
$X_{\dot{q}}$ $Y_{\dot{q}}$ $Z_{\dot{q}}$	Forward, side and vertical force due to pitch acceleration
$X_{\dot{r}}$ $Y_{\dot{r}}$ $Z_{\dot{r}}$	Forward, side and vertical force due to yaw acceleration

Table 4.2 Moment derivatives.

L_u M_u N_u	Rolling, pitching and yawing moment due to forward velocity
L_v M_v N_v	Rolling, pitching and yawing moment due to lateral velocity
L_w M_w N_w	Rolling, pitching and yawing moment due to vertical velocity
$L_{\dot{u}}$ $M_{\dot{u}}$ $N_{\dot{u}}$	Rolling, pitching and yawing moment due to forward acceleration
$L_{\dot{v}}$ $M_{\dot{v}}$ $N_{\dot{v}}$	Rolling, pitching and yawing moment due to lateral acceleration
$L_{\dot{w}}$ $M_{\dot{w}}$ $N_{\dot{w}}$	Rolling, pitching and yawing moment due to vertical acceleration
L_p M_p N_p	Rolling, pitching and yawing moment due to roll
L_q M_q N_q	Rolling, pitching and yawing moment due to pitch rate
L_r M_r N_r	Rolling, pitching and yawing moment due to yaw rate
$L_{\dot{p}}$ $M_{\dot{p}}$ $N_{\dot{p}}$	Rolling, pitching and yawing moment due to roll acceleration
$L_{\dot{q}}$ $M_{\dot{q}}$ $N_{\dot{q}}$	Rolling, pitching and yawing moment due to pitch acceleration
$L_{\dot{r}}$ $M_{\dot{r}}$ $N_{\dot{r}}$	Rolling, pitching and yawing moment due to yaw acceleration

Using the body axes system it is possible to give a precise meaning to each of the aero-derivatives introduced above, see Tables 4.1 and 4.2.

When the pilot moves a control in the cockpit he will cause changes to the pitch of the appropriate rotor blades. These control inputs will in turn generate off-trim forces and moments in a manner analogous to the effect of gusts. It is possible therefore to identify a set of *control derivatives* based on the following control deflections from trim, see Table 4.3:

$$A_1 \text{ or } \theta_{1_s} = \text{lateral cyclic pitch}$$

$$B_1 \text{ or } \theta_{1_c} = \text{longitudinal cyclic pitch}$$

$$\theta_c = \text{collective pitch}$$

$$\theta_{TR} = \text{tail rotor collective pitch}$$

Table 4.3 Control derivatives.

X_{A_1} Y_{A_1} Z_{A_1}	Forward, side and vertical force due to lateral cyclic pitch
X_{B_1} Y_{B_1} Z_{B_1}	Forward, side and vertical force due to longitudinal cyclic pitch
X_{θ_c} Y_{θ_c} Z_{θ_c}	Forward, side and vertical force due to collective pitch
$X_{\theta_{TR}}$ $Y_{\theta_{TR}}$ $Z_{\theta_{TR}}$	Forward, side and vertical force due to tail rotor pitch
L_{A_1} M_{A_1} N_{A_1}	Rolling, pitching and yawing moment due to lateral cyclic pitch
L_{B_1} M_{B_1} N_{B_1}	Rolling, pitching and yawing moment due to longitudinal cyclic pitch
L_{θ_c} M_{θ_c} N_{θ_c}	Rolling, pitching and yawing moment due to collective pitch
$L_{\theta_{TR}}$ $M_{\theta_{TR}}$ $N_{\theta_{TR}}$	Rolling, pitching and yawing moment due to tail rotor pitch

4.2.3 Normalized and non-dimensional derivatives

Although the full set of derivatives introduced above may be needed for an accurate mathematical representation of rotorcraft dynamic characteristics they are not all required when discussing typical helicopter handling qualities. Since the pilot is most interested in the attitude changes that occur as a result of control input and atmospheric disturbances mainly moment derivatives will be retained in the reduced set. Due to their widespread use, most of these derivatives have acquired a *common descriptor* based on their effect on the stability and control characteristics of a typical helicopter, see Table 4.4.

All the derivatives considered so far are affected by the aerodynamic characteristics of the aircraft. They will therefore be modified by changes in flight condition (density altitude, airspeed, rotor speed) and rotorcraft design (rotor radius, blade area). In order to make appropriate comparison it is common practice to make aero-derivatives *non-dimensional* in a similar manner to that used for lift and drag. In addition, as shall be seen later, when the equations of motion governing aircraft behaviour are manipulated it is sometimes convenient to *normalize* the derivatives using the mass and inertia properties of the helicopter. There is unfortunately no internationally agreed set of standard symbols used to distinguish between these various classes of aero-derivatives. A set appropriate to rotorcraft that has been developed from those used by Babister [4.2] is presented in Table 4.5.

Aero-derivatives help us understand the stability and control characteristics of

Table 4.4 Most commonly used aero-derivatives.

	Common descriptor		Common descriptor
X_u	Drag damping	L_p	Roll damping
Y_v	Side force	M_q	Pitch damping
Z_w	Heave damping	N_r	Yaw damping
L_v	Lateral static stability	L_{A_1}	Roll control power
M_u	Speed stability	M_{B_1}	Pitch control power
M_w	Angle of attack stability	$N_{\theta_{TR}}$	Yaw control power
N_v	Directional static stability	Z_{θ_c}	Heave control power
$L_{\theta_{TR}}$	Tail rotor roll	$Y_{\theta_{TR}}$	Tail rotor drift
M_{θ_c}	Pitch change with power	N_{θ_c}	Torque reaction

Table 4.5 Symbol set for various classes of aero-derivative.

Derivative type	Dimensional derivatives	Divisors	Normalized derivatives	Divisors	Non-dimensional derivatives
Force/linear velocity	$\mathring{X}_u\ \mathring{Y}_v\ \mathring{Z}_w$	mass	$X_u\ Y_v\ Z_w$	$\rho sA\Omega R$	$x_u\ y_v\ z_w$
Force/angular velocity	$\mathring{X}_p\ \mathring{Y}_q\ \mathring{Z}_r$	mass	$X_p\ Y_q\ Z_r$	$\rho sA\Omega R^2$	$x_p\ y_q\ z_r$
Moment/linear velocity	$\mathring{L}_u\ \mathring{M}_v\ \mathring{N}_w$	inertia	$L_u\ M_v\ N_w$	$\rho sA\Omega R^2$	$l_u\ m_v\ n_w$
Moment/angular velocity	$\mathring{L}_p\ \mathring{M}_q\ \mathring{N}_r$	inertia	$L_p\ M_q\ N_r$	$\rho sA\Omega R^3$	$l_p\ m_q\ n_r$
Force/control deflection	$\mathring{X}_{A_1}\ \mathring{Y}_{B_1}\ \mathring{Z}_{\theta_c}$	mass	$X_{A_1}\ Y_{B_1}\ Z_{\theta_c}$	$\rho sA\Omega^2 R^2$	$x_{A_1}\ y_{B_1}\ z_{\theta_c}$
Moment/control deflection	$\mathring{L}_{A_1}\ \mathring{M}_{B_1}\ \mathring{N}_{\theta_c}$	inertia	$L_{A_1}\ M_{B_1}\ N_{\theta_c}$	$\rho sA\Omega^2 R^3$	$l_{A_1}\ m_{B_1}\ n_{\theta_c}$

rotorcraft by providing a convenient manner through which to describe the factors that affect these characteristics. Equations of motion can be developed that tie these derivatives directly to the dynamic behaviour of rotorcraft. It is possible, therefore, to determine the particular derivatives that are key to shaping the pilot's perception of his aircraft. Likewise, the contribution made by the components of a typical helicopter to these derivatives can be identified and how the design choices made by rotorcraft manufacturers may affect the suitability of a helicopter for a particular role can be understood. Aero-derivatives are, therefore, fundamental to any study of aircraft stability and control.

4.3 EQUATIONS OF MOTION FOR A RIGID AIRFRAME

To fully understand the stability characteristics of an aircraft and why certain responses occur, it is necessary to determine the equations governing its motion and to define the parameters that dominate and influence them. The method used in this chapter to develop the necessary equations relies on a number of simplifying assumptions but applies to any rigid body (neglecting any structural distortion) which is subjected to *small* disturbances. By making such assumptions these essentially non-linear equations can be linearized and the task of analysing them is made much simpler.

Figure 4.2 shows a set of rectangular axes (O, x, y, z) fixed in the helicopter (body axes) with its origin at the body axes centre. Note that at this level of complexity the rotor dynamics are ignored. The rotor is assumed to be a fixed force and moment generating device. The components of velocity and force along the Ox, Oy and Oz axes are U, V, W and X, Y, Z respectively. The components of the rates of rotation about the same axes are p, q and r and the moments L, M and N. After Babister [4.2], consider an arbitrary point P at position x, y, z from the body axis centre, which has local components of velocity and acceleration u, v, w and a_x, a_y, a_z respectively.

The absolute velocity of the point P is now obtained by superimposing the velocity of the body axes centre onto the relative velocity of the point P. P is moving relative to the body axes centre, but this centre is also moving, so the absolute velocity of P is given by the sum of these two components. The body axes centre is moving with velocity U, V, W so denoting the absolute velocity of P by u', v', w' leads to:

$$u' = U - ry + qz \quad v' = V - pz + rx \quad w' = W - qx + py \tag{4.1}$$

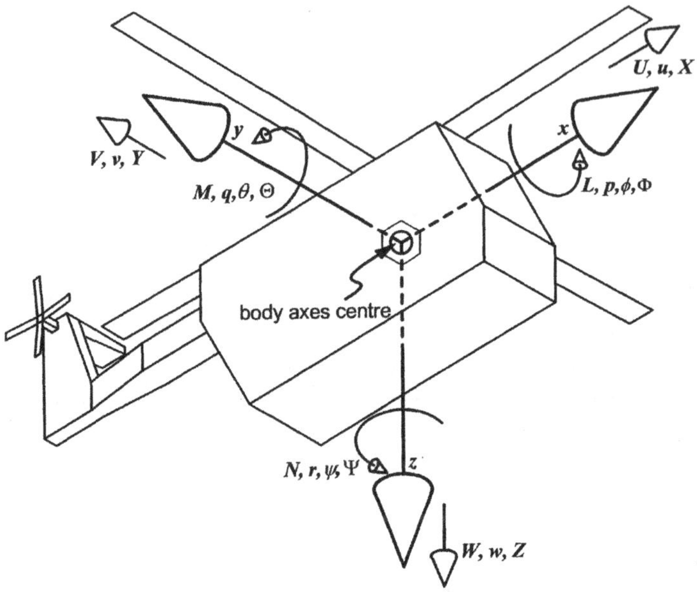

Fig. 4.2 Body axes system.

and similarly, the absolute accelerations are:

$$a'_x = \dot{u}' - rv' + qw' \quad a'_y = \dot{v}' - pw' + ru' \quad a'_z = \dot{w}' - qu' + pv' \tag{4.2}$$

The values of $\dot{u}'$, $\dot{v}'$, $\dot{w}'$ can be obtained by differentiation and since the airframe is rigid, $\dot{y} = \dot{x} = \dot{z} = 0$:

$$\dot{u}' = \dot{U} - y\dot{r} + z\dot{q} \quad \dot{v}' = \dot{V} - z\dot{p} + x\dot{r} \quad \dot{w}' = \dot{W} - x\dot{q} + y\dot{p}$$

Substituting for u', v', w' and for $\dot{u}'$, $\dot{v}'$, $\dot{w}'$, using Equations (4.1) and (4.2), leads to:

$$\left.\begin{aligned} a'_x &= \dot{U} - rV + qW - x(q^2 + r^2) + y(pq - \dot{r}) + z(pr + \dot{q}) \\ a'_y &= rU - \dot{V} - pW + x(pq + \dot{r}) - y(p^2 + r^2) + z(qr - \dot{p}) \\ a'_z &= -qU + pV + \dot{W} + x(pr - \dot{q}) + y(qr + \dot{p}) - z(p^2 + q^2) \end{aligned}\right\} \tag{4.3}$$

Having obtained expressions for the components of the absolute acceleration of the point P, consideration must now be given to the forces acting on the helicopter. To this end, consider an element of mass δm at the point P. Using the formula $F = ma$, the forces producing the components of acceleration a'_x, a'_y, a'_z are $a'_x\,\delta m$, $a'_y\,\delta m$, $a'_z\,\delta m$ respectively. The total force is obtained by summing the components of force at all such points over the whole helicopter using Equation (4.3):

$$\left.\begin{aligned} X &= m\{\dot{U} - rV + qW - d_x(q^2 + r^2) + d_y(pq - \dot{r}) + d_z(pr + \dot{q})\} \\ Y &= m\{rU + \dot{V} - pW + d_x(pq + \dot{r}) - d_y(p^2 + r^2) + d_z(qr - \dot{p})\} \\ Z &= m\{-qU + pV + \dot{W} + d_x(pr - \dot{q}) + d_y(qr + \dot{p}) - d_z(p^2 + q^2)\} \end{aligned}\right\} \tag{4.4}$$

where d_x, d_y, d_z are co-ordinates of the centre of gravity from the body axes centre.

The forces acting on the particle at point P will each have an associated moment about the axes. First consider the moments about the Ox axis; these will be rolling moments and will be denoted by L.

The elemental force $\delta m.a'_z$ has a moment about Ox given by:

$$\text{moment} = (\delta m.a'_z)y \quad \text{clockwise}$$

The elemental force $\delta m.a'_y$ has a moment about Ox given by:

$$\text{moment} = (\delta m.a'_y)z \quad \textit{anti}\text{clockwise}$$

Since the third elemental force $\delta m.a'_x$ has no moment about Ox, the total moment for such points is given by:

$$L = \Sigma\, \delta m(a'_z y - a'_y z)$$

Hence from Equation (4.2):

$$\left.\begin{aligned} L = \dot{p}\Sigma(y^2+z^2)\delta m + qr\Sigma(y^2-z^2)\delta m + (r^2-q^2)\Sigma yz\,\delta m + (pr-\dot{q})\Sigma xy\,\delta m \\ -(pq+\dot{r})\Sigma xz\,\delta m + (-qU+pV+\dot{W})\Sigma y\,\delta m - (rU+\dot{V}-pW)\Sigma z\,\delta m \end{aligned}\right\} \quad (4.5)$$

Moments and products of inertia are defined about the body axes centre as follows:

$$[I_{xx}]_b = \Sigma(y^2+z^2)\delta m = \text{moment of inertia about } x \text{ axis}$$
$$[I_{yy}]_b = \Sigma(x^2+z^2)\delta m = \text{moment of inertia about } y \text{ axis}$$
$$[I_{zz}]_b = \Sigma(x^2+y^2)\delta m = \text{moment of inertia about } z \text{ axis}$$
$$[I_{yz}]_b = \Sigma\, yz\,\delta m = \text{product of inertia about } y \text{ and } z \text{ axes}$$
$$[I_{xz}]_b = \Sigma\, xz\,\delta m = \text{product of inertia about } x \text{ and } z \text{ axes}$$
$$[I_{xy}]_b = \Sigma\, xy\,\delta m = \text{product of inertia about } x \text{ and } y \text{ axes}$$

So equation (4.5) can be rewritten as:

$$\left.\begin{aligned} L = \dot{p}[I_{xx}]_b + qr([I_{zz}]_b - [I_{yy}]_b) + (r^2-q^2)[I_{yz}]_b + (pr-\dot{q})[I_{xy}]_b \\ -(pq+\dot{r})[I_{xz}]_b + (-qU+pV+\dot{W})\Sigma y\,\delta m - (rU+\dot{V}-pW)\Sigma z\,\delta m \end{aligned}\right\} \quad (4.6)$$

Since the body axes centre is not necessarily at the centre of gravity the following transformation is required:

$$[I_{xx}]_b = I_{xx} + m(d_y^2 + d_z^2)$$
$$[I_{yy}]_b = I_{yy} + m(d_z^2 + d_x^2)$$
$$[I_{zz}]_b = I_{zz} + m(d_x^2 + d_y^2)$$
$$[I_{xy}]_b = I_{xy} + md_x d_y$$
$$[I_{xz}]_b = I_{xz} + md_x d_z$$
$$[I_{yz}]_b = I_{yz} + md_y d_z$$

Therefore Equation (4.6) becomes:

$$L = \dot{p}I_{xx} + qr\{I_{zz} - I_{yy}\} + (r^2-q^2)I_{yz} + (pr-\dot{q})I_{xy} - (pr+\dot{r})I_{xz} - Yd_z + Zd_y$$

By a similar analysis, equations for moments M and N can be determined:

$$\left.\begin{aligned} L &= \dot{p}I_{xx} + qr\{I_{zz} - I_{yy}\} - (pq + \dot{r})I_{xz} - (q^2 - r^2)I_{yz} - (\dot{q} - pr)I_{xy} - Yd_z + Zd_y \\ M &= \dot{q}I_{yy} - pr\{I_{xx} - I_{zz}\} - (r^2 - p^2)I_{xz} - (\dot{r} - pq)I_{yz} - (\dot{p} + qr)I_{xy} - Zd_x + Xd_z \\ N &= \dot{r}I_{zz} + qp\{I_{yy} - I_{xx}\} - (\dot{p} - qr)I_{xz} - (\dot{q} + pr)I_{yz} - (p^2 - q^2)I_{xy} - Xd_y + Yd_x \end{aligned}\right\} \quad (4.7)$$

where d_x, d_y, d_z are co-ordinates of the centre of gravity from the body axes centre and I_{xx}, I_{yy}, I_{zz}, I_{xy}, I_{yz}, I_{xz} are the moments and products of inertia about axes through the centre of gravity parallel to Ox, Oy, Oz. Linearizing the force and moment equations (Equations (4.4) and (4.7)) by means of the small perturbation theory and assuming that the CG is co-incident with the body axes centre leads to:

$$\left.\begin{aligned} X &= m\{\dot{u} + qW_e\} \\ Y &= m\{\dot{v} - pW_e + rU_e\} \\ Z &= m\{\dot{w} - qU_e\} \\ L &= I_{xx}\dot{p} - I_{xz}\dot{r} \\ M &= I_{yy}\dot{q} \\ N &= I_{zz}\dot{r} - I_{xz}\dot{p} \end{aligned}\right\} \quad (4.8)$$

Noting that the forces and moments arise from aerodynamic, gravitational and control sources and introducing the concept of derivatives, Equation (4.8) becomes:

$$m\dot{u} = u\mathring{X}_u + w\mathring{X}_w + q(\mathring{X}_q - mW_e) - \theta mg\cos\theta_e + B_1\mathring{X}_{B_1} + \theta_c\mathring{X}_{\theta_c}$$

$$m\dot{v} = v\mathring{Y}_v + p(\mathring{Y}_p + mW_e) + r(\mathring{Y}_r - mU_e)$$
$$+ \phi mg\sin\theta_e + \psi mg\cos\theta_e + A_1\mathring{Y}_{A_1} + \theta_{TR}\mathring{Y}_{\theta_{TR}}$$

$$m\dot{w} = u\mathring{Z}_u + w\mathring{Z}_w + q(\mathring{Z}_q + mU_e) - \theta mg\sin\theta_e + B_1\mathring{Z}_{B_1} + \theta_c\mathring{Z}_{\theta_c}$$

$$I_{xx}\dot{p} - I_{xz}\dot{r} = v\mathring{L}_v + p\mathring{L}_p + r\mathring{L}_r + A_1\mathring{L}_{A_1} + \theta_{TR}\mathring{L}_{\theta_{TR}}$$

$$I_{yy}\dot{q} = u\mathring{M}_u + w\mathring{M}_w + q\mathring{M}_q + B_1\mathring{M}_{B_1} + \theta_c\mathring{M}_{\theta_c}$$

$$I_{zz}\dot{r} - I_{xz}\dot{p} = v\mathring{N}_v + p\mathring{N}_p + r\mathring{N}_r + A_1\mathring{N}_{A_1} + \theta_{TR}\mathring{N}_{\theta_{TR}}$$

These equations are often expressed in matrix form as indicted below in the discussions on dynamic stability and control response (Section 4.11).

4.4 EQUATIONS OF FLAPPING MOTION

Development of the equations of flapping motion begins by considering a rotor system with a single hinge mounted a distance eR from the axis of rotation. The shaft rotates with a constant angular velocity, Ω, and the blade flaps with angular velocity, $\dot{\beta}$. If axes are fixed in the blade (see Fig. 4.3) then the dynamic situation can be described using [4.1]:

$$\left.\begin{aligned} L &= I_{xx}\dot{\omega}_1 + I_{xx}\omega_2\omega_3 \\ M &= I_{yy}\dot{\omega}_2 - I_{yy}\omega_3\omega_1 - m_b X_g R a_z \\ N &= (I_{xx} + I_{yy})\dot{\omega}_3 - (I_{xx} - I_{yy})\omega_1\omega_2 + m_b X_g R a_y \end{aligned}\right\} \quad (4.9)$$

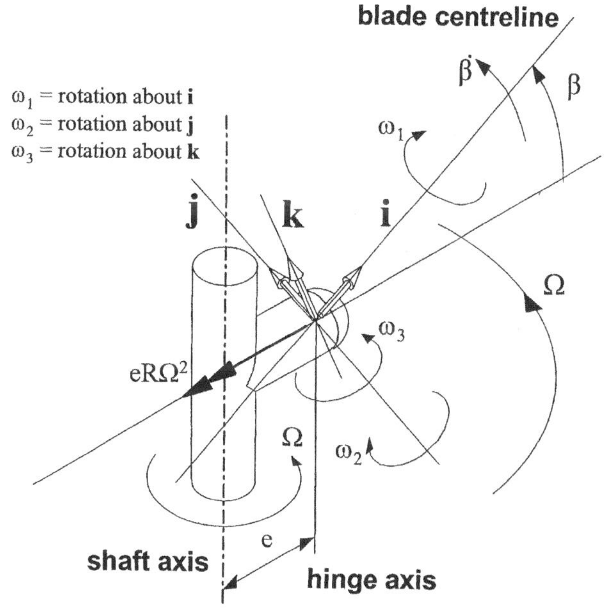

Fig. 4.3 Rotor axes system.

The axes are set to be parallel to the principal axes with their origin at the hinge, such that the **i**-axis lies along the blade span, the **j**-axis is perpendicular to the span and the **k**-axis completes the right-hand set. Although the blades are in reality highly flexible, it will be assumed, for simplicity, that they are rigid. If I_{xx} is the moment of inertia about **i** and I_{yy} is the moment of inertia about **j** then I_{zz}, the moment about **k**, is given by $I_{zz} = I_{xx} + I_{yy}$. Note that ω_1, ω_2 and ω_3 are the angular rates about the principal axes and a_x, a_y and a_z represent the acceleration of the hinge.

4.4.1 Pure flap motion

Analysis begins by assuming that the blade is constrained so that only flap motion is possible, see Fig. 4.4. From Fig. 4.4 it can be seen that the angular components of velocity, ω_1, ω_2 and ω_3 are given by:

$$\left.\begin{aligned} \omega_1 &= \Omega \sin\beta \\ \omega_2 &= -\dot{\beta} \\ \omega_3 &= \Omega \cos\beta \end{aligned}\right\} \tag{4.10}$$

Also the absolute acceleration of the axis system, a_0, is given by:

$$a_0 = a_x\mathbf{i} + a_y\mathbf{j} + a_z\mathbf{k}$$

$$a_0 = -\Omega^2 eR\cos\beta\mathbf{i} + \Omega^2 eR\sin\beta\mathbf{k}$$

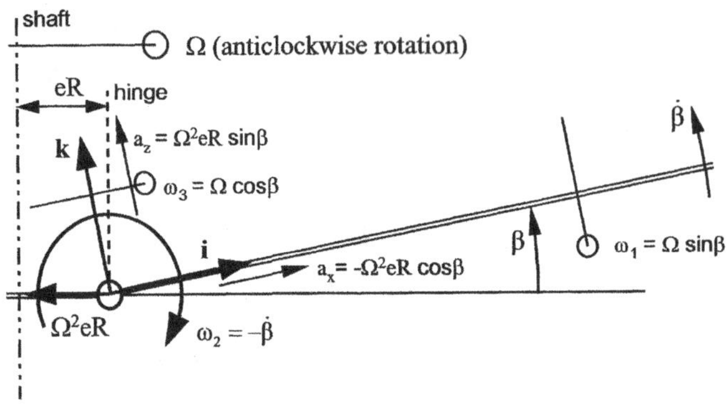

Fig. 4.4 Pure flap motion.

Hence:

$$\left.\begin{aligned} a_x &= -\Omega^2 eR\cos\beta \\ a_y &= 0 \\ a_z &= \Omega^2 eR\sin\beta \end{aligned}\right\} \tag{4.11}$$

Now substituting Equations (4.10) and (4.11) into Equation (4.9) and assuming small β:

$$\left.\begin{aligned} L &= 0 \\ -M &= I_{yy}\ddot{\beta} + \Omega^2(I_{yy} + m_b X_g eR^2)\beta \\ -N &= 2\Omega I_{yy}\dot{\beta}\sin\beta \end{aligned}\right\} \tag{4.12}$$

In other words if a rotating blade flaps upwards then in order to maintain it within the flapping plane no feathering moment is required but a rearwards moment of $2\Omega I_{yy}\dot{\beta}\sin\beta$ is necessary. This moment is equal and opposite to the Coriolis moment acting on the blade; it is usually so large that it is often relieved by a drag hinge. Therefore if the blade is hinged so that it is able to move in an in-plane sense then as the blade flaps up it will move forward or *lead*.

4.4.2 Pure lead/lag motion

Since a flapping blade is likely to move in the lead/lag sense it is necessary to determine whether the lead/lag motion will cause any rotation about the other two axes. It is now assumed that the flapping angle is zero and the blade moves forward on the drag hinge through an angle ξ, as in Fig. 4.5. In this case:

$$\left.\begin{aligned} \omega_1 &= 0 \\ \omega_2 &= 0 \\ \omega_3 &= \Omega + \dot{\xi} \end{aligned}\right\} \tag{4.13}$$

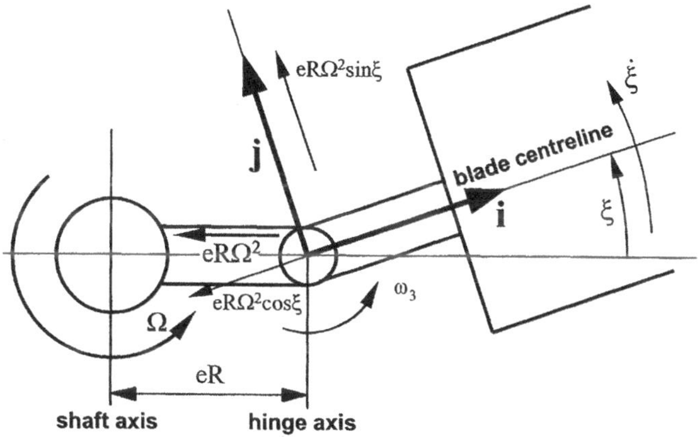

Fig. 4.5 Pure lead/lag motion.

and:

$$\left.\begin{aligned} a_x &= -\Omega^2 eR\cos\xi \\ a_y &= \Omega^2 eR\sin\xi \\ a_z &= 0 \end{aligned}\right\} \qquad (4.14)$$

and making the appropriate substitutions (Equations (4.13) and (4.14) into Equation (4.9):

$$\left.\begin{aligned} L &= 0 \\ M &= 0 \\ N &= I_{zz}\ddot{\xi} + m_b X_g eR^2\Omega^2\sin\xi \end{aligned}\right\} \qquad (4.15)$$

Thus when a rotating blade moves solely in a lead/lag sense no moments are generated that cause additional flapping or feathering motion.

4.4.3 Pure feather motion

As shall be seen later it is necessary to feather a rotor blade if changes in disk tilt and/or thrust are desired. Therefore consideration needs to be given to the possibility of feathering motion causing a flap and lead/lag response through mass and inertia effects. The blade is now constrained to feather through an angle θ whilst the flapping and lagging angles are zero, see Fig. 4.6.

Now from Fig. 4.6:

$$\left.\begin{aligned} \omega_1 &= \dot{\theta} \\ \omega_2 &= \Omega\sin\theta \\ \omega_3 &= \Omega\cos\theta \end{aligned}\right\} \qquad (4.16)$$

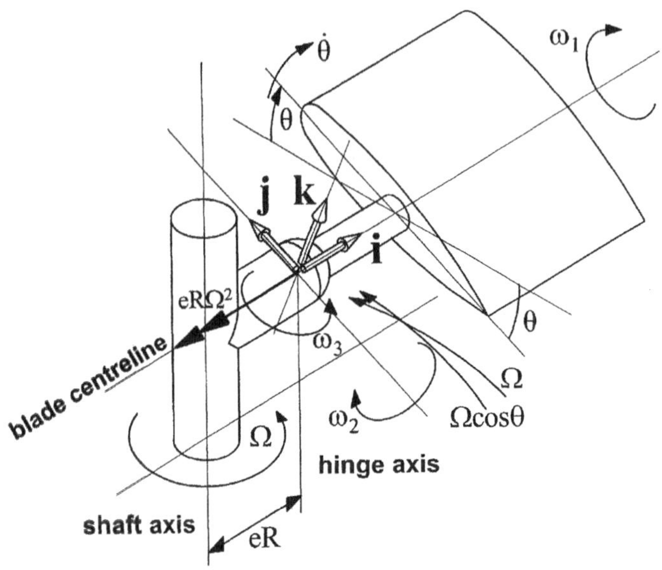

Fig. 4.6 Pure feather motion.

and:

$$\left.\begin{aligned} a_x &= \Omega^2 eR \\ a_y &= 0 \\ a_z &= 0 \end{aligned}\right\} \qquad (4.17)$$

As before, substituting Equations (4.16) and (4.17) into Equation (4.9):

$$L = I_{xx}\ddot{\theta} + I_{xx}\Omega^2 \sin\theta\cos\theta$$

$$M = I_{yy}\Omega\dot{\theta}\cos\theta - I_{yy}\Omega\dot{\theta}\cos\theta$$

$$N = -(I_{xx} + I_{yy})\Omega\dot{\theta}\sin\theta - (I_{xx} - I_{yy})\Omega\dot{\theta}\sin\theta$$

Hence:

$$\left.\begin{aligned} L &= I_{xx}\ddot{\theta} + I_{xx}\Omega^2\theta \\ M &= 0 \\ N &= -2I_{xx}\Omega\dot{\theta}\theta \end{aligned}\right\} \qquad (4.18)$$

The lagging moment due to feathering, $2I_{xx}\Omega\dot{\theta}\theta$, is usually extremely small when compared with the Coriolis flapping moment and can be neglected.

4.5 FLAP DYNAMICS

The above analysis and Equations (4.12), (4.15) and (4.18) show, to a first approximation, that the only dynamic linkage between flap, lead/lag and feather is the Coriolis effect that causes a flapping blade to lead or lag. Thus in the absence of a δ_3 hinge, lead/lag motion will not feedback to affect the flap response and it can be assumed that the feathering motion required to produce a flap response *aerodynamically* will not generate any dynamic effects that modify the flap behaviour. Consequently for the rotor the most fundamental dynamic equation is the flapping equation:

$$I_{yy}\ddot{\beta} + \Omega^2(I_{yy} + m_b X_g eR^2)\beta = -M \quad \text{or} \quad \ddot{\beta} + \Omega^2(1+\varepsilon)\beta = \frac{M_A}{I_{yy}} \tag{4.19}$$

To determine the nature of the flap dynamics it is necessary to consider the nature of the aerodynamic forcing. Three scenarios will be used to expose the basic dynamic characteristics of a typical rotor: a flap disturbance, the application of cyclic feathering and the application of a steady pitch/roll rate to the rotor. For simplicity all assume the helicopter is initially in a trimmed OGE hover.

4.5.1 Disturbed flapping motion at a constant feather angle

The rotor blades are assumed to be at a constant feather angle and the rotor is taken as rotating steadily at an angular velocity Ω. Since it is only the disturbed motion that is of interest the aerodynamic moment corresponding to the feather angle will be ignored. When the blade flaps up with angular velocity $\dot{\beta}$, there is a relative down velocity of $r\dot{\beta}$ at a point on the blade located r from the hinge, see Fig. 4.7. From Fig. 4.7 it can be seen that for small angles of inflow, flap and flap disturbance:

$$\Delta\alpha \approx \tan\Delta\alpha \approx \frac{-r\dot{\beta}}{\Omega(eR+r)}$$

Thus the change in elemental lift, δL, is given by:

$$\delta L = \frac{1}{2}\rho V^2 c\,\delta r\,C_L = -\frac{1}{2}\rho\Omega^2(r+eR)^2 ca\frac{r\dot{\beta}}{\Omega(r+eR)}\delta r = -\frac{1}{2}\rho\Omega r(r+eR)\dot{\beta}ca\,\delta r$$

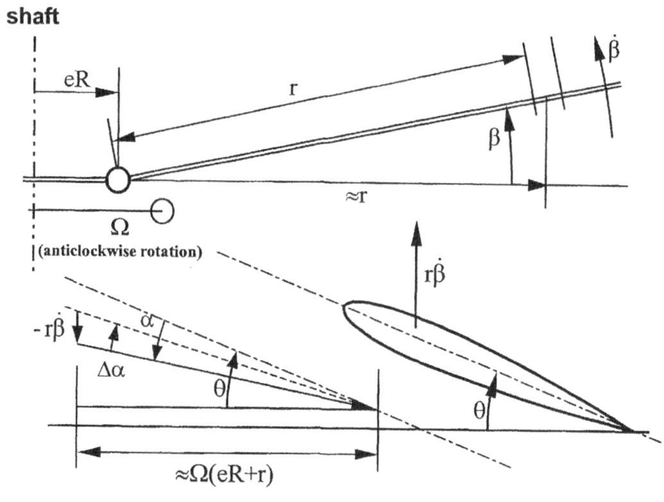

Fig. 4.7 Disturbed flapping at constant feather angle.

Now the aerodynamic moment is obtained by integrating the lift over the blade:

$$M_A = \int_0^{R(1-e)} r\,\mathrm{d}L = -\int_0^{R(1-e)} \frac{1}{2}\rho\Omega r^2(r+eR)\dot{\beta}ca\,\mathrm{d}r$$

$$= -\frac{1}{8}\rho\Omega ac\dot{\beta}R^4(1-e)^3(1+e/3)$$

Therefore:

$$\frac{M_A}{I_{yy}} = -\frac{\gamma}{8}\Omega\dot{\beta}(1-e)^3(1+e/3) \tag{4.20}$$

where:

$$\gamma = \frac{\rho acR^4}{I_{yy}} = \text{Lock's inertia number}$$

Therefore using Equations (4.19) and (4.20) the flapping equation becomes:

$$\ddot{\beta} + \frac{\gamma}{8}\Omega(1+e)^3(1+e/3)\dot{\beta} + \Omega^2(1+\varepsilon)\beta = 0 \tag{4.21}$$

Taking Laplace transforms:

$$s^2 + \frac{\gamma}{8}\Omega(1-e)^3(1+e/3)s + \Omega^2(1+\varepsilon) = 0$$

The above second-order equation relates directly to the standard form:

$$s^2 + 2\zeta\omega_n s + \omega_n^2 = 0$$

Hence:

$$\omega_n = \Omega\sqrt{1+\varepsilon} \quad \text{and} \quad \zeta = \frac{\gamma(1-e)^3(1+e/3)}{16\sqrt{1+\varepsilon}}$$

The motion described above is damped and harmonic in nature with a natural frequency, ω_n, of $\Omega\sqrt{1+\varepsilon}$. If the rotor has no flapping hinge offset then both e and ε will be zero. Therefore the natural frequency is exactly equal to the shaft frequency and consequently the flap mode is being forced at its resonant frequency; therefore it will display a phase shift of 90° between feather (the input) and flap (the output).

4.5.2 Flapping motion due to cyclic feathering

The rotor is subjected to a constant feather angle and a feather angle that is varied sinusoidally relative to the hub as expressed by the equation below:

$$\theta = \theta_0 - A_1\cos\psi - B_1\sin\psi \tag{4.22}$$

From Equation (4.19) it has already been shown that:

$$\ddot{\beta} + \Omega^2(1+\varepsilon)\beta = \frac{M_A}{I_{yy}}$$

If the independent variable is changed from time to blade azimuth angle by means of $\psi = \Omega t$, then:

$$\Omega^2 \frac{d^2\beta}{d\psi^2} + \Omega^2(1+\varepsilon)\beta = \frac{M_A}{I_{yy}} \tag{4.23}$$

In this scenario lift changes on the blade will arise from two sources: the change in angle of attack caused by the cyclic feathering and the flapping motion itself. It can be shown for an untwisted blade therefore that, since the moment due cyclic feathering is given by:

$$M_A = \int_0^{R(1-e)} r\,dL = \frac{1}{2}\rho ac\Omega^2(\theta_0 - A_1\cos\psi - B_1\sin\psi)\int_0^{R(1-e)} r(r+eR)^2\,dr$$

then the total flapping moment becomes:

$$\frac{M_A}{I_{yy}} = \frac{\gamma}{8}\Omega^2(1-e)^2\left[(\theta_0 - A_1\cos\psi - B_1\sin\psi)(1+2e/3) - (1-e)(1+e/3)\frac{d\beta}{d\psi}\right] \tag{4.24}$$

Thus using Equations (4.23) and (4.24):

$$\frac{d^2\beta}{d\psi^2} + \frac{\gamma}{8}(1-e)^3(1+e/3)\frac{d\beta}{d\psi} + (1+\varepsilon)\beta = \frac{\gamma}{8}(\theta_0 - A_1\cos\psi - B_1\sin\psi)(1-e)^2(1+2e/3) \tag{4.25}$$

Only the *steady state* solution is of interest here. Thus if the blade flapping is expressed as:

$$\beta = a_0 - a_1\cos\psi - b_1\sin\psi$$

Then solving Equation (4.25) gives:

$$a_1 = \frac{\gamma(1-e)^2(1+2e/3+e^2/3)}{8(\varepsilon^2+n^2)}[\varepsilon A_1 - nB_1]$$

$$b_1 = \frac{\gamma(1-e)^2(1+2e/3+e^2/3)}{8(\varepsilon^2+n^2)}[nA_1 + \varepsilon B_1]$$

$$a_0 = \frac{\gamma(1-e)^2(1+2e/3+e^2/3)}{8(1+\varepsilon)}\theta_0$$

$$n = \frac{\gamma}{8}(1-e)^3(1+e/3)$$

Now consider the case when the hinge offset is zero. Thus $\varepsilon = e = 0$ and:

$$a_1 = -B_1 \quad b_1 = A_1 \quad a_0 = \frac{\gamma}{8}\theta_0$$

Once again confirmation of the existence of a 90° phase shift is given since, for an anticlockwise rotating rotor, rearwards disk tilt is caused by an increase in feather over the starboard quarter and starboard disk tilt is generated by an increase in feather over the nose. Returning to the situation with non-zero hinge offset, examination of the equations for a_1 and b_1 indicate that cross-coupling will be present. This can be readily appreciated if the case of an increase in B_1 is considered; the magnitudes of both a_1 and b_1 will be affected. The cross-coupling can be expressed as a phase shift, ϕ, in terms of the azimuth angle between blade feather and blade flap response. For a blade with a uniform mass distribution:

$$X_g = \frac{(1-e)}{2} \quad \text{and} \quad I_{yy} = \frac{m_b}{3}(1-e)^2 R^2$$

So from Equation (4.19):

$$\varepsilon = \frac{m_b X_g e R^2}{I_{yy}} = \frac{3m_b(1-e)eR^2}{2m_b(1-e)^2 R^2} = \frac{3e}{2(1-e)}$$

Therefore:

$$\phi = \tan^{-1}\left[\frac{\gamma(1-e)^4(1+e/3)}{12e}\right]$$

Figure 4.8 shows the variation of ϕ with hinge offset and Lock number. The amount of cross-coupling, indicated by the reduction in ϕ from 90° increases with increasing hinge offset and reduces with increases in Lock number.

It should be noted that Lock number is air density dependent and therefore will reduce with altitude. The cross-coupling caused by hinge offset can be removed by mixing the cyclic feather demand such that when the pilot moves the cyclic stick forward, forward disk tilt is the *only* result. However, it should be appreciated that the mixing, if mechanically engineered, will only be exactly right at a single density altitude. During a ceiling climb, therefore, significant cross-coupling may occur as

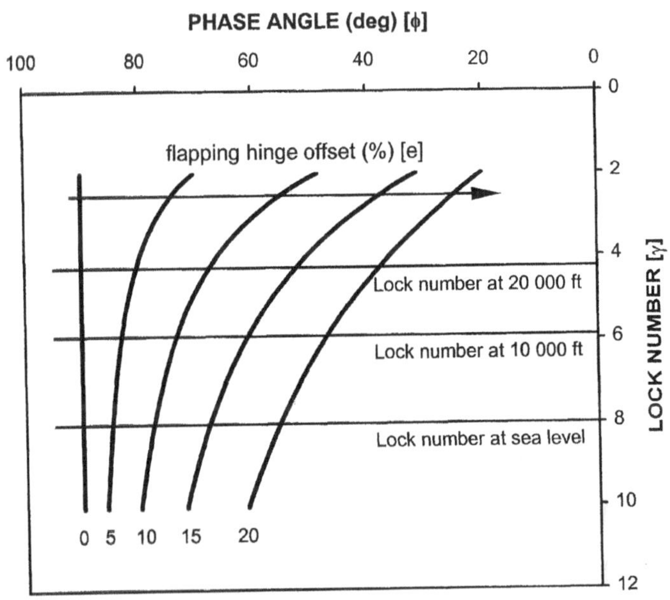

Fig. 4.8 Example of acceleration cross-coupling.

indicated on Fig. 4.8 using a sea level Lock number of 8. As the aircraft climbs the flap response will lead its sea level setting. Consequently, for an anticlockwise rotating rotor more and more roll to starboard will result as a nose down pitch demand is made at higher and higher altitude. This effect is called *acceleration cross-coupling* since it arises from moments generated by the rotor. Rate cross-coupling arising from a rotor under a steady pitch or roll rate will be covered next.

4.5.3 Flapping motion due to steady pitch or roll rate

The application of a steady pitch or roll rate (see Fig. 4.9) to a rotor system is an important scenario since it introduces not only the concept of *rate cross-coupling* but highlights the existence of *aerodynamic damping* resulting from the interaction between the main rotor and the fuselage. The following discussion will consider only the flapping motion due to a pitch rate; the situation when a steady roll rate is applied can be tackled in an analogous manner. The analysis begins by returning to the basic flapping equation (4.19):

$$I_{yy}\dot{\omega}_2 - I_{yy}\omega_1\omega_3 - m_b X_g R a_z = M$$

From Fig. 4.10, it can be seen that:

$$\left.\begin{aligned} \omega_1 &= q\sin\psi\cos\beta + \Omega\sin\beta \\ \omega_2 &= q\cos\psi - \dot{\beta} \\ \omega_3 &= -q\sin\psi\sin\beta + \Omega\cos\beta \end{aligned}\right\} \tag{4.26}$$

Therefore for a constant pitch rate:

$$\dot{\omega}_2 = \dot{q}\cos\psi - q\dot{\psi}\sin\psi - \ddot{\beta} = \dot{q}\cos\psi - q\Omega\sin\psi - \ddot{\beta} = -q\Omega\sin\psi - \ddot{\beta} \tag{4.27}$$

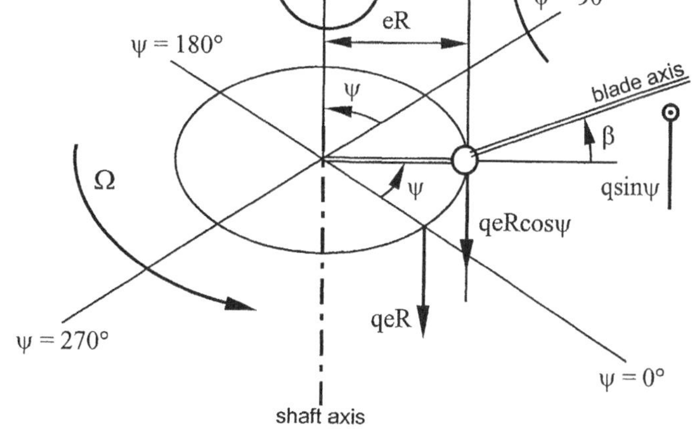

Fig. 4.9 Rotor hub under the influence of a steady pitch rate.

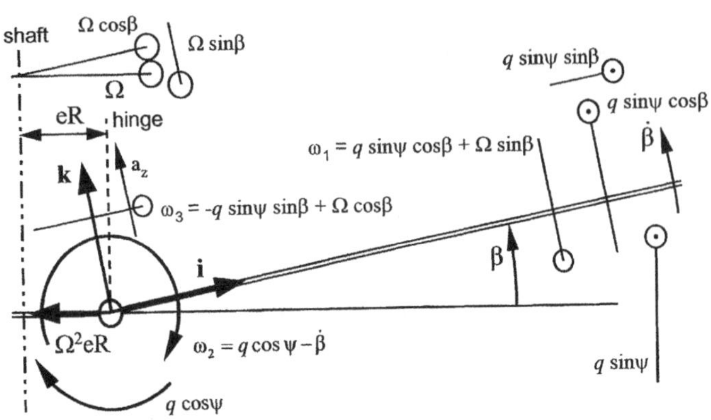

Fig. 4.10 Rotor axes system under the influence of a steady pitch rate.

In addition, using small angle approximations and assuming q^2 is small compared with Ω^2:

$$\left.\begin{aligned}\omega_3\omega_1 &= (-q\sin\psi\sin\beta + \Omega\cos\beta)(q\sin\psi\cos\beta + \Omega\sin\beta)\\ \omega_3\omega_1 &= -q^2\sin^2\psi\beta - q\Omega\sin\psi\beta^2 + q\Omega\sin\psi + \Omega^2\beta = q\Omega\sin\psi + \Omega^2\beta\end{aligned}\right\} \quad (4.28)$$

Thus substituting Equations (4.27) and (4.28) into Equation (4.9):

$$I_{yy}\dot{\omega}_2 - I_{yy}\omega_1\omega_2 - m_b X_g R a_z = M$$

$$-I_{yy}\ddot{\beta} - I_{yy}q\Omega\sin\psi - I_{yy}q\Omega\sin\psi - I_{yy}\Omega^2\beta - m_b X_g R a_z = M$$

$$\ddot{\beta} + 2q\Omega\sin\psi + \Omega^2\beta + \frac{m_b X_g R a_z}{I_{yy}} = \frac{M_A}{I_{yy}}$$

Now the vertical velocity of the hinge as a consequence of the pitch rate can be seen from Fig. 4.11 to equal:

$$V = \Omega eR\sin\psi\sin\theta - qeR\cos\psi = \Omega eR\sin\psi\theta - qeR\cos\psi \quad (4.29)$$

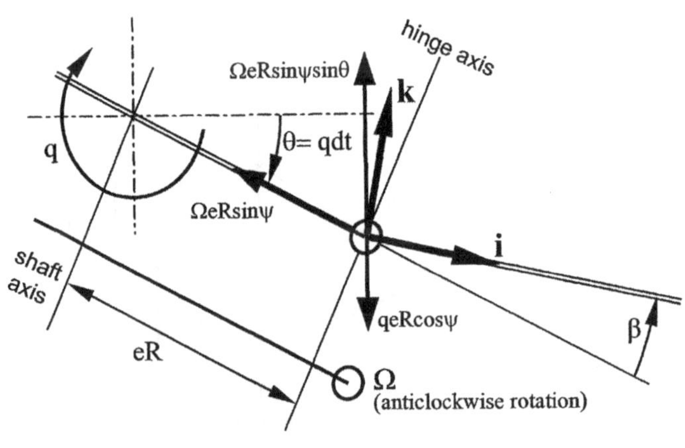

Fig. 4.11 Vertical velocity of rotor blade hinge.

Therefore the vertical acceleration, again as a consequence of the pitch rate, will be:

$$az = \frac{dV}{dt} = qeR\Omega \sin\psi + qeR\Omega \sin\psi \tag{4.30}$$

Substitution for the angular velocity components of the blade and the acceleration of the hinge given by Equations (4.11) and (4.30), leads to:

$$\ddot{\beta} + \Omega^2(1+\varepsilon)\beta = \Omega^2\frac{d^2\beta}{d\psi^2} + \Omega^2(1+\varepsilon)\beta = \frac{M_A}{I_{yy}} - 2\Omega q(1+\varepsilon)\sin\psi \tag{4.31}$$

Due to the combination of pitching and flapping, the change in blade incidence at a point r from the flapping hinge is given, for small flap angles, by:

$$\Delta\alpha = \frac{r\omega_2}{\Omega(r+eR)} = \frac{rq\cos\psi - r\dot{\beta}}{\Omega(r+eR)}$$

Therefore as before:

$$\delta L = \frac{1}{2}\rho V^2 s C_L = \frac{1}{2}\rho\Omega^2(r+eR)^2 cr\,\delta r\,a\left(\frac{q\cos\psi - \dot{\beta}}{\Omega(r+eR)}\right)$$

$$= \frac{1}{2}\rho\Omega(r+eR)cr\,\delta r\,a(q\cos\psi - \dot{\beta})$$

and:

$$M_A = \int_0^{R(1-e)} r\,dL = \frac{1}{2}\rho ac\Omega(q\cos\psi - \dot{\beta})\int_0^{R(1-e)} r^2(r+eR)dr$$

$$= \frac{1}{2}\rho ac\Omega(q\cos\psi - \dot{\beta})R^4(1-e)^3(1+e/3)$$

So:

$$\frac{M_A}{I_{yy}} = \frac{\gamma}{8}\Omega(q\cos\psi - \dot{\beta})(1-e)^3(1+e/3) = n\Omega(q\cos\psi - \dot{\beta}) \tag{4.32}$$

and on substitution of Equation (4.32) into Equation (4.31):

$$\frac{d^2\beta}{d\psi^2} + n\left(\frac{d\beta}{d\psi}\right) + (1+\varepsilon)\beta = n\frac{q}{\Omega}\cos\psi - 2\frac{q}{\Omega}(1+\varepsilon)\sin\psi \tag{4.33}$$

Once again only the steady state result is of interest. Thus if:

$$\beta = a_0 - a_1\cos\psi + b_1\sin\psi$$

then the solution to Equation (4.33) is given by:

$$a_1 = -\frac{qn(3\varepsilon+2)}{\Omega(\varepsilon^2+n^2)}$$

$$b_1 = -\frac{q[2\varepsilon(\varepsilon+1) - n^2]}{\Omega(\varepsilon^2+n^2)}$$

$$n = \frac{\gamma}{8}(1-e)^3(1+e/3)$$

Consider the case when the hinge offset is zero, $e = \varepsilon = 0$, so:

$$a_1 = -\frac{16q}{\gamma\Omega} \quad \text{and} \quad b_1 = -\frac{q}{\Omega}$$

Hence for a steady nose-up body pitch rate the rotor develops a nose-down tilt relative to the fuselage. The tilt is proportional to the pitch rate and is inversely proportional to Lock number. A moment is therefore generated via a forward tilt of the thrust vector that opposes the original nose-up moment. Thus the rotor develops *aerodynamic damping* which in the longitudinal sense will contribute, along with the tailplane, to the magnitude of M_q. Note also that a body pitch gives rise to lateral flapping such that a nose-up pitch produces a disk tilt to port. Hence *rate cross-coupling* occurs even when the hinge offset is zero, unlike with acceleration cross-coupling. It can be shown that for a uniform blade:

$$-\frac{a_1\Omega}{q} = \frac{4\gamma(1-e)^4(1+e/3)(5e+4)}{[144e^2 + \gamma^2(1-e)^8(1+e/3)^2]}$$

and

$$\cot\phi = \frac{96e(2+e) - \gamma^2(1-e)^8(1+e/3)^2}{4\gamma(1-e)^4(1+e/7)(5e+4)}$$

The variation of aerodynamic damping (proportional to $a_1\Omega/q$) and rate cross-coupling with hinge offset and Lock number are presented in Figs 4.12 and 4.13 respectively. Figure 4.12 indicates that at high Lock number the level of aerodynamic damping increases with hinge offset. Figure 4.13 shows that the tendency for the disk

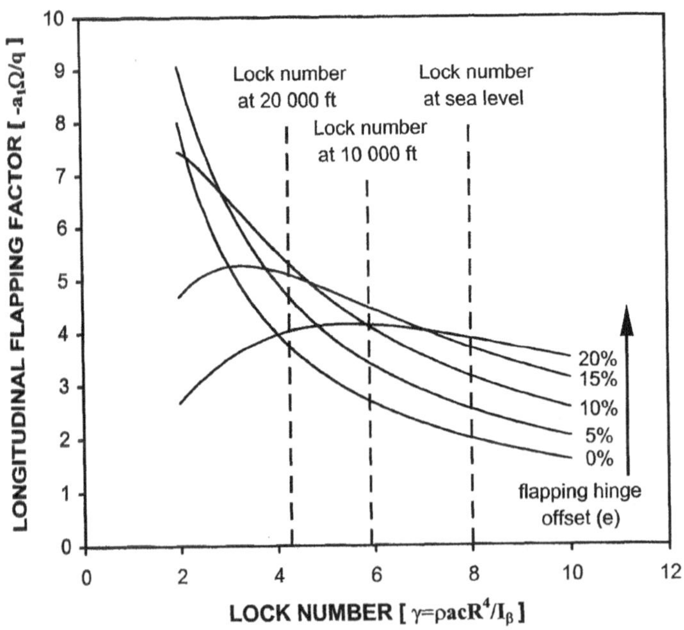

Fig. 4.12 Variation of aerodynamic damping with Lock number.

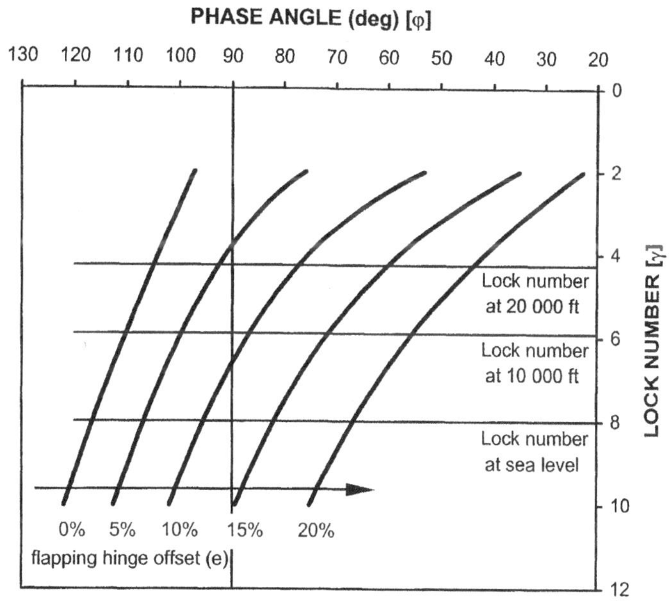

Fig. 4.13 Example of rate cross-coupling.

to tilt to port with a nose-up pitch rate changes with both Lock number and hinge offset. Recalling that Lock number is air density dependent allows an estimation of the variation of damping and cross-coupling with altitude. If a Lock number of 8 at sea level is assumed then it can be seen that at high altitude the aerodynamic damping will be greater for a rotor of modest hinge offset and the tendency of the aircraft to roll to the left may be replaced by a tendency to roll right, especially at high values of hinge offset. The increase in rotor damping with altitude may appear a little confusing but it should be remembered that the damping arises from the interplay between the tilt of the rotor due to aerodynamic forces and the inertia of rotor due to gyroscopic effects. Hence at high altitude the inertia forces will be stronger and the rotor will have a greater tendency to retain its position in space whilst the fuselage pitches nose up, thereby resulting in a greater relative nose-down disk tilt. Note that the increased off-axis response of a rotor with a high hinge offset at high altitude causes a reduction in aerodynamic damping.

4.5.4 Variation of control power with hinge offset

It has already been shown that a flapping hinge offset results in an increase in aerodynamic damping. Flight experience suggests that rotors with large hinge offsets also have a high degree of control sensitivity. In other words such a rotor causes a rapid acceleration in pitch and roll which combined with the high damping gives short time constants. Thus a rotor with a large hinge offset will display a crisper control response than one with a small offset; a steady pitch or roll rate being achieved rapidly after the application of step control input. The reason for this differing response lies in the fact that the hinge offset itself gives rise to additional forces and moments that

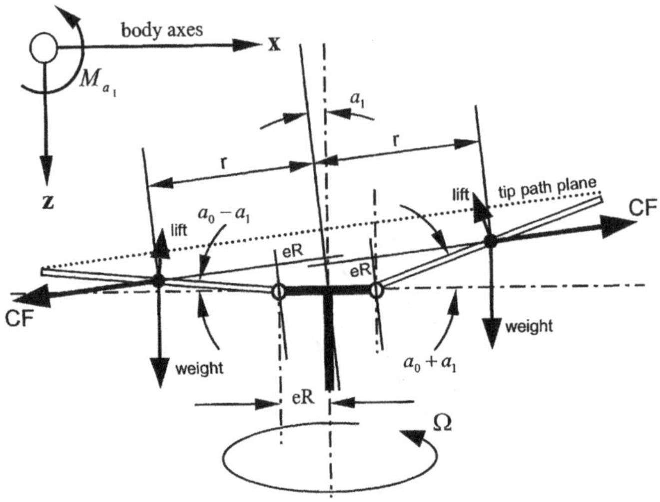

Fig. 4.14 Offset flapping hinges.

accelerate the aircraft in pitch or roll. From Fig. 4.14 it can be seen that as the blade flaps the lines of action of the centrifugal forces are no longer coincident; thus a couple is produced. This couple produces a moment at the hub that reinforces the disk tilt demanded by pilot control input: nose-up disk tilt produces a nose-up hub moment.

The hub moment arises from the separation between the lines of action of the centrifugal loads on opposite blades. In the longitudinal sense the separation is a function of the magnitude of a_1. Thus from Fig. 4.15 the maximum moment per blade is:

$$M = 2CF\,eR \tan a_1 \tag{4.34}$$

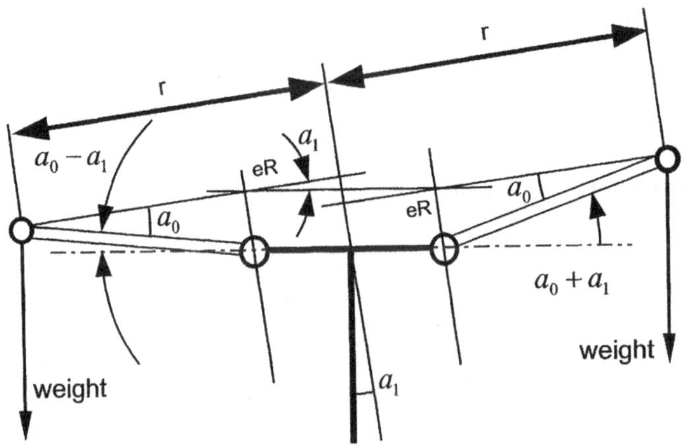

Fig. 4.15 Determination of hub moment.

Now:

$$CF = m_b r(\Omega \cos a_1)^2$$

where r is the radial location of the centre of mass from the axis of rotation.
For a blade with a uniform mass distribution:

$$CF = m_b\left(eR + \frac{1}{2}(1-e)R\cos a_0\right)\Omega^2\cos^2 a_1 \tag{4.35}$$

Hence using Equations (4.34) and (4.35):

$$M = eRmb\left(eR + \frac{1}{2}(1-e)R\cos a_0\right)\Omega^2\cos^2 a_1 \tan a_1$$

Assuming a small coning angle and disk tilt; $\tan a_1 = a_1$ and $\cos a_0 = \cos a_1 = 1$:

$$M = eR^2 m_b(1+e)\Omega^2 a_1$$

The average longitudinal moment is obtained by noting that the moment varies from the maximum value given above to zero. Thus since the longitudinal moment will only arise from longitudinal flapping and for b blades:

$$M = \frac{be}{2}(1+e)mbR^2\Omega^2 a_1 \tag{4.36}$$

Now the control power is defined as the moment generated for maximum control deflection, thus from Equation (4.36):

$$M_{a_1} = \frac{dM}{da_1} = \frac{1}{2}ebR^2 m_b(1+e)\Omega^2$$

Hence the larger the hinge offset the larger the additional moment. Thus it is now possible to describe the control response of a rotor with a hinge offset. When compared with a teetering rotor, a rotor with a significant hinge offset will display a larger control sensitivity since the additional moment described above will reinforce the moment achieved by tilting the thrust vector so giving a greater angular acceleration. The higher aerodynamic damping associated with hinge offset then causes the acceleration to decay rapidly. Thus the time to steady pitch or roll rate is inversely proportional to the size of flapping hinge offset, see Figs 4.16 and 17.

4.5.5 Semi-rigid rotors

The flap behaviour of a semi-rigid rotor can be modelled by approximating the rotor system to one with a hinge offset and a spring force. The size of the hinge offset, now referred to as the *effective hinge offset*, is chosen so that blade flap mode shape equates to the real blade under out-of-plane bending, see Fig. 4.18. Now the basic flapping equation is given by Equation (4.19) as:

$$I_{yy}\ddot{\beta} + \Omega^2(I_{yy} + m_b X_g eR^2)\beta = M_A$$

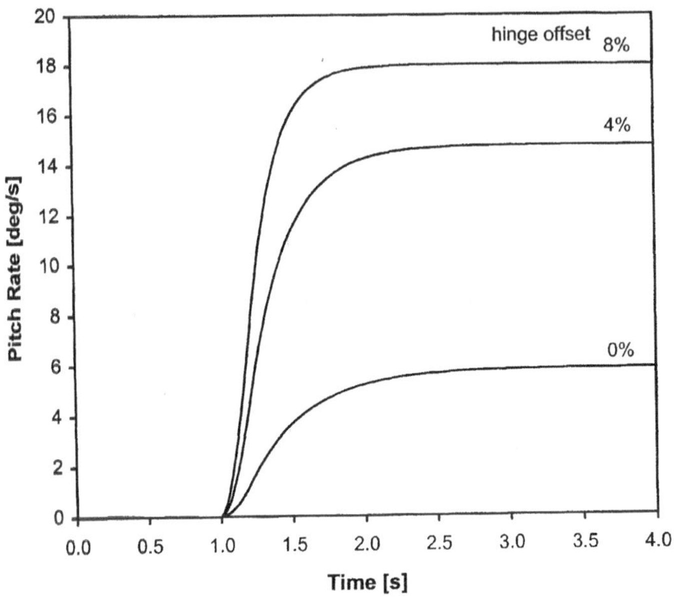

Fig. 4.16 Effect of hinge offset on pitch rate response.

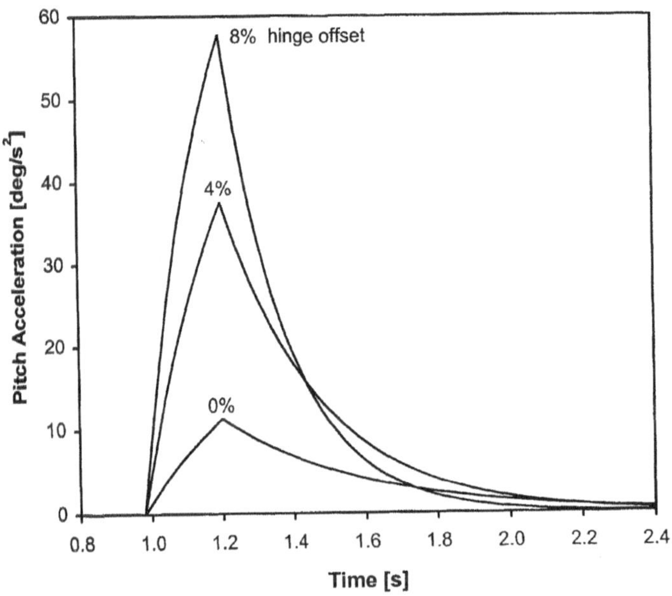

Fig. 4.17 Effect of hinge offset on pitch acceleration.

The spring produces a restoring moment proportional to the flap angle, thus Equation (4.19) becomes:

$$I_{yy}\ddot{\beta} + \Omega^2(I_{yy} + m_b X_g e R^2)\beta + K_\beta \beta = M_A \tag{4.37}$$

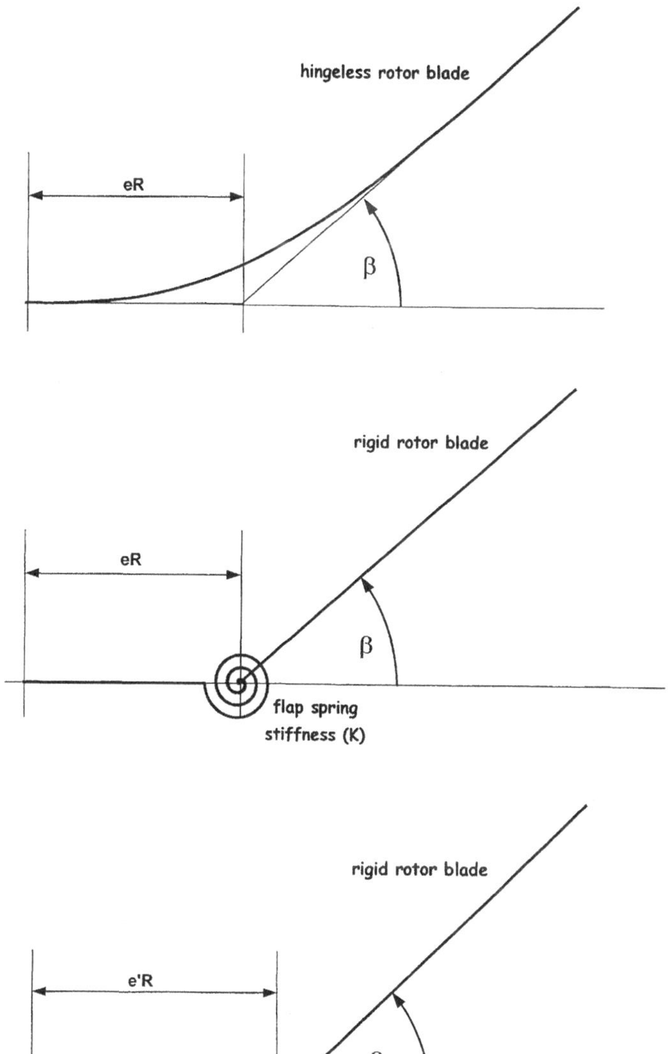

Fig. 4.18 Modelling hingeless rotors.

In order to determine the effect of this extra moment consider the case when the blade is disturbed in the flapping sense whilst in the hover. As before the aerodynamic moment is given, from Equation (4.20), by:

$$\frac{M_A}{I_{yy}} = -\frac{\gamma}{8}(1-e)^3(1+e/3)\Omega\dot{\beta}$$

Thus substituting into Equation (4.37):

$$I_{yy}\ddot{\beta} + \frac{\gamma}{8}(1-e)^3(1+e/3)I_{yy}\Omega\dot{\beta} + \Omega^2(I_{yy} + m_b X_g eR^2)\beta + K_\beta\beta = 0$$

or:

$$\ddot{\beta} + n\Omega\dot{\beta} + \Omega^2(1 + \varepsilon + \kappa)\beta = 0$$

where:

$$n = \frac{\gamma}{8}(1 - e)^3(1 + e/3) \quad \varepsilon = \frac{m_b X_g e R^2}{I_{yy}} \quad \kappa = \frac{K_\beta}{I_{yy}\Omega^2}$$

Taking Laplace transforms:

$$s^2 + n\Omega s + \Omega^2(1 + \varepsilon + \kappa) = 0$$

Comparing with the standard second-order characteristic equation:

$$\omega_n = \Omega\sqrt{1 + \varepsilon + \kappa} \quad \text{and} \quad \zeta = \frac{n}{2\sqrt{1 + \varepsilon + \kappa}}$$

The above equations indicate that the presence of the spring increases the natural frequency and reduces the flap damping. This is exactly the same effect as caused by increases in hinge offset, therefore a hingeless rotor may be adequately represented by a fully articulated rotor with a hinge offset greater than the geometric equivalent, as in Fig. 4.18.

4.6 ROTOR CONTRIBUTIONS TO STATIC AND DYNAMIC STABILITY

4.6.1 Changes to blade lift

The basic aerodynamics of rotor blades are similar to a conventional wing. At a given radial location the lift generated will depend on the flow velocity and incidence. The actual flow velocity and incidence will arise from the interaction of the rotational speed of the rotor, the inflow velocity, the airspeed, rate of climb or descent of the vehicle and the pitch setting of the blades. The variation of lift with changes in airspeed, vertical speed and blade pitch are best discussed by example. For simplicity this discussion will be *restricted* to the advancing and retreating blade tip ($r = R$ and $\psi = 90°$ and $270°$). Consider a hovering rotor that is subjected to the following:

(1) *Disturbance along the longitudinal axis.* Suppose the rotor is subjected to a disturbance equivalent to the rotor developing a forward airspeed component. When the disturbance appears the hovering rotor develops an asymmetry in tangential velocity and subsequent change in AOA and lift: more on the advancing side (A) and less on the retreating side (R).
(2) *Disturbance along the vertical axis.* If the rotor develops a sink rate the inflow velocity component is reduced and the AOA consequently increased. Since the rotor is in pure vertical flight the increase in lift, arising from the increase in AOA, will be equally distributed around the azimuth.
(3) *Change in blade pitch.* Suppose the rotor is disturbed such that the swash plate is moved instantly and the rotor attitude is initially unchanged. (This disturbance mode is an important case in considering the manoeuvre stability of a helicopter.) If the swash plate is tilted nose-up (positive direction) an increase in blade pitch

at $\psi = 90°$ and a reduction at $\psi = 270°$, will result. This will change the AOA in a similar sense resulting in more lift being generated at $\psi = 90°$ and less at $\psi = 270°$.

Now consider the same three disturbances applied to a rotor in trimmed forward flight with the cyclic pitch arranged to equalize the lift produced around the azimuth.

(1) *Disturbance along the longitudinal axis.* Suppose the rotor is subjected to a disturbance equivalent to the rotor developing a forward airspeed increment. The effect of the disturbance is similar to that observed in the hover: more lift is generated on the advancing side and less on the retreating side.
(2) *Disturbance along the vertical axis.* If the rotor develops a sink rate the inflow velocity component is reduced and the AOA consequently increased. Since the rotor is now in a combination of pure vertical flight and forward flight the increase in lift is not equally distributed around the azimuth due to the advancing/retreating effect.
(3) *Change in blade pich.* As before suppose the rotor is disturbed such that the swash plate is moved instantly and the rotor attitude is initially unchanged. If the swash plate is tilted nose-up an increase in blade pitch at $\psi = 90°$, and a reduction at $\psi = 270°$, will result. This will change the AOA in a similar sense resulting in more lift being generated at $\psi = 90°$ and less at $\psi = 270°$. The resulting change in the pitch attitude of the rotor is discussed below.

Calculations can be made to illustrate precisely how changes in airspeed, rate of climb or descent and pitch affect the lift generated on a blade. A rotor of 5.5 m radius, rotating at 35 rad/s and with a lift curve slope of 0.1/deg was used to produce the data contained in Tables 4.6 and 4.7.

The changes in lift described in these tables will cause the rotor blades to move about their flapping hinge, or bend within their flapping flexural element. An increase in lift will cause the blade to flap upwards and vice-versa. A rotor can be considered as a heavily damped system operating at (or close to) its natural frequency and therefore there will be approximately a 90° phase lag between input and output. The net result is that if an increase in lift reaches its maximum value at the point of maximum tangential velocity the blade will reach the point of maximum upwards flapping over the nose, that is approximately 90° later.

Table 4.6 Variations in lift: hover.

Disturbance	θ (deg)	Tangential speed (m/s)	v_i (m/s)	V (m/s)	α (deg)	L/S (N/m^2)	
nil	5.00	227.50	10.00	227.72	2.48	7887	
5 kts fwd	5.00	230.07	10.00	230.29	2.51	8157	A
5 kts fwd	5.00	224.93	10.00	225.15	2.45	7621	R
100 fpm descent	5.00	227.50	9.49	227.70	2.61	8291	
+ 0.5 deg pitch	5.50	227.50	10.00	227.72	2.98	9475	A
+ 0.5 deg pitch	4.50	227.50	10.00	227.72	1.98	6299	R

Table 4.7 Variations in lift: 120 kts.

Disturbance	θ (deg)	Tangential speed (m/s)	v_i (m/s)	V (m/s)	α (deg)	L/S (N/m^2)	
nil	3.52	289.23	10.00	289.40	1.54	7889	A
nil	8.12	165.77	10.00	166.07	4.67	7889	R
5 kts fwd	3.52	291.80	10.00	291.97	1.56	8121	A
5 kts fwd	8.12	163.20	10.00	163.51	4.62	7558	R
100 fpm descent	3.52	289.23	9.49	289.38	1.64	8403	A
100 fpm descent	8.12	165.77	9.49	166.04	4.84	8181	R
+ 0.5 deg pitch	4.02	289.23	10.00	289.40	2.04	10454	A
+ 0.5 deg pitch	7.62	165.77	10.00	166.07	4.17	7044	R

4.6.2 Rotor flapping and disk tilt

Consider a helicopter hovering in still air. If the longitudinal airspeed of the helicopter is changed from zero, by a gust or pilot inaction, the disk will tilt upward over the nose if the airspeed change is positive, that is forward, and vice-versa. This phenomena, the '*flap-back effect*', is very important in determining the static stability of the helicopter. Alternatively if the helicopter is allowed to sink the change in inflow will cause the rotor to generate more thrust thereby arresting the descent. In a similar manner a climb initiated by a gust will be damped out. Suppose a gust strikes the fuselage causing it to pitch up. This will have the initial effect of changing the pitch on the blades since the rotor will initially retain its position in space due to gyroscopic rigidity. Although the pilot has not applied any cyclic pitch a movement of the fuselage relative to the main rotor will have the same effect. Shortly after the fuselage pitches up the rotor will follow it and ultimately be re-aligned, at which point the lift will be equalized once again.

The response of the rotor to a change in forward airspeed, whilst in forward flight, is similar to that for the hover. The changes in lift on the advancing and retreating sides of the disk cause it to tilt up over the nose following an increase in airspeed and vice-versa (*flap-back*). The effects of changes in vertical speed and fuselage pitch attitude are different, however, with both situations giving rise to an unstable disk response. Suppose the helicopter develops a positive vertical speed (downwards). This will generate an increase in the angle of attack on both sides of the disk and hence an increase in lift. This increase is, however, not equal and the net result will be an upward tilt of the disk over the nose of the helicopter. The tilt back of the rotor will have the effect of increasing the AOA. If the action of a gust causes the helicopter to pitch nose-up this will increase the pitch of the advancing blade and reduce the pitch of the retreating blade as already described. The rotor will then flap in a 'nose-up' direction to achieve equality of lift, again as already mentioned. However due to the effect of the forward speed the attitude of the rotor relative to the fuselage at the point of lift equalization is further nose-up than the original disturbance. The tilt back angle required to equalize lift for a given pitch disturbance is determined in Table 4.8 and plotted in Fig. 4.19. For simplicity the rotor has a symmetric blade section and is initially at zero collective pitch (blade flapping is not considered). Note in this simple example that the blade pitch, measured relative to the horizon, is controlled solely by the swash plate tilt.

Table 4.8 Disk tilt response to a pitch disturbance (120 kts, rotor radius 6.5 m, RRPM 35 rad/s).

	ΩR (m/s)	V (m/s)	Total velocity vector angle (deg)	γ (deg)	θ (deg)	α (deg)	L/S (N/m^2)	
start	227.50	289.2	0.0	0.0	0.0	0.0	0	A
start	227.50	165.8	0.0	0.0	0.0	0.0	0	R
body tilt	227.50	289.2	0.0	0.0	10.0	10.0	102 475	A
body tilt	227.50	165.8	0.0	0.0	−10.0	−10.0	−33 663	R
disc tilt	227.50	288.5	7.9	10.0	10.0	2.1	21 709	A
disc tilt	227.50	167.1	−13.7	−10.0	−10.0	3.7	12 577	R
lift equal	227.50	288.4	8.4	10.7	10.0	1.6	15 940	A
lift equal	227.50	167.2	−14.7	−10.7	−10.0	4.7	15 939	R

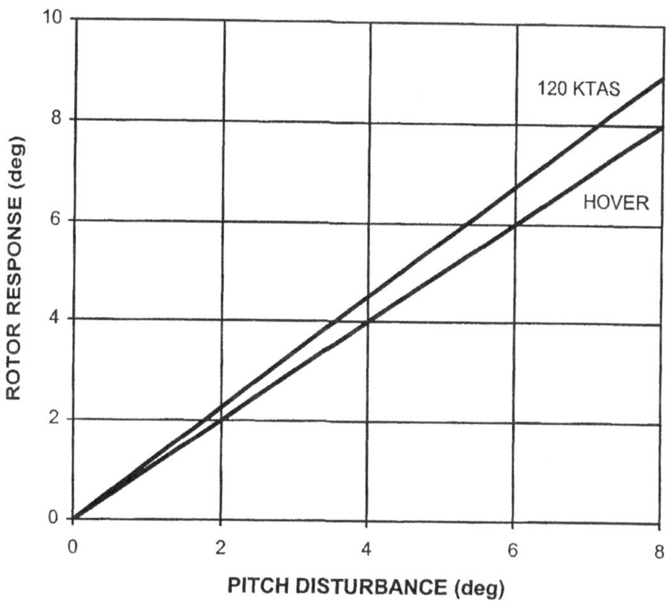

Fig. 4.19 Rotor response to a pitch attitude disturbance.

4.6.3 Main rotor contributions to speed and AOA stability

The foregoing discussion of the rotor response to changes in flight condition gives insight into the stability of a helicopter. Recalling that *static stability* is the initial response of the aircraft following a disturbance it is possible to lead from a consideration of *speed* and *AOA stability* to *static stability*. Before describing the stability of a conventional helicopter it will instructive to introduce the mechanisms by which control is effected. The basic method of control is through the generation of moments about the centre of gravity. In this case only pitching moments derived from the main rotor will be considered. On all helicopters a control moment is generated by changing the direction and magnitude of the thrust vector. Changes to the direction are achieved by disk tilt. An additional moment will arise on helicopters with a non-zero hinge offset, since as the disk tilts relative to the fuselage, centrifugal forces acting on the blades give rise to a couple which manifests itself as a hub moment which is proportional to the amount of disk tilt. Therefore, regardless of the type of rotor system, if a disk tilt is produced as a response to a disturbance a moment, pitching in this case, will be generated. Suppose the pilot trims the helicopter at a particular airspeed and then relaxes on the controls. Sometime later a gust strikes the aircraft causing the airspeed to increase. This increase will cause the rotor to flap-back, that is the disk will tilt rearwards. The movement of the disk relative to the fuselage will generate a nose-up pitching moment that will have the effect of reducing the airspeed. Therefore the main rotor generates a *stabilizing contribution* to the overall *speed stability* of the helicopter. Again suppose the pilot trims the helicopter, but this time the disturbance generates either a downwards component of airspeed or causes the fuselage to pitch upwards. In either case the result is an increase in the AOA on the advancing blade. Once again the rotor disk will tilt rearwards but this time further than the original AOA disturbance, thereby producing a net out-of-balance moment which generates a further disturbance. Therefore the main rotor generates a *destabilizing contribution* to the overall *AOA stability* of the helicopter.

4.7 LONGITUDINAL STATIC STABILITY

The main rotor, the fuselage and the horizontal stabilizer are assumed to be the only contributors to the longitudinal stability of the helicopter. Before going further it is worthwhile revising two important stability definitions:

(1) *Trim*. An aircraft is in *trim* when all the forces and moments acting on it are in balance. The aircraft is in a state of *equilibrium* and would continue in that condition unless acted upon by a gust, or affected by pilot action.
(2) *Static stability.* A body is statically stable if there is an initial tendency for it to return to its trim condition after an angular displacement or after a change in the transitional velocity. In helicopter parlance *static stability*, in forward flight, describes the response of the helicopter following a change in translation velocity (*speed stability*). *Manoeuvre stability* is the response to angular changes (*AOA stability*).

4.7.1 Hovering flight

Evidently a helicopter possesses neutral static stability with respect to angular displacements when the definition of static stability, given above, and the rotor response, detailed earlier, are considered. If a helicopter suffers an *angular* disturbance while hovering no direct *aerodynamic* moment arises which will restore it to its original attitude. The resultant rotor thrust always passes through the centre of gravity irrespective of the angular position of the helicopter (assuming no download on the horizontal stabilizer and fuselage). The response is equivalent to the neutral stability in roll displayed by a conventional fixed wing aircraft. In both cases it is the *subsequent* generation of a translational velocity that may give rise to a stabilizing response. In the fixed wing case the roll disturbance causes a lateral velocity to develop and the dihedral of the wings combined with this velocity produces a moment which will tend to return the aircraft to the trim condition. Similarly for a helicopter the angular displacement will result in a translational velocity due to the unbalanced horizontal component of the thrust vector. The positive speed stability of the rotor will then lead to the development of a moment tending to return the helicopter to the trim position. If a hovering helicopter is subjected to a disturbance in translational velocity then it is only the flap-back effect from the rotor that will tend to return the helicopter to its original attitude. At very low speeds contributions from the horizontal stabilizer and fuselage may be ignored.

4.7.2 Cross-coupling : collective to pitch attitude

Before describing the static and manoeuvre stability in forward flight it is important to understand the effect of changes in collective pitch, at constant airspeed and load factor, on the pitch attitude of the helicopter. The initial response of the helicopter will be dictated by the change in the magnitude, and direction, of the thrust vector and hub moment from the main rotor. As the pilot raises the collective lever the lift produced by all blades is increased thereby increasing the thrust and the coning angle. In forward flight the increased collective pitch will have a greater effect on the advancing side and consequently the amount of nose-down longitudinal flapping will be reduced. The net result of these changes is to reduce the size of the nose-down pitching moment generated by the main rotor. The helicopter will, therefore, pitch *nose up* as the collective lever is *raised.* Subsequent control activity and cyclic stick displacement from the level flight position will result from the effect of the relative airflow on the horizontal stabilizer and fuselage.

4.7.3 Climbing and descending flight

In a steady climb the cyclic stick is often held forward to counter the nose-up pitching moment from the tailplane as well as that from the main rotor. A rapid entry into flight idle glide (FIG) or autorotation will have the opposite effect. The amount of aft stick required during a rapid entry is of particular interest especially if there is a danger of infringing control margins. Sometimes even during a steady autorotation there is still concern over the amount of aft cyclic required to counter the net nose-down moment from the tailplane. In such cases the designer may arrange for a reduction in the effectiveness of the stabilizer, thereby reducing the upsetting moment. Measurement of the position of the cyclic stick in steady climbs or descents is part of a test technique called '*Trimmed Flight Control Positions*' (TFCP) which is discussed later (Section 5.2).

4.7.4 Forward flight

The response of a rotor to changes in speed is stable. Therefore a tail-less helicopter should possess a degree of positive static stability. Since, however, the fuselage contribution to static stability is variable and the addition of a horizontal surface will provide manoeuvre stability most helicopters are fitted with a tailplane at some location on the tail boom. Static (speed) stability, or otherwise, arises from the development of pitching moments following changes in the speed of the airflow approaching the helicopter. In summary, the contributors to static stability are:

(1) *The main rotor.* The main rotor provides a stable contribution to static stability. The size of the stabilizing moment increases with speed and rotor thrust.
(2) *The horizontal stabilizer.* The horizontal stabilizer produces a stabilizing contribution to static stability provided that it is downloaded. Therefore the inherent stability of the isolated main rotor (with speed) can be increased, or reduced, by the addition of a suitably sized downloaded, or uploaded, tailplane. The magnitude of the moment from the tailplane increases with speed.
(3) *The fuselage.* The contribution to static stability from the fuselage can be either stabilizing or destabilizing depending on the line of action of the lift and drag forces.

4.8 MANOEUVRE STABILITY

The response of a rotor to changes in AOA is unstable. Therefore a helicopter would possess negative manoeuvre stability without the addition of a surface to provide a stabilizing moment. A horizontal stabilizer is generally fitted to a helicopter to serve this purpose and thus provide a degree of manoeuvre stability. It will, however, also affect the static stability. Manoeuvre stability, or otherwise, arises from the development of pitching moments following changes in the angle of incidence of the airflow approaching the helicopter. Contributors to manoeuvre stability are:

(1) *The main rotor.* The main rotor provides an unstable contribution to manoeuvre stability. The size of the destabilizing pitching moment will increase with trim speed, above minimum power speed, and load factor.
(2) *The horizontal stabilizer.* The horizontal stabilizer provides a stabilizing contribution to manoeuvre stability. It does not matter whether the stabilizer is uploaded or downloaded when at the trim condition, since an increase in fuselage incidence will result in a nose-down pitching moment in both cases. The magnitude of the stabilizing moment increases with trim speed, but not load factor.
(3) *The fuselage.* The contribution to manoeuvre stability from the fuselage can be either stabilizing or destabilizing depending on the line of action of the lift and drag forces.

The destabilizing effect of aft rotor flapping (main rotor contribution to manoeuvre stability) is approximately proportional to the square of the forward speed as is the change in lift generated by the tailplane following an AOA change. Therefore, the size of stabilizer selected to provide stability at one trim speed is generally suitable for other speeds. However, since the magnitude of the destabilizing moment from the main rotor also increases with load factor, but the opposing moment from the horizontal stabilizer does not, it is highly likely that a helicopter will display manoeuvre instability if tested at both high load factor and high speed. A low set stabilizer may also suffer a variation in performance if heavily influenced by the downwash from the main rotor.

The position of the CG relative to the main rotor thrust vector and tailplane has an effect on the manoeuvre stability of a helicopter. If the CG is forward of the rotor, the increase in rotor thrust and aft tilt of the vector associated with an up gust may produce a stabilizing nose-down pitching moment, or at least a less severe nose-up moment. With an aft CG position, however, the situation is reversed and an increase in thrust and rearwards tilt of the thrust vector will generate an unstable pitching moment. The horizontal stabilizer will be required to provide a measure of manoeuvre stability, as described above. The overall stability will be weaker at higher weights and with aft CG positions since the destabilizing effect of the main rotor will be that much stronger. This is usually the reason a maximum allowable aft CG position is quoted in the operating manual. The forward limit is often established by the manufacturer either to prevent high oscillatory rotor loads or to provide adequate aft stick margin for a landing flare or rearward flight.

4.9 LONGITUDINAL DYNAMIC STABILITY AND CONTROL RESPONSE

Whilst static stability is concerned with an aircraft's initial motion following a perturbation, dynamic stability determines the aircraft's longer-term response to such a disturbance. An aircraft is dynamically stable if, following the removal of a disturbing force, it returns to its equilibrium position. Control response, on the other hand, is concerned with the response of the aircraft to a control input made by the pilot. This section examines the longitudinal dynamic stability and control response of a single main rotor helicopter in both the hover and forward flight. The analysis of the motions is simplified, as before, by assuming that there are no cross-coupling effects. This is rather less easily justified than for fixed wing aircraft, and care must be exercised in this respect. Before looking at the dynamic modes in detail, it is worth reminding ourselves of the derivatives which influence the longitudinal motion.

4.9.1 Longitudinal derivatives

In the longitudinal plane, the variation of X-force, Z-force and pitching moment, M, with respect to forward and vertical velocities, pitch rate and collective and longitudinal cyclic control movements must be considered.

4.9.1.1 Forward velocity derivatives

(1) *Forward force due to forward velocity* (X_u). The advancing blade sees an increase in forward speed as an increase in relative airspeed whilst the retreating blade see it as a decrease in relative airspeed. Assuming a phase angle of 90°, this causes the rotor disk to flap further back, which in turn causes the thrust vector to tilt rearwards resulting in a decrease in X-force. The flapback also results in an increase in rotor thrust and in H-force. Fuselage drag also increases with speed and at forward speeds in the range 35 to 50 kts this contribution can be equal to that of the main rotor. The overall effect of all these contributions is

to return the aircraft to its equilibrium position. To summarize, there are stabilizing contributions to X_u from the backward tilt of the main rotor; the change in magnitude of the thrust vector; the change in fuselage drag; and the change in rotor in-plane force.

(2) *Vertical force due to forward velocity* (Z_u). This derivative is zero in the hover as one would expect, and at low forward speeds an increase in forward speed will cause an increase in rotor lift. However, the amount of disk tilt is an important parameter in the determination of the rotor lift. At high forward speeds the tilt may be large and an increase in speed may then result in a decrease in rotor lift. The force Z_u is therefore zero at the hover, becomes negative (remember Z is positive downwards) and then positive. It may be undesirable for Z_u to be negative, however, and in this respect an aerodynamically clean fuselage is advantageous as a reduction in overall fuselage drag will result in a smaller disk tilt for a given speed.

(3) *Pitching moment due to forward speed* (M_u). An increase in forward speed causes the disk to flap back and hence tilts the thrust vector rearwards causing a nose-up pitching moment which gives a *stabilizing* (positive) contribution to M_u. A horizontal stabilizer also contributes significantly to the overall value of M_u, its effect depending on its setting angle and on the downwash changes resulting from the speed changes. The fuselage can also contribute to M_u, its contribution depending on the change in fuselage lift and drag with changes in speed and the distance of its centre of pressure from the CG. This derivative has a major effect on the dynamic motion of the helicopter and, although a positive value of M_u is necessary for static stability with respect to forward speed changes, if excessive, it will cause dynamic instability.

4.9.1.2 *Vertical velocity derivatives*

These are also known as the angle of attack derivatives as this is proportional to the vertical velocity for a fixed wing aircraft at a given forward speed. Angle of attack is not quite so clearly defined for a helicopter, but the derivatives are still known by either name. The derivatives are X_w (or X_α), Z_w (or Z_α), M_w (or M_α) and the 'downwash lag' derivative $M_{\dot{w}}$.

(1) *Forward force due to vertical speed* (X_w). The force X_w is usually small and insignificant and has little or no effect on either the static or dynamic stability characteristics of the aircraft.

(2) *Vertical force due to vertical speed* (Z_w). An increase in w means that the helicopter is moving vertically downwards, and this causes an increase in the blade angle of attack. This in turn produces an increase in blade lift as the inflow through the rotor decreases. The consequent increase in rotor thrust tends to nulify the increase in w, and Z_w is therefore the vertical damping derivative which is an important vertical response parameter, particularly in the hover and at low forward speeds. The force Z_w is always *stabilizing* (negative) and increases from its value in the hover to reach a limiting value in forward flight.

(3) *Pitching moment due to vertical speed* (M_w). The moment M_w is a measure of the static stability of the rotor with respect to changes in angle of attack. In the hover M_w is zero, but in forward flight an increase in w produces an effective increase in blade lift which is greater on the advancing blade than on the

retreating blade. This causes a backwards tilt of the rotor and a nose-up pitching moment. Consequently, the rotor contribution to M_w is positive and *destabilizing*. Its value increases approximately linearly with speed. Contributions to M_w also come from the fuselage and the horizontal stabilizer. The former is likely to be *destabilizing*, whereas the latter will normally be *stabilizing*. The tailplane will however be ineffective in the hover and at forward speeds when it is working in the downwash from the main rotor.

(4) *Pitching moment due to downwash lag* ($M_{\dot{w}}$). As the tailplane is a finite distance away from the main rotor and there will be a small delay between initiation of changes in rotor downwash due to angle of attack changes and their arrival at the tailplane. This is often ignored for simplicity.

4.9.1.3 Pitch rate derivatives

Assume that the helicopter is pitching nose-up with a constant angular velocity, q, and that the rotor is in equilibrium and pitching at the same rate. As the rotor may be regarded as a gyroscope it will be subjected to a precessing moment which would normally tend to tilt it to starboard. However, because of the response lag, the rotor actually tilts forward causing longitudinal forces and moments. This is the source of the aerodynamic damping discussed earlier. The derivatives X_q and Z_q are usually taken as negligible, so M_q is the only derivative worthy of note. M_q represents the change of pitching moment with changes in pitch rate. This is the '*pitch damping*' derivative. When the helicopter pitches nose-up there will be a favourable nose-down moment from the main rotor due to this aerodynamic damping. Thus, the rotor contribution to M_q is *stabilizing* (negative). Also, the angle of attack of the tailplane is changed producing more tailplane lift or less downforce and hence a favourable nose-down pitching moment providing a stabilizing contribution. There will also be a contribution from the fuselage that is usually stabilizing. Most helicopters will also require some form of augmentation of this damping if good handling qualities are to be achieved.

4.9.2 Control derivatives

Movement of the collective lever and fore/aft cyclic will also affect the motion of the helicopter in the longitudinal plane by altering the forces and moments acting. Collective and longitudinal cyclic pitch angles are denoted by θ_c and B_1 respectively, and their effect can be shown by the use of collective and cyclic control derivatives.

4.9.2.1 Collective pitch derivatives

When the collective pitch, θ_c, is increased, each blade experiences an increase in lift and the total rotor thrust is increased. The pitch increase also leads to an increase in flapback of the disk in forward flight and a subsequent nose-up pitching moment. This helps to explain the derivatives:

(1) *Forward force due to collective* ($X_{\theta c}$). The forward force due to collective is usually negligibly small.

(2) *Vertical force due to collective* ($Z_{\theta c}$). An increase in collective always produces

an increase in thrust (−ve Z) so is always negative. It is known as the collective control power derivative, or the *heave control power*.

(3) *Pitching moment due to collective* ($M_{\theta c}$). As outlined above, the increased flapback and thrust combine to produce a nose-up pitching moment in forward flight so the derivative is positive. As the disk does not flapback in the hover, $M_{\theta c}$ is zero in this flight regime, provided the horizontal stabilizer is unloaded and the tail rotor is conventional (not canted).

4.9.2.2 *Cyclic pitch derivatives*

Remembering that only fore/aft cyclic are being considered, any such pitch change will result in a change in the disk tilt, also fore/aft, and of the thrust vector. Hence, a pitching moment will be generated, nose-down for forward stick deflection and nose-up for rearward stick. The corresponding cyclic derivatives are:

(1) *Forward force due to longitudinal cyclic* (X_{B1}). The forward force due to longitudinal cyclic is usually negligible.
(2) *Vertical force due to longitudinal cyclic* (Z_{B1}). The vertical force due to longitudinal cyclic is negligible.
(3) *Pitching moment due to longitudinal cyclic* (M_{B1}). Pitching moments are generated as described above, and M_{B1} is known as the pitch control power derivative, or the '*longitudinal cyclic control power*'.

4.9.3 Longitudinal motion of the helicopter

To analyze the longitudinal dynamic stability characteristics of the helicopter it is necessary to consider the equations governing its motion. In order to study these equations, simplifying assumptions are required to ease the computational task. The matrices describing the equations of motion often include several acceleration dependent derivatives such as $M_{\dot{u}}$. Assuming all these are negligible and the centre of gravity is aligned with the main rotor drive shaft such that $d_x = d_y = d_z = 0$, then:

$$\mathbf{M}_m \dot{\mathbf{x}} = \mathbf{M}_a \mathbf{x} + \mathbf{M}_c \mathbf{u}$$

where:

$$\dot{\mathbf{x}} = \begin{bmatrix} \dot{u} \\ \dot{w} \\ \dot{q} \\ \dot{\theta} \end{bmatrix} \quad \mathbf{x} = \begin{bmatrix} u \\ w \\ q \\ \theta \end{bmatrix} \quad \mathbf{u} = \begin{bmatrix} \theta_c \\ B_1 \end{bmatrix} \quad \mathbf{M}_c = \begin{bmatrix} \mathring{X}_{\theta_c} & \mathring{X}_{B_1} \\ \mathring{Z}_{\theta_c} & \mathring{Z}_{B_1} \\ \mathring{M}_{\theta_c} & \mathring{M}_{B_1} \\ 0 & 0 \end{bmatrix}$$

$$\mathbf{M}_m = \begin{bmatrix} m & 0 & 0 & 0 \\ 0 & m & 0 & 0 \\ 0 & 0 & I_{yy} & 0 \\ 0 & 0 & 0 & 1 \end{bmatrix} \quad \mathbf{M}_a = \begin{bmatrix} \mathring{X}_u & \mathring{X}_w & \mathring{X}_q - mW_e & -mg\cos\theta_e \\ \mathring{Z}_u & \mathring{Z}_w & \mathring{Z}_q + mU_e & -mg\sin\theta_e \\ \mathring{M}_u & \mathring{M}_w & \mathring{M}_q & 0 \\ 0 & 0 & 1 & 0 \end{bmatrix}$$

Modern control theory when applied to multi-input/multi-output (MIMO) systems states that the Characteristic Equation (CE) can be obtained from the following:

$$\mathbf{det}(s\mathbf{I} - \mathbf{A}) = 0$$

where **A** is the result of pre-multiplying $\mathbf{M}_a$ by $\mathbf{M}_m^{-1}$. The CE when solved will show the nature of the controls fixed response of the helicopter to a disturbance. So:

$$CE = \left| \begin{bmatrix} s & 0 & 0 & 0 \\ 0 & s & 0 & 0 \\ 0 & 0 & s & 0 \\ 0 & 0 & 0 & s \end{bmatrix} - \begin{bmatrix} X_u & X_w & (X_q - W_e) & -g\cos\theta_e \\ Z_u & Z_w & (Z_q + U_e) & -g\sin\theta_e \\ M_u & M_w & M_q & 0 \\ 0 & 0 & 1 & 0 \end{bmatrix} \right| = 0$$

Therefore:

$$CE = \begin{vmatrix} s - X_u & -X_w & W_e - X_q & g\cos\theta_e \\ -Z_u & s - Z_w & -U_e - Z_q & g\sin\theta_e \\ -M_u & -M_w & s - M_q & 0 \\ 0 & 0 & -1 & s \end{vmatrix} = 0$$

This determinant will be of the form, $As^4 + Bs^3 + Cs^2 + Ds + E = 0$, where the coefficients in the polynomial can be expressed in terms of aerodynamic derivatives, see Bramwell [4.1]. The quartic may be solved by numerical computer methods when the values of the coefficients are known. Consequently the characteristic equation can be evaluated and factorized. For helicopters, in most cases, the equation factorizes to:

$$(T_1 s + 1)(T_2 s + 1)(s^2 + 2\zeta\omega_n s + \omega_n^2) = 0$$

The three modes of motion implied by this equation are summarized below:

(1) *Vertical velocity mode.* The vertical velocity mode, described by $(T_1 s + 1) = 0$ is a stable, heavily damped subsidence in vertical velocity. The motion is decoupled from speed and pitch and has a time constant of the order of 1 to 2 seconds.
(2) *Forward speed mode.* The forward speed mode, described by $(T_2 s + 1) = 0$ is a stable, heavily damped subsidence in speed. The motion is coupled with pitch attitude and pitch rate. It has a short time constant of the order of 0.5 second.
(3) *Pitching oscillation.* The stability of the pitching oscillation is both speed and flight condition dependent. In the climb or at high speed the oscillation can be unstable, possibly degenerating to an exponential divergence at high speed. The oscillation couples with the forward speed mode and is mainly due to rotor flapping caused by speed changes.

These basic equations of motion govern the aircraft in all flight regimes but differing values of the derivatives account for differences in the behaviour and will, of course, give different characteristic equations to solve. It should be remembered that **all** modes will be excited following a disturbance or pilot input. The various dynamic modes described above can be separated into long-term modes and short-term modes. The long-term modes characterize the dynamic stability of the helicopter, whereas the short-term modes affect the pilot's perception of the aircraft during manoeuvres, that is its control response.

4.9.4 Dynamic stability in the hover

Some of the derivatives can be assumed to approximate to zero in the hover case (X_w, Z_u, Z_q, Z_w, and M_w) and the characteristic equation typically solves to give two real roots and a pair of complex roots. One real root can be shown to be $(s - Z_w) = 0$, so that $s = Z_w$. This represents a heavily damped subsidence such that if a helicopter is disturbed, by a vertical gust for example, the subsequent heave motion is quickly damped out. The motion is a pure convergence with no oscillation and confirms that the vertical motion is completely decoupled from the pitching and fore/aft motions, a prediction arising from examination of the Z-force equation in the hover. The other real root represents the forward speed mode. In the hover the pitching oscillation or *falling leaf mode* masks this mode.

The physical description of the motion associated with the complex root is handled easily. Assume that the hovering helicopter experiences a small horizontal velocity disturbance, the relative airspeed change causes the rotor to tilt backwards and exert a nose-up pitching moment on the helicopter. A nose-up attitude then develops and the rearward component of the thrust vector decelerates the aircraft until its forward motion is stopped. At this point the disturbing disk tilt and rotor moment vanish but the helicopter is left in a nose-up attitude and backward motion begins. This causes the rotor to flap forwards and exert a nose-down moment. The thrust vector tilts forward and the rearward motion is stopped but the helicopter is now left in a nose-down attitude which accelerates it forward and the cycle begins again. The motion is generally unstable and its amplitude increases steadily, bearing out the analytical solution. The instability is entirely due to the characteristic backward flapping of the rotor with forward speed. However, the pitch damping derivative M_q will affect the rate of divergence. Making M_q more negative will reduce the real part of the complex root and hence increase the time to double amplitude of the motion. It can never make the motion stable, however, and at best only neutral stability can be achieved.

Bramwell [4.1], discusses the implications of positive M_u on the motion, that is the possibility of making the rotor flap forward with speed, but this only leads to a pure divergence which is even more undesirable. He also quotes the results of investigations into the effects of configuration changes on the dynamic stability of the hovering helicopter but it appears that no reasonable change will significantly improve it. In particular, CG position has little effect on the stability but simply affects the fuselage attitude adopted, even for blades with offset hinges or hingeless rotors. Although a hub moment can be exerted in these cases and it is therefore no longer necessary for the thrust vector to act through the CG, thrust changes due to forward speed and pitch rate are typically zero in the hover. The motion can be approximated to a neutral oscillation whose frequency depends on M_u and M_q:

$$\text{falling leaf frequency} = \omega = \sqrt{\frac{-M_u g}{M_q}}$$

4.9.5 Dynamic stability in forward flight

The characteristic equation in forward flight resolves into four roots but it is not so easy to generalize as in the case with a conventional fixed wing aircraft. For the latter, the characteristic equation resolves into pairs of complex conjugate roots representing two oscillatory motions, one of short period and high damping – the SPPO – and the other of long period which is lightly damped – the phugoid. In the case of a helicopter the characteristic equation solves into four roots but for a particular helicopter two pairs of complex roots may be found at one flight condition, two real and a pair of complex roots at a second condition and four real roots at a third. The reason for this is the large variation in the value of the derivatives over the flight envelope. In the longitudinal case M_w has the greatest influence, with M_q, M_u and Z_w also playing a part in the overall result. Although rotorcraft do not, strictly speaking, exhibit the SPPO and phugoid motions described for fixed wing aircraft, there are certain analogies which can be drawn.

The phugoid motion of a fixed wing aircraft is an oscillation involving changes in height and speed at approximately constant incidence. A disturbance producing, for example, an increase in lift causes the aircraft to climb slowly as lift now exceeds weight. The climb (which is at constant incidence) results in a decrease in speed and consequent loss of lift and eventually leads to a situation where lift and weight are again equal but the aircraft continues to slow down. A descent begins as weight exceeds lift and the consequent increase in speed produces an increase in lift and the climb starts again. The oscillatory motion experienced by a helicopter is influenced by the respective values of the speed stability, M_u, the angle of attack stability, M_w, the pitch damping, M_q and the pitching moment of inertia, I_{yy}.

Consider the motion following a disturbance that causes the helicopter to adopt a nose-down attitude and start to descend. Initially, rotorcraft with neutral angle of attack stability ($M_w = 0$) will be considered. The component of aircraft weight acting along the flight path accelerates the helicopter but as speed increases the rotor disk flaps back (*speed stability*) and a consequent nose-up pitching moment and angular acceleration occurs. This soon produces an angular velocity such that the pitch damping causes the fuselage to rotate to a greater angle of attack than the rotor, and this effectively neutralizes the thrust vector tilt due to static stability. The flight path is still downwards, however, and the component of weight acting along it ensures that the airspeed continues to increase so that the preceding steps are repeated. The angular velocity continues to increase with consequent increase in fuselage angle of attack. In turn, the thrust continues to increase until it is sufficient to level off the glide path. At this point the helicopter has reached its maximum forward speed, maximum nose-up pitch rate, and maximum fuselage angle of attack. The thrust now exceeds the weight and the aircraft begins to climb. The component of weight acting along the flight path now starts to slow down the helicopter. The rotor disk flaps forward as the rearward tilt due to speed stability is exceeded by the forward tilt due to pitch damping. The nose-down damping moment reduces the pitch rate and the fuselage angle of attack until they reach their trim values. The helicopter is still climbing, however, so the speed continues to decrease and the rotor flaps forward. The resulting nose-down pitching moment starts a similar sequence of events (with opposite signs of course) and this will be repeated until the oscillation eventually either damps out (stable) or grows worse (unstable). It is possible to characterize the long-term mode as [4.1]:

$$\omega_n = \sqrt{\frac{-Z_u g}{U_e}}$$

$$\zeta = \frac{X_u}{2}\sqrt{\frac{-U_e}{Z_u g}}$$

The *frequency* of the long-term mode is therefore inversely proportional to the trim speed, U_e. An increase in trim speed thus reduces the frequency resulting in a larger period for the oscillation. An increase in Z_u (related to the lift coefficient), on the other hand, has the opposite effect. The *damping* of the long-term mode is affected by the same factors, but in the opposite sense, so an increase in trim speed will increase the damping in addition to reducing the frequency of the response. The drag of the helicopter (related to X_u) also affects the 'phugoid' frequency. A helicopter carrying large external stores (higher drag coefficient) is therefore likely to exhibit a more heavily damped long-term response. A less simplified relationship for ζ shows that the pitch damping, M_q, adds to the damping of the long-term mode whereas M_u reduces it [4.3].

4.9.6 Short-term or pitch (subsidence) mode

Instead of the SPPO, there are usually two aperiodic motions, one with a short time constant and one with a longer time constant, the former being masked by the latter. Just as the dynamic stability of the helicopter is directly related to its long-term modes the control response is characterized by the short-term modes. Whenever a pilot makes a control input the helicopter is excited dynamically and if left to its own devices will exhibit all the modes discussed above. However when the pilot wishes to manoeuvre the aircraft he will only be concerned with the response in the short term and therefore the short-term dynamic modes along with the control derivatives can be used to predict the handling qualities of the helicopter. In the matrix equations this means that for pitch subsidence everything but the third row and column can be ignored, so the equations of motion presented earlier reduce to:

$$\dot{q} = M_q.q + M_{B_1}.B_1 + M_{\theta_c}.\theta_c$$

With no collective input this becomes:

$$\dot{q} = M_q.q + M_{B_1}.B_1$$

$$\frac{q}{B_1} = \frac{M_{B_1}}{(s - M_q)}$$

which describes a classic first-order type of response. Thus the time constant of the pitch subsidence mode is dependent solely on the value of M_q, the pitch damping derivative.

4.9.7 Effect of aero-derivatives on dynamic stability modes

The dynamic response of a helicopter is governed by the values of the aerodynamic and control derivatives that make up the characteristic equation and stability matrices. Most of the important derivatives are speed dependent so it will be instructive to examine the variation in the dynamic modes with airspeed as well as determining the effect of modifying single derivative values.

4.9.7.1 *Effect of airspeed*

Some idea of the likely behaviour can be obtained by examining the aero-derivative and control matrices. Below are a set of matrices for three flight cases: hover, 60 KTAS and 120 KTAS.

$$\mathbf{A}_{\text{hov}} = \begin{bmatrix} -0.0253 & 0.0215 & 0.6675 & -9.7838 \\ 0.0276 & -0.3120 & 0.0135 & -0.7215 \\ 0.0482 & 0.0051 & -1.8955 & 0 \\ 0 & 0 & 0.9986 & 0 \end{bmatrix}$$

$$\mathbf{B}_{\text{hov}} = \begin{bmatrix} 6.9418 & -9.2861 \\ -93.9176 & -0.0021 \\ 0.9554 & 26.4011 \\ 0 & 0 \end{bmatrix}$$

$$\mathbf{A}_{60} = \begin{bmatrix} -0.0243 & 0.0392 & -0.6705 & -9.8014 \\ -0.0467 & -0.7285 & 30.8626 & -0.4200 \\ 0.0280 & 0.0248 & -2.2189 & 0 \\ 0 & 0 & 0.9995 & 0 \end{bmatrix}$$

$$\mathbf{B}_{60} = \begin{bmatrix} 4.6289 & -8.0560 \\ -107.3895 & -21.2286 \\ 10.7004 & 27.6889 \\ 0 & 0 \end{bmatrix}$$

$$\mathbf{A}_{120} = \begin{bmatrix} -0.0460 & 0.0385 & 2.7192 & -9.8052 \\ 0.0221 & -0.9008 & 61.5403 & 0.3205 \\ 0.0299 & 0.0380 & -2.6060 & 0 \\ 0 & 0 & 0.9989 & 0 \end{bmatrix}$$

$$\mathbf{B}_{120} = \begin{bmatrix} 3.8024 & -7.0223 \\ -135.2501 & -49.3052 \\ 20.9344 & 30.9867 \\ 0 & 0 \end{bmatrix}$$

The dynamic stability of the helicopter can be assessed by studying the eigenvalues of the **A** matrix. The eigenvalues equate to the solutions of the differential equations that underpin the matrix itself. Recalling that a negative real part is indicative of a convergent response and that a pair of complex roots imply an oscillatory motion we are in a position to describe the dynamic modes:

(1) *Hover*. The eigenvalues of the $\mathbf{A}_{\text{hov}}$ are:

$$0.0548 + 0.4805i, \quad 0.0548 - 0.4805i, \quad -0.3142, \quad -2.0282$$

These eigenvalues imply an unstable second-order dynamic mode and two stable first-order responses. The second-order mode has a natural frequency of 0.484 rad/s (period of 13.0 s) and a relative damping value of −0.1133 (T_2 of 12.6 s). The first-order responses have time constants of 3.18 s (−0.3142) and 0.49 s (−2.0282).

(2) 60 *KIAS*. The eigenvalues of the $\mathbf{A}_{60}$ matrix are:

$$0.0735 + 0.3822i,\ 0.0735 - 0.3822i,\ -0.4725,\ -2.6460$$

These eigenvalues imply an unstable second-order dynamic mode and two stable first-order responses. The second-order mode has a natural frequency of 0.389 rad/s (period of 16.2 s) and a relative damping value of −0.1885 (T_2 of 9.4 s). The first-order responses have time constants of 2.12 s (−0.4725) and 0.38 s (−2.6460).

(3) 120 *KIAS*. The eigenvalues of the $\mathbf{A}_{120}$ matrix are:

$$0.1995 + 0.3784i,\ 0.1995 - 0.3784i,\ -0.4191,\ -3.5326$$

These eigenvalues imply an unstable second-order dynamic mode and two stable first-order responses. The second-order mode has natural frequency of 0.428 rad/s (period of 14.7 s) and a relative damping value of −0.4664 (T_2 of 3.5 s). The first-order responses have time constants of 2.39 s (−0.4191) and 0.28 s (−3.5326).

The period of the oscillatory mode is fairly long and this would be called the *long-term dynamic mode* which characterizes the *dynamic stability*. Therefore this helicopter is dynamically unstable and the degree of instability increases with speed.

4.9.7.2 *Effect of increased pitch damping*

The effect of an increase in pitch damping, achieved by a direct increase in the value of the pitching moment due to pitch rate (M_q), can be easily shown by increasing the appropriate value in the **A** matrix. Suppose that the value of M_q is doubled from −2.505 to −5.212:

$$\mathbf{A}_{120} = \begin{bmatrix} -0.0460 & 0.0385 & 2.7192 & -9.8052 \\ 0.0221 & -0.9008 & 61.5403 & 0.3205 \\ 0.0299 & 0.0380 & -5.2120 & 0 \\ 0 & 0 & 0.9989 & 0 \end{bmatrix}$$

$$\mathbf{B}_{120} = \begin{bmatrix} 3.8024 & -7.0223 \\ -135.2501 & -49.3052 \\ 20.9344 & 30.9867 \\ 0 & 0 \end{bmatrix}$$

The effect of this change on the dynamic response of the helicopter can be seen from the change in the eigenvalues:

(1) *Before*. 0.1995 + 0.3784*i*, 0.1995 − 0.3784*i*, −0.4191 and −3.5326, which result in the following engineering parameters:

$$\zeta = -0.4664,\ \omega_n = 0.428\ \text{rad/s},\ T_1 = 2.39 \text{ and } T_2 = 0.28$$

(2) *After.* $0.0501 + 0.2916i$, $0.0501 - 0.2916i$, -0.5411 and -5.718, which result in the following:

$$\zeta = -0.1693,\ \omega_n = 0.296\ \text{rad/s},\ T_1 = 1.85 \text{ and } T_2 = 0.17$$

This suggests that increasing M_q will reduce the frequency of the long-term mode and shorten the time constant for the short-term mode (control response).

4.9.7.3 *Effect of change in speed stability*

In a similar manner the effect of a change in the speed stability can be demonstrated by a change to the value of the pitching moment due to speed (M_u). In order to isolate this effect it is necessary to eliminate the influence that the pitching moment due to vertical speed (M_w) may have on the ensuing long-term dynamics. This is achieved by setting M_w to zero. Consider the following three cases:

(1) *Standard speed stability* ($M_w = 0$)

$$\zeta = -0.0769,\ \omega_n = 0.324\ \text{rad/s},\ T_1 = 1.06\ \text{s},\ T_2 = 0.38\ \text{s}$$

(2) *Half standard speed stability*

$$\mathbf{A}_{120} = \begin{bmatrix} -0.0460 & 0.0385 & 2.7192 & -9.8052 \\ 0.0221 & -0.9008 & 61.5403 & 0.3205 \\ 0.0150 & 0 & -2.6060 & 0 \\ 0 & 0 & 0.9989 & 0 \end{bmatrix}$$

$$\zeta = -0.0115,\ \omega_n = 0.233\ \text{rad/s},\ T_1 = 1.08\ \text{s},\ T_2 = 0.37\ \text{s}$$

(3) *Twice standard speed stability*

$$\mathbf{A}_{120} = \begin{bmatrix} -0.0460 & 0.0385 & 2.7192 & -9.8052 \\ 0.0221 & -0.9008 & 61.5403 & 0.3205 \\ 0.0600 & 0 & -2.6060 & 0 \\ 0 & 0 & 0.9989 & 0 \end{bmatrix}$$

$$\zeta = -0.1440,\ \omega_n = 0.449\ \text{rad/s},\ T_1 = 1.04\ \text{s},\ T_2 = 0.37\ \text{s}$$

The effect of M_u can now be seen clearly. Only the long-term mode is affected significantly and it is evident that stronger speed stability causes a more divergent, and higher frequency, long-term response. The moment arising from flying at an off-trim speed is greater if the speed stability is stronger. This larger moment causes the helicopter to return towards trim more aggressively thereby producing a more divergent response.

4.10 LATERAL/DIRECTIONAL STATIC STABILITY

When addressing the question of lateral and directional stability it is convenient to assume that the motion can be analyzed separately from the longitudinal motion.

Cross-coupling terms are therefore ignored to obtain a reasonable feel for the subject but it must be remembered that in the real case cross-coupling will occur and these effects will almost certainly modify the results obtained. Here only the lateral/directional static stability of the helicopter shall be considered which is dominated by the derivatives L_v, L_p, N_v, and N_r. Major contributions to these derivatives come from the main rotor, the tail rotor, the fin and the fuselage. The control displacements, required to maintain steady side-slipping flight at constant heading which are related to the corresponding static stabilities, are also described. The lateral cyclic displacement is related to the *lateral* (*rolling*) *static stability*, (L_v) and the yaw pedal displacement is related to the *directional* (*yawing*) *static stability*, (N_v). In order that the control movements should be in the conventional sense it is necessary that both L_v and N_v are stabilizing. It is worth remembering the sign convention for control deflections; positive control displacement produces negative aircraft response (lateral cyclic stick movements to the left and left push of the yaw pedals are taken as positive).

4.10.1 Directional (yawing) static stability

Contributions to the directional stability of a helicopter arise from the tail rotor, the fin, the fuselage and the main rotor. Suppose the helicopter was in a sideslip to starboard, then for positive static stability there must be a yawing moment also to starboard which tends to align the aircraft with the relative wind direction. The change of yawing moment, N, due to sideslip velocity, v, is the derivative N_v, so for static stability, this must have a positive value.

4.10.1.1 Tail rotor contribution to N_v

The tail rotor always provides a stabilizing contribution to N_v, which arises from the change in tail rotor thrust with change in sideslip velocity. In steady forward flight, the thrust produced depends on the blade angle of attack. Consider a blade element at radius, r, from the tail rotor hub. In forward flight at speed U_e, the blade element experiences a velocity in the plane of rotation of $U_e \pm \Omega_{tr} r$ (where Ω_{tr}, is the tail rotor rotational speed) and a velocity perpendicular to the plane of rotation, v_{tr}, due to inflow. Now the angle of attack of the blade element is given by:

$$\alpha = \theta_{tr} - \phi$$

where θ_{tr} is the tail rotor collective pitch angle and ϕ is the inflow angle given, on the advancing side, by:

$$\phi = \tan^{-1}\left[\frac{v_{tr}}{U_e + \Omega_{tr} r}\right]$$

In positive side-slipping flight with sideslip velocity v, the relative air flow direction now makes an angle β with the plane of rotation, given by $\beta = \tan^{-1}(v/U_e)$. The inflow through the tail rotor is now increased by the value of the sideslip velocity and the inflow angle is now given by:

$$\phi_s = \tan^{-1}\left[\frac{v_{tr} + v}{U_e + \Omega_{tr} r}\right]$$

This changes the blade element angle of attack to:

$$\alpha_s = \theta_{tr} - \phi_s$$

The angle of attack has therefore been reduced by the sideslip (assuming tail rotor collective remains constant) which in this case will result in a reduction in tail rotor thrust, ΔT_{tr}. Now the fuselage torque reaction, Q_{MR}, is initially balanced by the tail rotor thrust moment. So:

$$Q_{MR} = T_{tr} l_{tr}$$

where T_{tr} = tail rotor thrust and l_{tr} = tail rotor moment arm. The reduction in tail rotor thrust caused by the positive sideslip means that this balance is no longer maintained (assuming no rapid change in main rotor torque) and an out-of-balance yawing moment N_{tr}, is produced that is given by:

$$N_{tr} = Q_{MR} - (T_{tr} - \Delta T_{tr}) l_{tr} = \Delta T_{tr} l_{tr}$$

This yawing moment will cause the helicopter to yaw to starboard into the direction of the sideslip. Hence, the tail rotor contribution to N_v is positive and therefore stabilizing. This applies to both tractor and pusher tail rotors.

4.10.1.2 *Fin contribution to N_v*

In forward flight with no sideslip or rotor wake interference a symmetrical fin produces no net sideforce. However, should the helicopter sideslip to the right, say, the relative airflow is such that the fin now has an effective angle of attack equal to the sideslip angle. The net sideforce, Y_F, produced by the fin comprises components of both the fin lift, L_F, and drag D_F. This force will give rise to a yawing moment, N_F, given by:

$$N_F = Y_F l_F$$

where l_F is the moment arm. If the fin lift and drag coefficients and the effective fin area, S_F, are known, then:

$$N_F = [\tfrac{1}{2}(U_e^2 + v^2)\rho S_F (C_L \cos\beta + C_D \sin\beta)] l_F$$

The direction of this moment is such that it will always tend to yaw the helicopter into the direction of the sideslip, thus it is stabilizing. The size of the moment for a given fin design will depend on the forward speed, U_e, and lateral velocity, v.

4.10.1.3 *Fuselage contribution to N_v*

The body lift and drag forces produced by the fuselage of a helicopter during sideslip will result in a net sideforce. This sideforce will produce a yawing moment which will be stabilizing or destabilizing depending on the relative positions of the force and the CG of the aircraft. In forward flight with no sideslip, a symmetrical fuselage would produce no net sideforce. However, in the same manner as the fin, the fuselage in a sideslip presents an angle of attack to the relative airflow equal to the sideslip angle. Fuselage *lift* and *drag* forces will then be produced (although the lift will usually be very small) parallel with and perpendicular to the relative airflow. Components of these forces will give rise to the net sideforce, Y_{fus}, which will in turn produce a yawing moment, N_{fus}. The point of action of the force in relation to the CG will determine whether or not this moment is stabilizing. The moment is given by:

$$N_{\text{fus}} = Y_{\text{fus}} l_{\text{fus}} = [\tfrac{1}{2}\rho(U_e^2 + v^2) S_{\text{fus}} C_D \sin\beta] l_{\text{fus}}$$

Note that the component of the fuselage lift force has been neglected. The magnitude of the fuselage contribution to N_v for a given fuselage shape will increase with increasing forward speed, sideslip velocity and fuselage drag area.

4.10.1.4 *Main rotor contribution to N_v*

As the helicopter sideslips, the main rotor will flap away from the direction of the airflow as has been described above (a_1 and b_1 effects). The tilting of the thrust vector will produce a horizontal thrust component that manifests itself as a sideforce at the rotor hub. If the main rotor is tilted forward with respect to the z-axis of the helicopter then a yawing moment will be generated. This moment will provide a destabilizing contribution to N_v but it will be small in relation to the others.

4.10.2 Lateral (rolling) static stability

Just as the mainplane provides the main contribution to the lateral static stability of a fixed wing aircraft, so the main rotor provides the major contribution in a helicopter. Other contributions come from the tail rotor, fin, fuselage and horizontal stabilizer. Also, as with fixed wing aircraft, no direct restoring moment arises as a result of a disturbance in bank angle (it has zero *stiffness* about the roll axis). The lateral static stability is, in fact, provided by the sideslipping motion that occurs subsequent to a change in bank angle. Suppose a disturbance in bank angle occurs and is followed by a sideslip to starboard, a rolling moment to port is required to restore equilibrium: L_v must therefore be negative for stability.

4.10.2.1 *Main rotor contribution to L_v*

Sideslip induces the characteristic flapback effect with the rotor flapping away from the relative wind direction. This causes a tilt of the thrust vector that will produce a restoring moment about the CG. The magnitude of the rolling moment will be dependent on the size of the flapping angle and the height of the rotor hub above the CG. Offset hinges will also increase its magnitude.

4.10.2.2 *Other contributions to L_v*

The contributions to lateral static stability from the tail rotor, the fin and the fuselage all arise as a result of the sideforces produced on these components during a sideslip. The magnitude of each contribution will depend on the individual forces and the distances of their lines of action from the rolling axis. Normally all the contributions are stabilizing. The horizontal stabilizer can also contribute to L_v in a similar fashion to that of the main and tail planes of a fixed wing aircraft. If a helicopter is fitted with symmetrical tail surfaces either side of the tail boom then as the aircraft rolls the downgoing side of the stabilizer encounters the relative airflow at an angle which effectively increases its angle of attack and hence its lift force. The upgoing side will experience the opposite effect and a decrease in angle of attack and lift. The imbalance provides a moment that acts to stop the roll and, in the subsequent sideslip, any dihedral will further increase this restoring moment.

4.11 LATERAL/DIRECTIONAL DYNAMIC STABILITY AND CONTROL RESPONSE

As with the longitudinal axis the lateral/directional dynamic stability of a helicopter is evaluated by observing the longer-term response to a disturbance. The aircraft will be dynamically stable if, following the removal of a disturbance force, it has a tendency to return to the trimmed attitude. Equally control response is again concerned with the response of the helicopter to control inputs made by the pilot. The analysis of the motions is simplified by assuming that there are no cross-coupling effects. Once again the reader should be reminded that this assumption is of questionable validity and care must be taken when reading-across the theoretical results, mentioned below, to the behaviour of an actual aircraft. Before looking at the dynamic modes in detail, it is worth reminding ourselves of the aero-derivatives that influence the lateral/directional motion.

4.11.1 Lateral/directional derivatives

4.11.1.1 Sideslip derivatives

Mechanisms for the generation of sideforce, rolling moment and yawing moment as a result of a lateral velocity disturbance have already been discussed.

4.11.1.2 Roll rate derivatives

(1) *Sideforce due to changes in roll rate* (Y_p). The sideforces produced at the fin, the fuselage and the tail rotor due to changes in the relative airflow direction (and that due to the main rotor flapping) in a roll are generally negligible. This derivative is therefore normally ignored.

(2) *Rolling moment due to roll rate* (L_p). The main rotor contribution to roll rate damping is analogous to the pitch rate damping (M_q) discussed above. This *aerodynamic damping* effect is a function of Lock number and the size of the hinge offset. Smaller contributions are also provided by the tail rotor, the fin and the fuselage.

(3) *Yawing moment due to roll rate* (N_p). Contributions to N_p depend largely on the size and position of the tail rotor and fin. A high tail rotor, for example, would have a considerable effect, see Fig. 4.20.

4.11.1.3 Yaw rate derivatives

(1) *Sideforce due to changes in yaw rate* (Y_r). It is present but considerable self-cancelling occurs due to effects fore and aft of CG. Y_r is therefore usually small enough to be negligible.

(2) *Rolling moment due to yaw rate* (L_r). The yaw rate produces a change in the direction of the relative airflow to the fin and tail rotor in particular. The sideforces produce a sizeable rolling moment if their lines of action are sufficiently

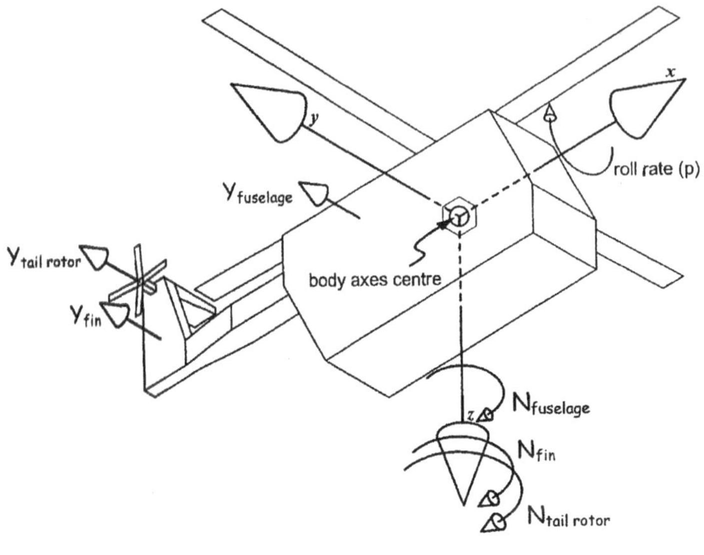

Fig. 4.20 Yawing moment due to roll rate.

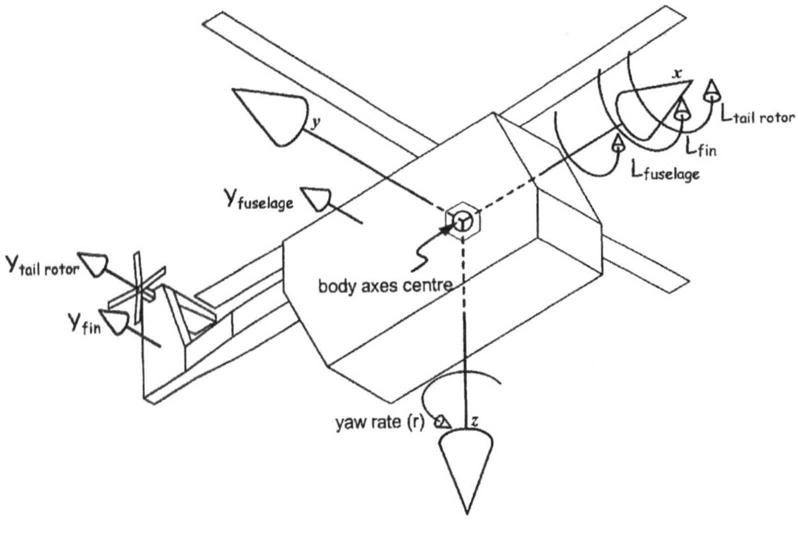

Fig. 4.21 Rolling moment due to yaw rate.

high above the rolling axis, see Fig. 4.21. If the rotor shaft is at an appreciable angle to the vertical, the main rotor may also provide a contribution to the rolling moment.

(3) *Yawing moment due to yaw rate* (N_r). If a helicopter yaws to starboard the tail rotor appears to be sideslipping to port. A blade element then experiences relative airflow from a direction that will effectively increase its angle of attack. There will be an associated increase in thrust and this will produce a damping moment opposing the yaw rate. The effect is present in both the hover and

forward flight. A starboard yaw rate also produces relative airflow to both fin and fuselage which gives rise to a net sideforce from each surface. Both associated moments make stabilizing contributions to N_r.

4.11.2 Control derivatives

The controls influencing the lateral and directional motion of the helicopter are assumed to be only lateral cyclic pitch and tail rotor collective pitch. Although movement of the collective lever changes torque and hence the tail rotor thrust required for trim, this will not be considered in our discussion so collective may be assumed effectively constant during lateral control inputs. The derivatives are denoted by L_{A_1}, $L_{\theta_{tr}}$, N_{A_1}, $N_{\theta_{tr}}$.

4.11.3 Lateral/directional motion of the helicopter

Details of the general equations governing lateral and directional motion are given earlier. It is common practice for lateral/directional motion to assume that all acceleration dependent derivatives are negligible and that the centre of gravity is on the main rotor drive shaft such that $d_x = d_y = d_z = 0$. Thus:

$$\mathbf{M}_{\mathrm{m}}\dot{\mathbf{x}} = \mathbf{M}_{\mathrm{a}}\mathbf{x} + \mathbf{M}_{\mathrm{c}}\mathbf{u}$$

where:

$$\dot{\mathbf{x}} = \begin{bmatrix} \dot{v} \\ \dot{p} \\ \dot{r} \\ \dot{\phi} \end{bmatrix} \quad \mathbf{x} = \begin{bmatrix} v \\ p \\ r \\ \phi \end{bmatrix} \quad \mathbf{u} = \begin{bmatrix} A_1 \\ \theta_{\mathrm{tr}} \end{bmatrix} \quad \mathbf{M}_{\mathrm{c}} = \begin{bmatrix} \mathring{Y}_{A_1} & \mathring{Y}_{\theta_{\mathrm{tr}}} \\ \mathring{L}_{A_1} & \mathring{L}_{\theta_{\mathrm{tr}}} \\ \mathring{N}_{A_1} & \mathring{N}_{\theta_{\mathrm{tr}}} \\ 0 & 0 \end{bmatrix}$$

$$\mathbf{M}_{\mathrm{m}} = \begin{bmatrix} m & 0 & 0 & 0 \\ 0 & I_x & -I_{xz} & 0 \\ 0 & -I_{xz} & I_z & 0 \\ 0 & 0 & 0 & 1 \end{bmatrix}$$

$$\mathbf{M}_{\mathrm{a}} = \begin{bmatrix} \mathring{Y}_v & (\mathring{Y}_p + mW_{\mathrm{e}}) & (\mathring{Y}_r - mU_{\mathrm{e}}) & mg\cos\theta_{\mathrm{e}} \\ \mathring{L}_v & \mathring{L}_p & \mathring{L}_r & 0 \\ \mathring{N}_v & \mathring{N}_p & \mathring{N}_r & 0 \\ 0 & 1 & 0 & 0 \end{bmatrix}$$

To proceed further it is necessary to convert the matrix equation into the form:

$$\dot{\mathbf{x}} = \mathbf{A}\mathbf{x} + \mathbf{B}\mathbf{u}$$

This is achieved by pre-multiplying $\mathbf{M}_\mathrm{m}$, $\mathbf{M}_\mathrm{a}$ and $\mathbf{M}_\mathrm{c}$ by the inverse of $\mathbf{M}_\mathrm{m}$, ($\mathbf{M}_\mathrm{m}^{-1}$). Noting that $r = \dot{\psi}$ and including the effect of ψ on lateral acceleration:

$$\mathbf{M}_\mathrm{m}^{-1} = \begin{bmatrix} \frac{1}{m} & 0 & 0 & 0 & 0 \\ 0 & \frac{I_{zz}}{(I_{xx}I_{zz} - I_{xz}^2)} & \frac{I_{xz}}{(I_{xx}I_{zz} - I_{xz}^2)} & 0 & 0 \\ 0 & \frac{I_{xz}}{(I_{xx}I_{zz} - I_{xz}^2)} & \frac{I_{xx}}{(I_{xx}I_{zz} - I_{xz}^2)} & 0 & 0 \\ 0 & 0 & 0 & 1 & 0 \\ 0 & 0 & 0 & 0 & 1 \end{bmatrix}$$

Hence:

$$\mathbf{A} = \begin{bmatrix} Y_v & (Y_p + W_\mathrm{e}) & (Y_r - U_\mathrm{e}) & g\cos\theta_\mathrm{e} & g\sin\theta_\mathrm{e} \\ L_v & L_p & L_r & 0 & 0 \\ N_v & N_p & N_r & 0 & 0 \\ 0 & 1 & 0 & 0 & 0 \\ 0 & 0 & 1 & 0 & 0 \end{bmatrix} \quad \mathbf{B} = \begin{bmatrix} Y_{A_1} & Y_{\theta_\mathrm{tr}} \\ L_{A_1} & L_{\theta_\mathrm{tr}} \\ Y_{N_1} & N_{\theta_\mathrm{tr}} \\ 0 & 0 \\ 0 & 0 \end{bmatrix}$$

where:

$$Y_v = \frac{\mathring{Y}_v}{m} \quad Y_p = \frac{\mathring{Y}_p}{m} \quad Y_r = \frac{\mathring{Y}_r}{m} \quad Y_{A_1} = \frac{\mathring{Y}_{A_1}}{m} \quad Y_{\theta_\mathrm{tr}} = \frac{\mathring{Y}_{\theta_\mathrm{tr}}}{m}$$

$$L_v = \frac{I_{zz}\mathring{L}_v}{(I_{xx}I_{zz} - I_{xz}^2)} + \frac{I_{xz}\mathring{N}_v}{(I_{xx}I_{zz} - I_{xz}^2)} \qquad L_p = \frac{I_{zz}\mathring{L}_p}{(I_{xx}I_{zz} - I_{xz}^2)} + \frac{I_{xz}\mathring{N}_p}{(I_{xx}I_{zz} - I_{xz}^2)}$$

$$L_r = \frac{I_{zz}\mathring{L}_r}{(I_{xx}I_{zz} - I_{xz}^2)} + \frac{I_{xz}\mathring{N}_r}{(I_{xx}I_{zz} - I_{xz}^2)} \qquad L_{A_1} = \frac{I_{zz}\mathring{L}_{A_1}}{(I_{xx}I_{zz} - I_{xz}^2)} + \frac{I_{xz}\mathring{N}_{A_1}}{(I_{xx}I_{zz} - I_{xz}^2)}$$

$$L_{\theta_\mathrm{tr}} = \frac{I_{zz}\mathring{L}_{\theta_\mathrm{tr}}}{(I_{xx}I_{zz} - I_{xz}^2)} + \frac{I_{xz}\mathring{N}_{\theta_\mathrm{tr}}}{(I_{xx}I_{zz} - I_{xz}^2)} \qquad N_v = \frac{I_{xx}\mathring{N}_v}{(I_{xx}I_{zz} - I_{xz}^2)} + \frac{I_{xz}\mathring{L}_v}{(I_{xx}I_{zz} - I_{xz}^2)}$$

$$N_p = \frac{I_{xx}\mathring{N}_p}{(I_{xx}I_{zz} - I_{xz}^2)} + \frac{I_{xz}\mathring{L}_p}{(I_{xx}I_{zz} - I_{xz}^2)} \qquad N_r = \frac{I_{xx}\mathring{N}_r}{(I_{xx}I_{zz} - I_{xz}^2)} + \frac{I_{xz}\mathring{L}_r}{(I_{xx}I_{zz} - I_{xz}^2)}$$

$$N_{A_1} = \frac{I_{xx}\mathring{N}_{A_1}}{(I_{xx}I_{zz} - I_{xz}^2)} + \frac{I_{xz}\mathring{L}_{A_1}}{(I_{xx}I_{zz} - I_{xz}^2)} \qquad N_{\theta_\mathrm{tr}} = \frac{I_{xx}\mathring{N}_{\theta_\mathrm{tr}}}{(I_{xx}I_{zz} - I_{xz}^2)} + \frac{I_{xz}\mathring{L}_{\theta_\mathrm{tr}}}{(I_{xx}I_{zz} - I_{xz}^2)}$$

As before the characteristic equation (CE) can be obtained from the following:

$$\det(s\mathbf{I} - \mathbf{A}) = 0$$

So:

$$\text{CE} = \begin{vmatrix} s - Y_v & -Y_p - W_e & U_e - Y_r & -g\cos\theta_e & g\sin\theta_e \\ -L_v & s - L_p & -L_r & 0 & 0 \\ -N_v & -N_p & s - N_r & 0 & 0 \\ 0 & -1 & 0 & s & 0 \\ 0 & 0 & -1 & 0 & s \end{vmatrix} = 0$$

Thus for lateral/directional motion, $As^5 + Bs^4 + Cs^3 + Ds^2 + Es = 0$, which usually becomes:

$$(T_1 s + 1)(T_2 s + 1)(s^2 + 2\zeta\omega_n s + \omega_n^2)s = 0$$

The four modes of motion are summarized below:

(1) *Heading mode.* The heading mode is represented by the root $s = 0$ that indicates that the aircraft has neutral yaw angle stability.
(2) *Yawing mode.* The yawing mode, which is equivalent to the fixed wing, *spiral mode*, is represented by $(T_1 s + 1) = 0$. The mode is independent of roll and lateral translation and is an exponential motion that can be either convergent or divergent. The time constant is moderately long being typically between 5 and 20 seconds.
(3) *Rolling mode.* The roll mode, described by $(T_2 s + 1) = 0$, is a damped subsidence in pure roll. The motion has a short time constant of the order of 1 to 2 seconds.
(4) *Lateral/directional oscillation.* The Lateral/Directional Oscillation (LDO) or *Dutch roll*, is an oscillation in roll and yaw, which like the pitching oscillation can be flight condition dependent. Typically the oscillation is unstable in the hover and in a climb.

4.11.4 Dynamic stability in the hover

If the heading mode is ignored as it has little effect on the other motions a quartic equation results. In the longitudinal case, not only is forward speed zero, but some of the derivatives are also zero. This is not strictly true in the lateral/directional case as the rolling and yawing motions are cross-coupled, being represented by the derivatives L_r and N_p. However, if the tail rotor is assumed to be located on the roll axis then L_r can be taken as negligible and a situation analogous to that for longitudinal motion occurs.

One of the real roots can be shown to be $(s - N_r) = 0$ so that $s = N_r$. This root confirms that the yawing motion is independent of the rolling and sideways motions, and represents a damped subsidence in yaw. Any disturbance generating a yaw rate will be damped out by the moment $N_r . r$ and the aircraft will be left pointing in a new direction This root also characterizes the yaw control response in the hover. The other real root is usually a large negative root representing a heavily damped roll subsidence.

The complex root represents a divergent oscillation involving changes in bank angle, heading and sideways velocity and is often referred to as the falling leaf mode. A disturbance in bank angle causes the helicopter to move sideways; this motion causes the rotor to flap back and eventually stop the sideways velocity but the aircraft is left with some bank angle and the motion reverses direction. At the same time, the sideways velocity causes a change in tail rotor thrust and sideforce which then produces a yawing motion. The falling leaf motion can be considered as an undamped oscillation and the frequency of the mode is shown to be given by:

$$\omega = \sqrt{\frac{L_v g}{L_p}}$$

4.11.5 Dynamic stability in forward flight

The characteristic equation in forward flight again resolves into two real roots and a complex pair. The large, negative, real root again represents a heavily damped roll subsidence. The small real root now corresponds to the *spiral mode*, and the complex pair represent an oscillation, the lateral/directional oscillation (LDO), analogous to the *Dutch roll* mode of a fixed wing aircraft.

4.11.5.1 The spiral mode

As forward speed increases, the directional static stability, N_v, increases and the yaw subsidence in the hover becomes a spiral mode. The mode is dependent on the value of the factor:

$$\dot{\phi} = p = \frac{g}{U_e} \frac{[N_r L_v - L_r N_v]}{[L_p N_v - N_p L_v]} \phi$$

Usually L_r, N_v and N_p are positive whilst L_v, L_p and N_r are negative. Since N_p depends on the height of the tail rotor above the roll axis it is unlikely to be the dominant term therefore $(L_p N_v - N_p L_v)$ will usually be negative and for spiral stability $N_r L_v > L_r N_v$.

Now N_v is a measure of the helicopter's tendency to '*yaw into wind*' and L_v is its tendency to '*roll wings level*'. In order for the aircraft to display positive spiral stability it must 'roll wings level' rather than 'yaw into wind'. Therefore L_v (the dihedral effect) stabilizes the spiral mode. It should also be noted that the mode depends on the trimmed forward speed, U_e. However the effect of changes in speed are difficult to determine theoretically since the values of all the important derivatives change with speed.

4.11.5.2 The lateral/directional oscillation (Dutch roll mode)

In forward flight the directional stability of the helicopter increases and the oscillation found in the hover becomes more like the Dutch roll of a fixed wing aircraft as the helicopter weathercocks with little sideways translation. It is worth noting, however, that in addition to roll and yaw, helicopters exhibit a significant amount of pitching motion during this oscillation. The LDO frequency and damping are given approximately by:

$$\omega = \sqrt{Y_v N_r + U_e N_v} \qquad \zeta = \frac{-(Y_v + N_r)}{2\sqrt{Y_v N_r + U_e N_v}}$$

Now N_v is positive and Y_v, and N_r are negative, therefore the LDO will always be stable provided there is no sign change in these derivatives. The value and sign of N_v is subject to change due to modifications in the contributions from the various parts of the helicopter. For example, changes in CG position can result in the fuselage contribution becoming negative, also main rotor wake interference may reduce or increase the value of N_v for the tailplane. Note also that if the overall value of N_v were negative the destabilizing effect would increase with forward speed, U_e. Interestingly Prouty [4.3], adopts a slightly different approach which ultimately shows that the lateral stability, L_v, can destabilize the LDO if it is excessive.

Prouty treats the LDO as an oscillation in roll and yaw but with the aircraft CG maintaining a straight flight path. In other words the LDO and spiral mode are completely decoupled. In mathematical terms this assumption means that the sideforce aero-derivatives are zero and that the bank angle and gravity effects are ignored. These assumptions change the 5 × 5 matrix presented above to the following 3 × 3:

$$\begin{bmatrix} \dot{v} \\ \dot{p} \\ \dot{r} \end{bmatrix} = \begin{bmatrix} 0 & 0 & -U_e \\ L_v & L_p & L_r \\ N_v & N_p & N_r \end{bmatrix} \cdot \begin{bmatrix} v \\ p \\ r \end{bmatrix}$$

Generating a characteristic equation for the LDO, using $\det(s\mathbf{I} - \mathbf{A}) = 0$:

$$\mathrm{CE} = \begin{vmatrix} s & 0 & U_e \\ -L_v & s - L_p & -L_r \\ -N_v & -N_p & s - N_r \end{vmatrix} = 0$$

$$= s[(s - L_p)(s - N_r) - L_r N_p] + U_e[L_v N_p + N_v(s - L_p)] = 0$$

$$\mathrm{CE} = s^3 - s^2(L_p + N_r) + s(L_p N_r - L_r N_p + U_e N_v) + U_e(L_v N_p - N_v L_p) = 0 \quad (4.38)$$

Assuming that $N_r \ll L_p$ and making use of the *Bairstow* assumption, which states that for *lightly damped* systems represented by the cubic $c_3 s^3 + c_2 s^2 + c_1 s + c_0 = 0$, then $c_2 s^2 + c_0 = 0$ and the cubic becomes:

$$c_2 s^2 + \left(c_1 - c_3 \frac{c_0}{c_2} \right) s + c_0 = 0$$

Thus Equation (4.38) becomes:

$$CE = s^2 - \left[N_r - \frac{N_p L_r}{L_p} + \frac{U_e L_v N_p}{L_p^2} \right] s + U_e \left[N_v - \frac{L_v N_p}{L_p} \right] = 0$$

Consequently:

$$\omega_{\mathrm{LDO}} = \sqrt{\frac{U_e}{L_p}(N_v L_p - L_v N_p)} \qquad \zeta_{\mathrm{LDO}} = \frac{\dfrac{1}{L_p}\left[N_p L_r - L_p N_r - \dfrac{U_e L_v N_p}{L_p} \right]}{2\sqrt{\dfrac{U_e}{L_p}(N_v L_p - L_v N_p)}}$$

Using typical values for the constituent aero-derivatives it can be seen that if L_v is excessive it is possible for ζ_{LDO} to become negative, indicating an unstable dynamic mode.

4.11.6 Short-term or the roll (subsidence) mode

In most aircraft, both fixed and rotary wing, the Dutch roll tends to have a reasonably long period and the spiral mode a long time constant. Damping in roll tends to occur rapidly with time constants of the order of 0.2 second. Hence the roll subsidence mode can be decoupled from the other motions. In the matrix equations this means that for roll subsidence everything but the second row and column can be ignored, so the equations of motion presented earlier reduce to:

$$\dot{p} = L_p . p + L_{A_1} . A_1 + L_{\theta_{tr}} . \theta_{tr}$$

With no pedal input this becomes:

$$\dot{p} = L_p . p + L_{A_1} . A_1$$

$$\frac{p}{A_1} = \frac{L_{A_1}}{(s - L_p)}$$

which describes a classic first order type of response. Thus the time constant of the roll subsidence mode is dependent solely on the value of L_p, the roll damping derivative. As forward speed increases the value of L_p will change. Typically the motion will remain heavily damped but the time constant will increase from that in the hover. It should be noted that a similar relationship can be generated for yaw control in the hover:

$$\dot{r} = N_r . r + N_{\theta_{tr}} . \theta_{tr}$$

$$\frac{r}{\theta_{tr}} = \frac{N_{\theta_{tr}}}{(s - N_r)}$$

4.11.7 Effect of aero-derivatives on dynamic stability modes

The lateral/directional dynamic response of a helicopter is governed by the values of the aerodynamic and control derivatives that make up the characteristic equation and stability matrices. As was the case with the longitudinal motion most of the important derivatives are speed dependent so it is instructive to examine the variation in the dynamic modes with airspeed as well as determining the effect of modifying single derivative values.

4.11.7.1 Effect of airspeed on dynamic modes

An estimate of the likely behaviour can be made by examining the aero-derivative and control matrices. Below are a set of matrices for three flight cases: hover, 60 KTAS and 120 KTAS. Note that the yaw angle response has been removed and that the state vector has been re-ordered to $[v, p, \phi, r]^T$.

$$\mathbf{A}_{\text{hov}} = \begin{bmatrix} -0.0394 & -0.6653 & 9.7665 & 0.1365 \\ -0.0624 & -3.1111 & 0 & -0.0278 \\ 0 & 1 & 0 & 0.0837 \\ 0.0391 & -0.5046 & 0 & -0.3742 \end{bmatrix}$$

$$\mathbf{B}_{\text{hov}} = \begin{bmatrix} -9.7060 & 5.6145 \\ -47.0134 & -1.1236 \\ 0 & 0 \\ -8.4807 & -15.1333 \end{bmatrix}$$

$$\mathbf{A}_{60} = \begin{bmatrix} -0.1320 & 0.9829 & 9.7929 & -30.4417 \\ -0.0455 & -3.0478 & 0 & -0.2221 \\ 0 & 1 & 0 & 0.0535 \\ 0.1016 & -0.4899 & 0 & -1.1085 \end{bmatrix}$$

$$\mathbf{B}_{60} = \begin{bmatrix} -9.7054 & 5.7541 \\ -47.0742 & -1.1515 \\ 0 & 0 \\ -8.3750 & -15.5094 \end{bmatrix}$$

$$\mathbf{A}_{120} = \begin{bmatrix} -0.2317 & -1.4240 & 9.8014 & -61.1382 \\ -0.0524 & -2.9711 & 0 & -0.4099 \\ 0 & 1 & 0 & -0.0121 \\ 0.1514 & -0.4664 & 0 & -1.7557 \end{bmatrix}$$

$$\mathbf{B}_{120} = \begin{bmatrix} -9.8901 & 8.3638 \\ -47.4028 & -1.6738 \\ 0 & 0 \\ -8.3069 & -22.5440 \end{bmatrix}$$

The dynamic modes can be described as:

(1) *Hover.* The eigenvalues of the $\mathbf{A}_{\text{hov}}$ matrix are:

$$-0.0484 + 0.3835i, \quad -0.0484 - 0.3835i, \quad -0.2381, \quad -3.1898$$

These eigenvalues imply three stable modes, one second-order (oscillatory) response and two first-order responses. The second-order mode (LDO) has a natural frequency of 0.387 rad/s giving an observed period of 15.4 s and a relative damping value of 0.125 ($T_{1/2}$ of 14.3 s). The first-order responses have time constants of 4.20 s (−0.2381) and 0.31 s (−3.1898).

(2) 60 *KIAS.* The eigenvalues of the $\mathbf{A}_{60}$ matrix are:

$$-0.5329 + 1.7540i, \quad -0.5329 - 1.7540i, \quad -0.0508, \quad -3.1717$$

Again these eigenvalues imply three stable modes. In this case the second-order mode has a natural frequency of 1.833 rad/s giving an observed period of 3.6 s and a relative damping value of 0.291 ($T_{1/2}$ of 1.3 s). The first-order responses have time constants of 19.69 s (−0.0508) and 0.32 s (−3.1717).

(3) *120 KIAS.* The eigenvalues of the $\mathbf{A}_{120}$ matrix are:

$$-0.8882 + 2.9488i, \quad -0.8882 - 2.9488i, \quad -0.0528, \quad -3.1292$$

Here the second-order mode has a natural frequency of 3.08 rad/s giving an observed period of 2.1 s and a relative damping value of 0.288 ($T_{1/2}$ of 0.8 s). The first-order responses have time constants of 18.94 s (−0.0528) and 0.32 s (−3.1292).

Thus the *Lateral/Directional Oscillation* (*LDO*) is stable throughout the speed range although the period and damping of the oscillatory mode varies with airspeed. The helicopter would appear, therefore, *dynamically stable* to the pilot with the frequency increasing with airspeed. This result agrees with detailed analysis given above. Also, since the *spiral mode* is characterized by the first order mode with the longer time constant, note that for the example helicopter the mode is stable with the rate of convergence being greater at 120 KIAS than at 60 KIAS. The *roll control response* is, therefore, characterized by the other, shorter, first-order mode and for the example helicopter there appears to be very little change in this mode with airspeed.

4.11.7.2 *Effect of increased roll damping*

The effect of an increase in roll damping, achieved by a direct increase in the numerical value of the rolling moment due to roll rate (L_p), can be shown by increasing the appropriate value in the **A** matrix. Consider the 60 knot case discussed earlier: and suppose the value of L_p is increased from −3.0478 to −6.0956:

$$\mathbf{A}_{60} = \begin{bmatrix} -0.1320 & 0.9829 & 9.7929 & -30.4417 \\ -0.0455 & -6.0956 & 0 & -0.2221 \\ 0 & 1 & 0 & 0.0535 \\ 0.1016 & -0.4899 & 0 & -1.1085 \end{bmatrix}$$

$$\mathbf{B}_{60} = \begin{bmatrix} -9.7054 & 5.7541 \\ -47.0742 & -1.1515 \\ 0 & 0 \\ -8.3750 & -15.5094 \end{bmatrix}$$

The effect of this change on the dynamic response can be seen from the change in the eigenvalues:

(1) *Before.* −0.5329 + 1.7540*i*, −0.5329 − 1.7540*i*, −0.0508, −3.1717, resulting in the following engineering parameters:

$$\zeta = 0.291, \ \omega_n = 1.833 \text{ rad/s}, \ T_1 = 19.69 \text{ and } T_2 = 0.32$$

(2) *After.* −0.5892 + 1.7384*i*, −0.5892 − 1.7384*i*, −0.0183, −6.1393, resulting in the following:

$$\zeta = 0.321,\ \omega_n = 1.836\,\text{rad/s},\ T_1 = 54.54 \text{ and } T_2 = 0.163$$

This suggests that increasing L_p hardly affects the LDO but will shorten the time constant of the roll mode. In addition, the spiral mode will have reduced stability since it will take much longer for the helicopter to return to wings level flight following a pulse input of lateral cyclic. It is interesting to note that a doubling of the rate damping has halved the time constant thereby improving the control predictability by reducing the time to achieve a steady roll rate.

4.11.7.3 *Effect of increased yaw damping*

As with the roll damping example, the effect of increasing N_r can be seen by changing the numerical value of the appropriate matrix element. Suppose the value of N_r is increased from −1.1085 to −4.4340:

$$\mathbf{A}_{60} = \begin{bmatrix} -0.1320 & 0.9829 & 9.7929 & -30.4417 \\ -0.0455 & -3.0478 & 0 & -0.2221 \\ 0 & 1 & 0 & 0.0535 \\ 0.1016 & -0.4899 & 0 & -4.4340 \end{bmatrix}$$

$$\mathbf{B}_{60} = \begin{bmatrix} -9.7054 & 5.7541 \\ -47.0742 & -1.1515 \\ 0 & 0 \\ -8.3750 & -15.5094 \end{bmatrix}$$

The effect of this change on the dynamic response can be seen from the eigenvalues of the modified matrix: −3.9231, −2.5885, −0.8745, −0.2277, resulting in the following modal characteristics: $T_1 = 0.255$, $T_2 = 0.386$, $T_3 = 1.144$ and $T_4 = 4.392$. These results suggest that increasing N_r has a major effect on both the LDO and the spiral mode. The LDO has degenerated into two convergent first-order responses implying that it has been completely suppressed. In addition, the time constant of the spiral mode has been reduced thereby implying significant strengthening of the spiral stability.

4.11.7.4 *Effect of increased lateral static stability*

As with longitudinal stability the static derivatives, in this case L_v and N_v, have a significant effect on the dynamic modes. Before studying the effect of increased dihedral effect (L_v) it would be instructive to surmise the likely results. If a helicopter has increased lateral static stability and all other derivatives are left unchanged then for a given amount of lateral velocity (sideslip) it will generate a larger moment away from the direction of the sideslip. This will reduce the time taken for the aircraft to roll wings-level following a pulse on lateral cyclic and will probably cause a more oscillatory LDO since the roll attitude will change more readily for the same variations in lateral velocity.

Changing the numerical value of the appropriate matrix element can check these deductions. Suppose L_v is changed from -0.0455 to -0.1820:

$$\mathbf{A}_{60} = \begin{bmatrix} -0.1320 & 0.9829 & 9.7929 & -30.4417 \\ -0.1820 & -3.0478 & 0 & -0.2221 \\ 0 & 1 & 0 & 0.0535 \\ 0.1016 & -0.4899 & 0 & -1.1085 \end{bmatrix}$$

$$\mathbf{B}_{60} = \begin{bmatrix} -9.7054 & 5.7541 \\ -47.0742 & -1.1515 \\ 0 & 0 \\ -8.3750 & -15.5094 \end{bmatrix}$$

The effect of this change on the dynamic response can be seen from the eigenvalues of the modified matrix: $-0.3670 + 1.9618i$, $-0.3670 - 1.9618i$, -0.1454 -3.4079, resulting in the following: $\zeta = 0.184$, $\omega_n = 1.996$ rad/s, $T_1 = 6.83$ and $T_2 = 0.29$. These results confirm the deductions above. An increase in lateral static stability reduces the time constant of the spiral mode from almost 20 s to below 7 s and causes, approximately, a 40% reduction in the damping of the LDO.

4.11.7.5 *Effect of increased directional static stability*

If a helicopter has increased directional static stability and all other derivatives are left unchanged then for a given amount of lateral velocity (sideslip) it will generate a larger yawing moment towards the direction of the sideslip. This will serve to increase the time taken for the aircraft to roll wings-level following a pulse on lateral cyclic because there will be a stronger tendency to 'turn-into-wind'. In addition, one would expect the LDO to be less oscillatory since there should be smaller variations in lateral velocity for the same size of input.

Once again changing the numerical value of the appropriate matrix element can check these deductions. Suppose N_v is changed from 0.1016 to 0.4064:

$$\mathbf{A}_{60} = \begin{bmatrix} -0.1320 & 0.9829 & 9.7929 & -30.4417 \\ -0.0455 & -3.0478 & 0 & -0.2221 \\ 0 & 1 & 0 & 0.0535 \\ 0.4064 & -0.4899 & 0 & -1.1085 \end{bmatrix}$$

$$\mathbf{B}_{60} = \begin{bmatrix} -9.7054 & 5.7541 \\ -47.0742 & -1.1515 \\ 0 & 0 \\ -8.3750 & -15.5094 \end{bmatrix}$$

The effect of this change on the dynamic response can be seen from the modified eigenvalues: $-0.5837 + 3.4925i$, $-0.5837 - 3.4925i$, -0.0184, 3.1025, resulting in the following engineering parameters: $\zeta = 0.165$, $\omega_n = 3.541$ rad/s, $T_1 = 54.35$ and $T_2 = 0.32$. These results certainly confirm our deductions regarding the spiral mode. The time constant for this mode has increased by almost threefold from over 19 s to nearly 55 s.

Chapter 5
Stability and Control Testing

5.1 ASSESSING FLIGHT CONTROL MECHANICAL CHARACTERISTICS

When evaluating control aspects of rotorcraft the first aspect to consider is the characteristics of the control system. As the pilot must control the swashplate through the flight controls a deficiency in their operation will affect all areas of flight. Even excellent aircraft handling qualities can be masked by poor flight control mechanical characteristics (FCMC). Testing can be divided into quantitative aspects which normally take place on the ground and qualitative aspects which are conducted in flight. As quantitative testing is concerned with the measurement of forces and displacements it can be conducted more easily and safely on the ground. For reversible systems, however, the amount of ground testing that can be undertaken is limited by the requirement to have the rotors turning and by the need to evaluate realistic flight forces. Qualitative testing is concerned with the effect that the control system characteristics have on the conduct of role tasks. This section will only cover conventional control systems with displacement controllers.

In any assessment that includes FCMC a comprehensive control reference system is defined. This establishes the test conditions in case the controls of an aircraft are modified at some stage subsequent to the assessment. The reference system records the range of movement for all the controls. As the distances through which controls move and the forces needed to move them vary depending on where the measurement is made, it is necessary to define this point which is termed the control reference point (CR Point). The definition is usually given in specification documents such as the Ministry of Defence Standard 00-970 [5.1]. Having defined the CR Point for each control the next stage is to determine a position within the throw of each control from which displacements are measured. The position is known as the control reference position (CR Posn). The CR Posn can be defined anywhere along the control throw although it is usual to set the position at the mid-point for the cyclic axes and the yaw pedals. The convention for the collective is to use the minimum pitch position. By measuring the distance to the cyclic CR Posn from three points on the aircraft structure it is possible to reference the control envelope to the cockpit as a whole. As the pedals and collective only move in one direction, or follow an arc, a single measurement normally suffices to define the location of the CR Posns for these types of controllers.

The cyclic control envelope is the area described by the maximum cyclic displacement (measured at the CR Point) in all directions. The envelope can only be measured fully with rotors stopped. Of course the assessment of the acceptability of the envelope can only be made in flight. Determining if the control envelope is satisfactory is not as simple as it may seem at first. If items within the cockpit restrict the range of control

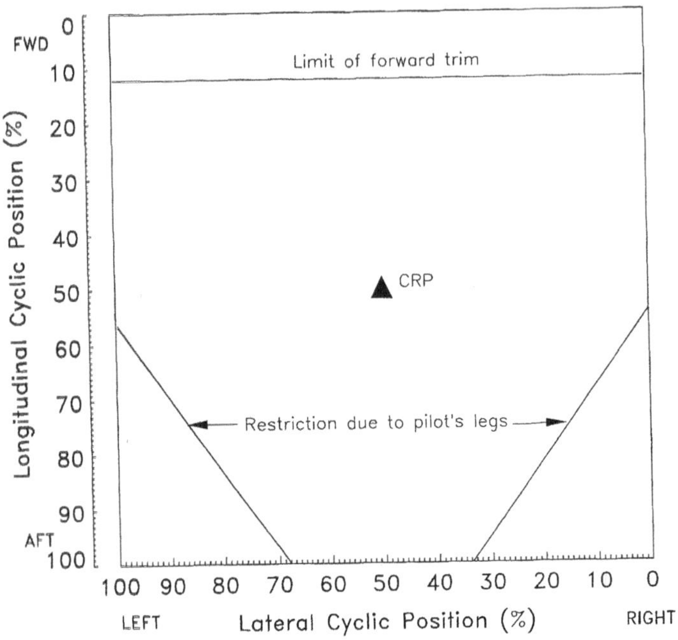

Fig. 5.1 Cyclic stick displacement envelope.

movement it is not necessarily a deficiency; it is only significant if the restriction is encountered during flight manoeuvres. Of course it is important that the pilot is able to reach the *required* envelope of all the controls from the normal seated position. A typical presentation of a cyclic envelope is shown in Fig. 5.1. The envelope may be affected by the operation of mechanical interlinks and is measured with the collective and yaw pedals at both ends of their displacement range. The position of the pilot's seat can affect the amount of the envelope that can be used and this is placed in the most restrictive position – fully forward and normally fully up. Where the displacement that can be achieved using the trim system is less than the full displacement then this too is presented on the envelope plot. Problems are sometimes experienced with the envelopes of the other controls; for example, to lower the collective lever fully some pilots have to lean over to an uncomfortable degree. This highlights the importance of recording the assessing pilot's anthropometric data and seating position.

5.1.1 Evaluating control forces

Measuring the force gradient for a control is achieved using either a spring balance or a force gauge placed at the relevant CR Point. The force is applied in a slowly increasing manner to note the precise force at which the control starts to move; this will give the value of the breakout force plus the static friction $(B+F)$. Having overcome $B+F$, the standard convention is then to use only sufficient force to keep the control moving. The force required is recorded at regular intervals of displacement

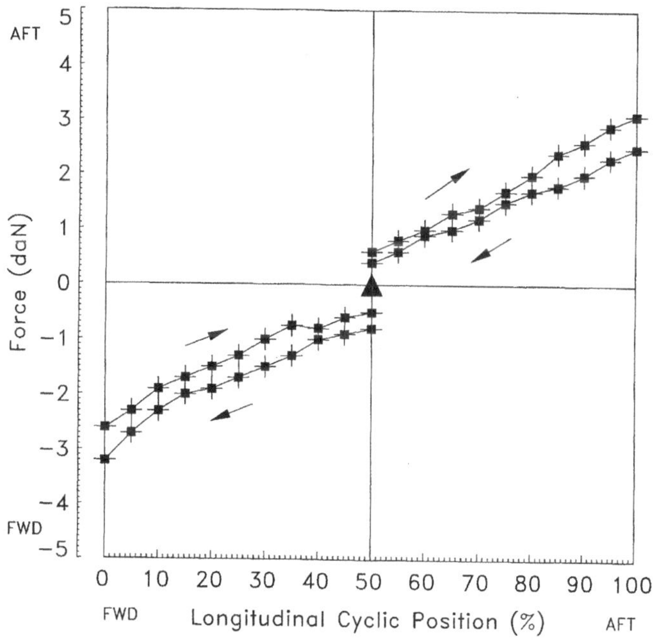

Fig. 5.2 Force/displacement plot: longitudinal cyclic stick.

to allow the production of a force/displacement plot as shown in Fig. 5.2. When the limit of displacement has been reached the force is reduced gradually until the control moves back towards the trim point. Note on Fig. 5.2 that having reached full displacement the control will not start to move back towards the trim position until the force has been reduced by a value equal to twice the static friction. If a control friction device is fitted it is usually placed in the fully OFF position when measuring the forces. If there is no trim system then it is still necessary to measure and document the value of the sliding friction. An example of a collective force/displacement plot is shown in Fig. 5.3. Pilots are very sensitive to the value of $B + F$ as the majority of control inputs are small being used to make minor adjustments to the helicopter's flight path. If $B + F$ is set too high it will make precise control difficult and some pilots deliberately fly holding a trim force during tasks such as instrument approaches to overcome this problem. On the other hand too weak a force will lead to inadvertent small control inputs due to vibration or turbulence. Where an adjustable friction device is fitted this is assessed to determine if it can be used to set precise amounts of friction. For obvious reasons it should not be possible to lock the control by setting excess friction.

The designer of the control system will set the value of the cyclic force gradient to allow the pilot to move the control rapidly, but at the same time sufficient force gradient must be provided to prevent unintentionally large inputs. In the case of aircraft with high cyclic control power the need to provide protection may predominate and the force gradient may be quite high. This will have a significant effect on handling qualities during manoeuvres where the control has to be moved faster than can be achieved with a beeper trim system, or where trimming is not desirable, such as during a flyaway from the hover following a single engine failure. In this case the need to

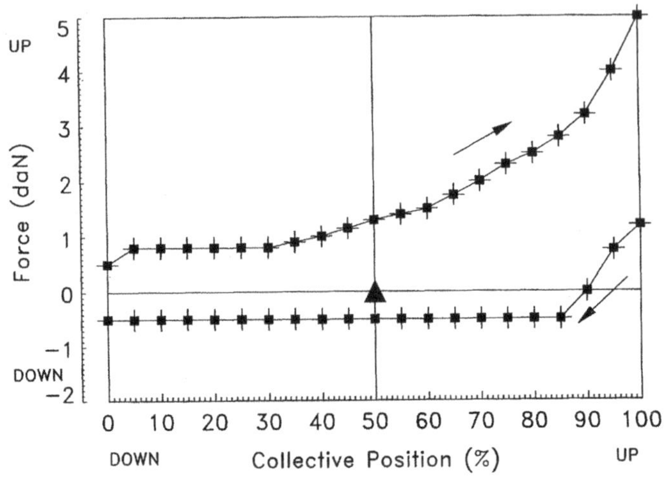

Fig. 5.3 Force/displacement plot: collective lever.

overcome a high spring feel force will make the flyaway more difficult to achieve and may cause an excessive height loss. On the other hand too shallow a force gradient will also lead to problems with the pilot making excessively rapid inputs; the pilot may then feel constrained into making only gentle inputs to avoid this problem and thus may not achieve the full agility of the helicopter. It is also necessary to assess the maximum force that a beeper trim system can produce, known as the *limit control force*, to determine if the pilot could retain control with a trim runaway. Holding the control at one end of its available range and then trimming fully in the opposite direction stimulates this.

Where a force trim release button is fitted the sudden release of the force will often cause the control to 'jump' as the pilot changes the amount of force he is applying. In the case of the cyclic this is known as *stick jump*. To the pilot this can become irritating, as it is not possible to make small changes to the datum trim position. To test control jump, the control is displaced from the trim position, the release operated and the reaction of the control noted. The amount of displacement used for this test is a realistic amount that an operational pilot might employ before operating the release. After ground tests, an airborne assessment is made to determine if any control oscillations cause an undesirable aircraft response.

Associated with the force gradient tests are the determination of the centring characteristics. This is the tendency of the cyclic to return towards the trim point when any displacing force is removed. *Positive* centring indicates a return *towards* the trim point while *absolute* centring is achieved if the control returns *precisely* to the original trim position. When conducting this test the control is displaced from trim and then allowed to return slowly towards trim. Where centring is positive but not absolute there will be a trim control displacement band (TCDB). This occurs as a consequence of a weak breakout force that is insufficient to overcome sliding friction. If the cyclic is moved to any position within this band and released it will stay at that position.

The implications of a trim control displacement band will vary according to the flight task and will not always have a detrimental effect on aircraft control. For example, when instrument flying a large TCDB would make it difficult to control aircraft attitude but a small band would allow the pilot to make small attitude changes without having to re-trim. A small TCDB may also make hovering easier again by obviating the need to make constant trim inputs. It is important, however, to understand that a TCDB will have implications for AFCS functions which use cyclic position as an input. For example, an attitude command/attitude hold system would not return the aircraft to the original attitude after a pilot input if a TCDB was present. The TCDB can be reduced by reducing the amount of friction in the system or by increasing breakout force.

It should be remembered that force gradients are not always linear and sometimes soft stops are incorporated such as the intermediate and maximum pitch stops in the Aerospatiale AS 341 Gazelle collective control. These soft stops can be incorporated for a number of reasons such as cueing the pilot to the limit of the power-on collective lever movement or defining the surge-free range of lever displacement. The cueing properties of the stop and the force required to overcome it are assessed during manoeuvres that require large and rapid control movements. For example, if the force required to overcome a stop is too high it might have serious implications during an engine-off landing if it interfered with the application of collective pitch at touchdown.

Since harmonization of control forces is an important aspect that will affect the aircraft's handling qualities, the force gradient in any axis should be matched with the breakout force. For instance, if there is a high breakout force and a shallow force gradient it will be difficult to make precise inputs which require the control to be moved through the trim point. This may make manoeuvres such as hovering and NOE flight difficult. Harmonization of the forces of all the flight controls and between the lateral and longitudinal axes of the cyclic is also important for good control. It is the pilot's perception of harmony between the forces that is the important issue not necessarily the actual values. This is due to the physiology of the human body that makes it easier to generate higher forces with the legs than with the arms; similarly it is easier to push the cyclic to the left than pull it to the right. Specifications such as ADS-33E [5.2] encompass this idea of 'perceived' or 'apparent' harmony by merely stating that '*forces, displacements, and sensitivities ... shall be compatible*', without specifying values. Where the control forces are poorly harmonized manoeuvring is more difficult particularly during manoeuvres that require extensive use of all controls such as pirouettes [5.2] or downwind quickstops.

5.1.2 Freeplay

Freeplay will inevitably exist in any mechanical control system due to wear and backlash between the various components. Freeplay that exists between the flight controls and the blade pitch change linkage is termed *total system freeplay* while wear or backlash between the flight controls and the spring feel unit is termed *trim system freeplay*.

Measuring total system freeplay (for an irreversible system) is performed with rotors stopped and an observer stationed by the rotor head. The pilot makes control inputs in the axis under investigation and the observer confirms that a blade pitch change has taken place. The process is repeated using gradually smaller inputs until there is no resultant blade pitch change. Results can be confirmed with blades turning by observing the tip path plane and observing when there is no response to the control.

Total system freeplay is always undesirable as it will delay the aircraft response to inputs and therefore adversely affect handling qualities during tasks such as precision hovering which involve small, high-frequency control inputs.

Determining trim system freeplay is rather more straightforward and involves measuring the distance through which the control can be moved without needing to overcome a force. Although generally undesirable, some pilots like the opportunity that trim system freeplay affords to make small control adjustments without having to overcome $B+F$. Although it is different to a TCDB, a band of trim system freeplay does share the characteristic that control displacements within the band do not require operation of the trim system as there will be no force to hold off.

5.1.3 Assessing mass balance and control dynamics

The mass balance characteristics of the cyclic and collective are assessed to determine the tendency of the control to move due to the influence of gravity or other accelerations. Clearly this will usually only be a problem where there is no force feel system fitted or the force gradient is very shallow. The assessment is made initially with no control friction set to establish a base-line condition, and then repeated with 'normal' amounts of friction. Any manoeuvres which produce forces on the controls such as pull-ups/push-overs and steep turns can be employed. Poor mass balance characteristics can increase the pilot's workload as it prevents the control being released for other than brief periods or requires any adjustable friction devices to be set to possibly undesirably high levels.

Identifying the control dynamics is an essential part of any FCMC assessment. This consists of evaluating the effect of control 'raps' and releases from an off-trim condition as well as assessing the effect of any biomechanical feedback and the mass balancing of controls. Although some limited testing can take place on the ground the majority of tests need to be conducted in-flight. Release-to-trim (RTT) tests of the cyclic and control raps of the cyclic and collective are approached with caution in case the control dynamics lead to a divergent aircraft response. For obvious reasons the tests are not conducted with the aircraft at the edges of the cleared flight envelope. The procedure involves two crew members; the handling pilot gives a countdown before each RTT or rap while the other crew member positions both hands near the control, ready to suppress any divergent response. A series of incremental RTTs or raps in each axis are made and the control dynamics recorded. For a well-damped response it is usually sufficient to record the number of overshoots but for less damped responses time histories may be needed. An example of poor dynamics includes a collective control where if a rap is made and only a small amount of friction has been set the subsequent aircraft heave motion will cause a collective displacement in the opposite direction; this forces a divergent oscillation which may be difficult to suppress.

Part of this process of assessing control dynamics involves looking for biomechanical feedback. This is a problem related to FCMC that can degrade handling qualities and is the process whereby the motion of the aircraft causes the pilot's arm to move resulting in an unintended control input. Biomechanical feedback is most likely to occur when there is little $B+F$ in the control system and the control has poor mass balance characteristics. The pilot's seating position may also be a factor if it prevents the pilot achieving a position where the arm can be braced to prevent inadvertent movements caused by aircraft motions. This coupling between the pilot and the control may be a problem when flying in turbulence as gusts will cause an aircraft disturbance that will subsequently result in a control input. If this is combined with poor control dynamics the result may be very serious.

5.1.4 Trim system evaluation

The final part of the FCMC assessment is to measure the beeper trim system characteristics and then to determine their suitability for the intended role. For a 'beeper' trim system using a trim motor the two important aspects are firstly the trim rate and secondly any lag in the system. Measurement of the rate is made on the ground by operating the trim and obtaining the displacement against time from the data replay. If no instrumentation is available then displacement against time is measured using a stopwatch and tapes fitted to the control. To conduct the airborne evaluation flight tasks are selected which require accurate trimming and other tasks that require more rapid displacement of the control. Too rapid a trim rate is distracting and frustrating for the pilot conducting the precision task while too slow a rate requires the pilot to hold the trim force for too long after a larger displacement. When measuring trim lag the trim control is moved in one direction and then trimmed in the opposite direction to measure the time it takes for backlash in the system to be taken up and the flight control to be moved.

5.2 ASSESSING STATIC STABILITY

5.2.1 Longitudinal static stability

The static stability of a helicopter will manifest itself to the pilot as the amount of forward stick required to maintain an airspeed greater than trim. The pilot will also 'expect' to perceive some change in control position as he trims the helicopter through its speed range in level flight. There are, therefore, two different test techniques:

(1) *Apparent static stability tests.* The helicopter is trimmed, in level flight, at a series of airspeeds from minimum to maximum. The control position data obtained is often referred to as the '*Trimmed Flight Control Positions*' (TFCPs). As suggested earlier TFCPs can also be assessed in steady climbing and descending flight.

(2) *Collective fixed static stability tests.* The helicopter is trimmed at an airspeed and the collective fixed at '*Power For Level Flight*' (PFLF). The pilot then attempts to hold an off-trim speed, either greater or less, and accepts the ensuing climb or descent. The stick position data obtained in this test is directly related to the strength of the speed stability (M_u) *provided* the rate of climb or descent is not excessive and the control power is constant.

The difference in the results obtained from these tests will depend on the pitch response of the helicopter to changes in collective pitch. In addition to the control position data the variation of pitch attitude with airspeed is also noted. A large change in attitude with speed may be used by the pilot to compensate for poor cyclic stick position cues. If, on the other hand, the variation in attitude is very small the attitude hold function of an AFCS would not be very effective as an airspeed hold.

5.2.1.1 *Collective fixed test results*

The variation of longitudinal cyclic pitch (B_1) with airspeed from a trimmed condition is relatively easy to estimate theoretically. The linearized form of the longitudinal equations of motion for a helicopter, with the centre of gravity situated at the body axes origin, can be written as follows:

$$m[\dot{u} - rV_e + qW_e] = X_u . u + X_w . w + X_q . q - mg\theta \cos\theta_e + X_{B_1} . B_1 + X_{\theta_c} . \theta_c$$

$$m[\dot{w} - qU_e + qV_e] = Z_u . u + Z_w . w + Z_q . q - mg\theta \sin\theta_e + Z_{B_1} . B_1 + Z_{\theta_c} . \theta_c$$

$$I_{yy}\dot{q} - I_{xz}\dot{r} - I_{xz}\dot{p} = M_u . u + M_w . w + M_q . q + M_{B_1} . B_1 + M_{\theta_c} . \theta_c$$

When performing speed stability testing the pilot endeavours to achieve the desired off-trim speed with the aircraft wings level and with no pitch, roll or yaw rate evident. Therefore:

$$\dot{u} = \dot{w} = p = q = \dot{p} = \dot{q} = \dot{r} = \theta_c = 0$$

Thus the equations of motion reduce to:

$$0 = X_u . u + X_w . w - mg\theta \cos\theta_e + X_{B_1} . B_1$$

$$0 = Z_u . u + Z_w . w - mg\theta \sin\theta_e + Z_{B_1} . B_1$$

$$0 = M_u . u + M_w . w + M_{B_1} . B_1$$

If it is assumed that the trimmed pitch attitude is small, such that $\cos\theta_e = 1$ and $\sin\theta_e = \theta_e$, then:

$$0 = X_u . u + X_w . w - mg\theta + X_{B_1} . B_1$$

$$0 = Z_u . u + Z_w . w - mg\theta\theta_e + Z_{B_1} . B_1$$

$$0 = M_u . u + M_w . w + M_{B_1} . B_1$$

Thus:

$$w = -\frac{1}{M_w}[M_u . u + M_{B_1} . B_1]$$

and assuming that the pitch attitude change is small:

$$0 = Z_u . u + Z_w . w - mg\theta\theta_e + Z_{B_1} . B_1$$

$$= Z_u . u - \frac{Z_w}{M_w}[M_u . u + M_{B_1} . B_1] + Z_{B_1} . B_1$$

$$0 = M_w Z_u . u - Z_w M_u . u - Z_w M_{B_1} . B_1 + M_w Z_{B_1} . B_1$$

$$= u[M_w Z_u - Z_w M_u] - B_1[Z_w M_{B_1} - M_w Z_{B_1}]$$

$$\frac{B_1}{u} = \frac{[M_w Z_u - Z_w M_u]}{[Z_w M_{B_1} - M_w Z_{B_1}]}$$

Since there is little change in vertical force with changes in longitudinal cyclic:

$$\frac{dB_1}{du} = \frac{M_w Z_u - Z_w M_u}{Z_w M_{B_1}}$$

this equation will only be valid if the pitch attitude required to hold the off-trim speed is not large when compared with the trim speed. Once again it should be appreciated that the presence of M_{B_1} (longitudinal cyclic control power) in the above equation means that the test cannot be used to evaluate the *magnitude* of the static stability (M_u) since the amount of stick deflection required is dependent on the control power. In addition it should be noted that the above equation contains a contribution from M_w. Only small excursions from the trim speed can, therefore, be tested since if large rates of descent or climb are experienced the results will be corrupted by M_w effects.

5.2.1.2 *Trimmed flight control positions tests*

Determining the theoretical variation of longitudinal cyclic pitch, and pitch attitude, with trimmed airspeed is more difficult. The control deflection required and the aircraft attitude is dictated by the requirement to achieve equilibrium of both forces and moments whilst maintaining level flight. Basically equations arising from considering equilibrium of X-force, Z-force and M-moment are obtained. At each trim condition these equations are solved simultaneously to give values for longitudinal flapping, relative to the shaft (a_{1_s}), pitch attitude (θ) and thrust. These are then used to determine the longitudinal cyclic pitch (B_1) and cyclic stick position required. On a conventional helicopter the M_{θ_c} effect on cyclic position will be the same as those seen during steady climbs and descents at the same collective pitch and off-trim airspeed. A forward longitudinal control displacement or force should be required in order to initiate and maintain an increased forward airspeed [5.1]. Pitch attitude variation with airspeed will typically be a compromise between maintaining a near level attitude for crew comfort and profile drag considerations, whilst achieving pitch attitudes relatable to airspeeds for instrument flight.

5.2.1.3 *Practical considerations*

Provided only small speed changes are made during collective fixed static stability (CFSS) testing, and the same range of airspeeds are evaluated during a TFCP test, the effect of changes in collective pitch or power (M_{θ_c}) can be evaluated by direct comparison. Consider Fig. 5.4 which shows the typical case of a helicopter with a nose-up trim change with increased power (collective). Above minimum power speed the pilot must increase collective pitch to generate sufficient thrust to maintain level flight at higher speeds. He must now apply extra forward cyclic, to counter the nose-up cross-coupling effect over and above that required to overcome the basic pitch-up tendency with increased speed (flap-back effect). Static stability test results would, therefore, show 'increased stability' when TFCP results were compared with data obtained from a collective fixed test. Note that the speed stability, or M_u effect, can also be clearly seen by comparing the collective fixed test data with a horizontal line from the trim point ($M_u = 0$). The non-linear nature of the M_{θ_c} effect can be explained with reference to a typical power curve where the power decrement required to maintain a lower speed in level flight reduces as the minimum power speed is approached.

CFSS tests are normally accomplished by establishing a trim condition (airspeed/

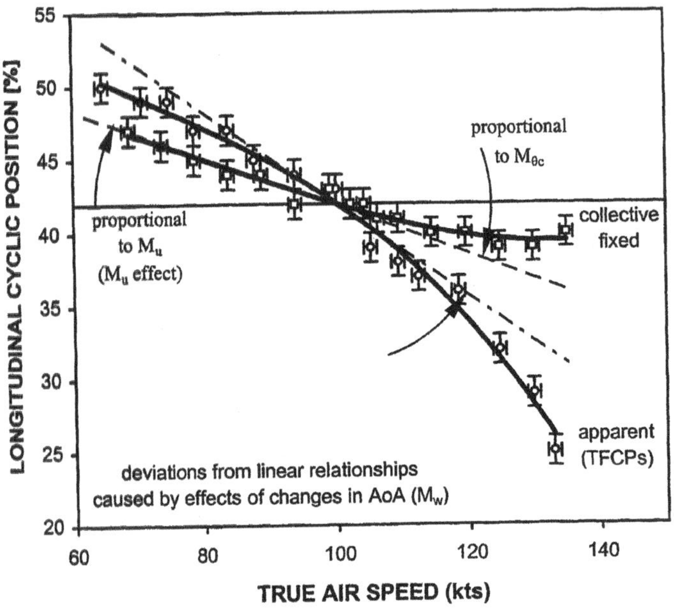

Fig. 5.4 Longitudinal static stability test data.

power combination) with zero control forces. Then without changing the collective position, trim setting or rotor speed, the helicopter is stabilized at incremental airspeeds both faster and slower than the trim airspeed using cyclic only. Directional control inputs are made to maintain ball centred or zero sideslip flight. An airspeed range either side of trim is assessed using 2 knot then 5 knot increments. Ideally the helicopter is kept within 1000 feet of the specified test altitude. Airspeeds faster or slower than trim are flown alternately to achieve this. Excursions from the test altitude band can be corrected using the collective control. To achieve this the collective trim position or power/torque for level flight at the datum speed is noted. Collective pitch is then used as required to regain the desired altitude and, without re-trimming the cyclic, is adjusted back to its initial position before the test is continued.

Although, as discussed, the collective to pitch coupling can be discerned by comparison between TFCP testing in level flight and collective fixed data, the coupling can also be evaluated by performing trimmed climbs and descents at constant airspeed. As well as highlighting any problems with excessive collective-to-pitch cross-coupling these tests will identify the possibility of an encroachment of control margins. Discontinuities in the collective position versus longitudinal control position data may suggest handling problems caused by, for example, the stalling of a down-loaded tailplane at high negative angles of attack which may occur at high ROC and low IAS. TFCPs are normally evaluated at a number of key role-relatable airspeeds such as climbing/endurance/autorotation speed, cruise speed and maximum level flight speed. Since M_{θ_c} increases with airspeed it is normal to evaluate the lower speeds first.

In summary, during static (speed) stability tests the aim is to identify the M_u effect (collective fixed) and the M_{θ_c} effect (TFCPs in climbs/descents and collective fixed versus TFCPs in level flight). If, however, during these tests the angle of attack changes markedly from its value at the trim point (level flight at the trim speed) it is possible that M_w effects will corrupt the results. Typical advice is therefore to only test modest deviations from trim (± 15 kts for collective fixed static stability tests and ± 1000 fpm for TFCPs in climbs/descents).

5.2.2 Lateral and directional static stability

5.2.2.1 *Steady heading sideslips*

The control deflections required to maintain an unaugmented helicopter in a steady sideslip are related to its lateral and directional static stabilities. In a steady sideslip the rates of roll and yaw will be zero, therefore the rolling and yawing aerodynamic moments must have been balanced by control deflections generating control moments. The linearized form of the lateral/directional equations of motion for a helicopter, with the centre of gravity situated at the body axes origin, can be written as follows:

$$m[rU_e + \dot{v} - pW_e] = Y_v . v + Y_p . p + Y_r . r + mg\phi \cos\theta_e + Y_{A_1} . A_1 + Y_{\theta_{tr}} . \theta_{tr}$$

$$I_{xx}\dot{p} - I_{xz}\dot{r} = L_v . v + L_p . p + L_r . r + L_{A_1} . A_1 + L_{\theta_{tr}} . \theta_{tr}$$

$$I_{zz}\dot{r} - I_{xz}\dot{p} = N_v . v + N_p . p + N_r . r + N_{A_1} . A_1 + N_{\theta_{tr}} . \theta_{tr}$$

Now when performing a SHSS the pilot endeavours to achieve the off-trim condition by co-ordinated movement of both the lateral cyclic and the yaw pedals without developing any sustained pitch, roll or yaw rate. Once 'on-condition' the lateral velocity will be constant, indicated by a fixed sideslip angle. So:

$$\dot{v} = \dot{p} = \dot{r} = p = r = 0$$

Thus the equations of motion reduce to:

$$0 = Y_v . v + mg\phi \cos\theta_e + Y_{A_1} . A_1 + Y_{\theta_{tr}} . \theta_{tr}$$

$$0 = L_v . v + L_{A_1} . A_1 + L_{\theta_{tr}} . \theta_{tr}$$

$$0 = N_v . v + N_{A_1} . A_1 + N_{\theta_{tr}} . \theta_{tr}$$

If it is assumed that the roll attitude remains small, such that $\phi \approx 0$, then:

$$0 = Y_v . v + Y_{A_1} . A_1 + Y_{\theta_{tr}} . \theta_{tr}$$

$$0 = L_v . v + L_{A_1} . A_1 + L_{\theta_{tr}} . \theta_{tr}$$

$$0 = N_v . v + N_{A_1} . A_1 + N_{\theta_{tr}} . \theta_{tr}$$

which leads to:

$$0 = v(Y_v . L_{\theta_{tr}} - Y_{\theta_{tr}} . L_v) + A_1(Y_{A_1} . L_{\theta_{tr}} - Y_{\theta_{tr}} . L_{A_1})$$

$$\frac{A_1}{v} = \frac{(Y_v . L_{\theta_{tr}} - Y_{\theta_{tr}} . L_v)}{(Y_{\theta_{tr}} . L_{A_1} - Y_{A_1} . L_{\theta_{tr}})}$$

If the rolling moment due to tail rotor collective (pedal) is negligible, such that $L_{\theta_{tr}} = 0$, then:

$$\frac{A_1}{v} = -\frac{L_v}{L_{A_1}}$$

also:

$$0 = v(Y_v . N_{A_1} - Y_{A_1} . N_v) + \theta_{tr}(Y_{\theta_{tr}} . N_{A_1} - Y_{A_1} . N_{\theta_{tr}})$$

$$\frac{\theta_{tr}}{v} = \frac{(Y_v . N_{A_1} - Y_{A_1} . N_v)}{(Y_{A_1} . N_{\theta_{tr}} - Y_{\theta_{tr}} . N_{A_1})}$$

If the yawing moment due to lateral cyclic is negligible, such that $N_{A_1} = 0$, then:

$$\frac{\theta_{tr}}{v} = -\frac{N_v}{N_{\theta_{tr}}}$$

For positive lateral static stability, the slope of the control deflection versus sideslip graph will be negative. That is, in order to increase sideslip to starboard the pilot must apply an increased right lateral cyclic stick deflection. For positive directional static stability, the slope of the pedal deflection versus sideslip graph will be positive. That is, in order to increase sideslip to starboard the pilot must apply increased left pedal. Note that in each case the magnitude of the control deflection required to maintain a given sideslip angle depends, on the degree of stability *and* the amount of control power. Useful techniques to determine the relationship between stability and control power are turns on one control (TO1C) which are described later.

As PEs are often present in sideslipping flight the test technique consists of stabilizing in a steady heading sideslip at constant EAS, recording roll attitude, lateral cyclic and yaw pedal positions. It should be noted that steady heading sideslip (SHSS) tests will only indicate the amount of control displacement required to counter the rolling and yawing moments generated by the sideslip. The amount of control displacement should increase as lateral velocity increases and for stability should be in the same direction as the sideslip for lateral cyclic and in opposition for the yaw pedals. Whether, for example, a small control displacement is due to weak stability or high control power must be determined by some other test technique such as turns on one control, which are described later. Figures 5.5 and 5.6 illustrate the presentation of typical results. From these plots it can be seen that at each speed the control movement is in the correct sense and that increasing control deflection is required for increasing lateral velocity.

Tests are normally made in level flight at the minimum power speed, $V_{NE} - 20$ kts and one intermediate speed. In climbing and autorotative flight the speed will normally be the recommended climbing speed and the speed for minimum rate of descent in autorotation respectively. If no sideslip angle indicator is fitted an approximation of sideslip can be obtained by employing one of the methods described below:

(1) *SHSS technique – GSDI fitted aircraft.* The helicopter is flown into wind with zero roll angle and slipfall central (GSDI drift needle reading zero) and aircraft heading is noted. The helicopter is then yawed to change heading by the sideslip angle required while maintaining the aircraft track into wind. The GSDI drift needle should indicate the *change* in sideslip angle and, if the correct flight path has been maintained, this value will give a working approximation of sideslip angle.

(2) *SHSS technique – non-GSDI fitted aircraft.* The helicopter is flown down a line feature (a runway) into wind and aircraft heading is noted. The helicopter is then yawed to change heading with zero roll angle and slipfall central by the sideslip angle required while maintaining aircraft track down the line feature.

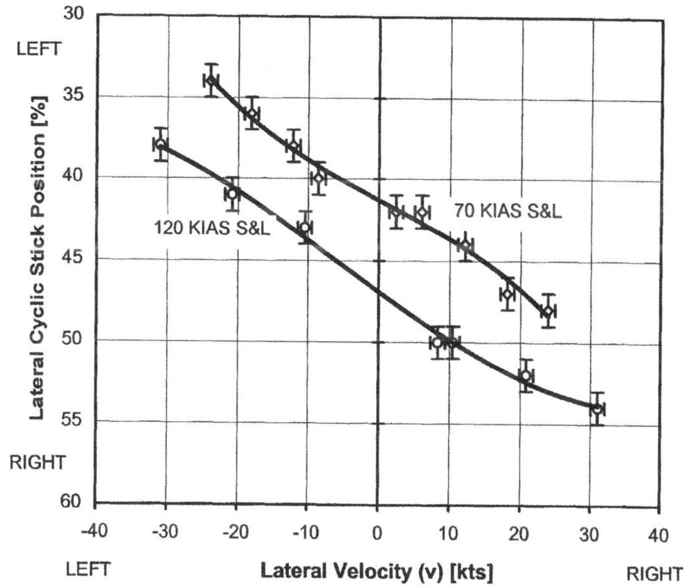

Fig. 5.5 SHSS test data – lateral static stability.

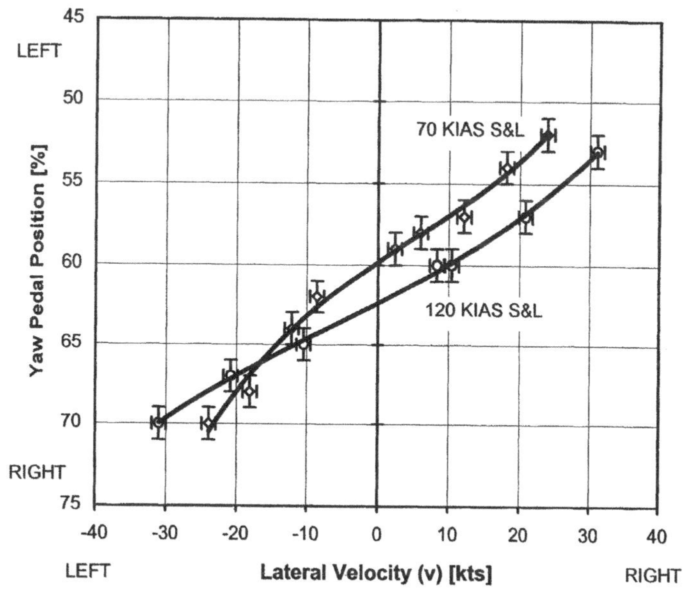

Fig. 5.6 SHSS test data – directional static stability.

Note the actual heading change when established in a steady-state condition and this approximates to the sideslip angle.

An approximate 'calibration' of slip-ball displacement to sideslip angle can also be obtained during either of the above techniques. Also, if a suitable line feature is selected and there is no wind (or very light wind down the line feature) it may be possible to estimate inherent sideslip by noting the line feature heading and then comparing this to the heading required to track down the line feature with wings level and ball centred. The difference between the line feature heading and the heading required to track the line feature will equate to the inherent sideslip value for the test airspeed.

Before starting, the ASI PECs with sideslip are obtained for the test speeds. This is particularly important at high IAS to avoid exceeding V_{NE}. Steady sideslips are flown at constant EAS to eliminate inconsistencies due to pressure errors at high angles of sideslip. If ASI PECs with sideslip are not available, it should be possible to obtain satisfactory steady sideslip results by using the following method:

(1) Stabilize the aircraft wings level with inherent sideslip (ball centred) at the IAS which gives the EAS for the test condition.
(2) Smoothly but rapidly yaw the aircraft to the required sideslip angle controlling the bank with lateral cyclic to maintain a SHSS, and note the new IAS immediately.
(3) Maintain this IAS and record the test parameters.
(4) Yaw the aircraft smartly back to wings level and the inherent sideslip value (ball centred) and note that the IAS returns to the original value.

5.2.2.2 *Turns on one control*

As discussed above the standard test for lateral and directional static stability is the steady heading sideslip (SHSS). The problem with this test technique is that it only indicates *cockpit* stability, that is how the stability appears to the pilot. A technique that may be of use in determining the relationship between stability and control power is that of *turns on one control* or '*TO1C*'. Although a co-ordinated turn normally involves both cyclic and pedal movements and possibly collective control, consider what happens if an attempt is made to turn the helicopter using only one control, cyclic stick or pedals, while the other is fixed.

From a condition of steady, trimmed level flight, consider the effects of initiating a turn to starboard using *yaw pedals* only. Firstly, as right pedal is pushed forward the helicopter will yaw to the right and sideslip to the left. The sideslip will cause the rotor disk to flap away from the relative airflow and this will produce a rolling moment to starboard. The subsequent bank to starboard will reduce the sideslip to the left and eventually cause a sideslip to the right. Finally, the helicopter will settle into a steady turn with the yaw pedal deflection adjusted as necessary to maintain the roll attitude of the aircraft. It will then normally have adopted an attitude in which it is yawing to starboard, sideslipping to starboard, banked to starboard. The amount of sideslip required to generate the roll response will indicate the strength of L_v since the test has not involved any contribution from L_{A_1}. This, of course, assumes that the rolling moment due to pedal $L_{\theta_{tr}}$ is negligible. Consideration of the stabilized turn leads to:

$$\theta_{tr} = -\frac{1}{N_{\theta_{tr}}}(N_r \, . \, r + N_v \, . \, v)$$

Assume a starboard turn (*r positive*) with starboard sideslip (*v positive*). The turn is opposed by the yaw damping (N_r being *negative*) and assisted by the directional stability

(N_v being *positive*). Thus, $N_r . r$ *is negative* and $N_v . v$ is *positive* so the sign of the term $(N_r . r + N_v . v)$ depends on the relative magnitudes of the two terms. As $N_{\theta_{tr}}$ is *negative by defintion*, if $(N_r . r + N_v . v)$ is negative then θ_{tr} will be *positive* and vice versa. The pedal deflection required can therefore be summarized as follows:

- The yaw pedals are deflected to yaw the helicopter *into* the turn (right pedal forward, θ_{tr} is *negative*), then $N_r . r > N_v . v$ and the yaw damping term is dominant.
- The yaw pedals are deflected to yaw the helicopter *out of* the turn (left pedal forward, θ_{tr} is *positive*), then $N_r . r < N_v . v$ and the directional stability term is dominant.
- No pedal movement from the trim position is necessary to maintain the turn (θ_{tr} is zero), then $N_r . r = N_v . v$ and neither term is dominant.

Thus it is possible during the stabilized portion of a turn on one control-pedal (TO1C-P) to determine the relative magnitudes of the directional stability and the yaw damping. Equally the strength of the L_v effect can be gauged by the readiness with which the helicopter responds in roll to the sideslip generated by a TO1C-P. Assuming there is little or no contribution from $L_{\theta_{tr}}$, strong lateral stability will be present if the helicopter rolls smartly to the right following the application of right pedal and the onset of left sideslip. Pedal-only turns are generally more difficult to perform than turns on cyclic alone (see later discussion) as some sideslip must be generated before the aircraft responds, as there will be a lag between yaw pedal application and the helicopter rolling into a turn. Consequently the input is usually in the form of a steady ramp with the rate of application varied to establish whether this has an effect on the subsequent response of the aircraft. Care must be exercised, as it is relatively easy to exceed sideslip limits during this test. Longitudinal cyclic is used as necessary to maintain the airspeed.

Now consider the effects of initiating a turn to starboard using *lateral cyclic* only. Firstly, as right cyclic is applied the helicopter will roll to the right and commence sideslipping to the right as the aircraft descends. The sideslip will cause the tail rotor and fin to generate a yaw moment starboard. Eventually, the helicopter will settle into a steady turn with the lateral cyclic deflection adjusted as necessary to maintain the roll attitude of the aircraft. The amount of sideslip required to generate the yaw response will indicate the strength of N_v since the test has not involved any contribution from $N_{\theta_{tr}}$. This, of course, assumes that the yawing moment due to lateral cyclic N_{A_1} is negligible. Consideration of the stabilized turn yields:

$$A_1 = -\frac{1}{L_{A_1}}(L_r . r + L_v . v)$$

In a turn to starboard (*r positive*) with starboard sideslip (*v positive*), $L_r . r$ will be positive (act to starboard) whilst $L_v . v$ will be negative (act to port). As L_{A_1} is negative, by definition A_1 will be positive if $(L_r . r > L_v . v)$ and negative if $(L_r . r < L_v . v)$. The control deflection can thus be summarized as:

- The cyclic is deflected *out of* the turn, that is towards the left (A_1 positive), therefore $L_r . r > L_v . v$ and the yaw rate contribution is dominant.
- The cyclic is deflected *into* the turn (A_1 negative), thus $L_r . r < L_v . v$ and the sideslip contribution is dominant.

- Once established in the turn no cyclic movement from the position required for trim is necessary therefore neither term dominates as $L_r \, . \, r = L_v \, . \, v$.

With the aircraft established in a steady constant bank angle turn inspection of a sideslip gauge, skid-ball or suitably mounted string will confirm the strength of N_v. Weak directional stability will be evident if large values of sideslip are observed. Alternatively, strong directional stability will be evident if little sideslip is recorded.

Turns on one control-cyclic (TO1C-C) testing begins by stabilizing the helicopter at the required balanced flight condition and recording the trimmed control positions, aircraft heading, sideslip and bank angle. The rate of application and the magnitude of the lateral cyclic displacement is chosen to achieve a bank angle which is related to the requirements of the role, although the bank angle chosen should be approached incrementally (20° is a good initial condition). At the desired bank angle longitudinal cyclic is used as necessary to maintain constant IAS. The cyclic is then returned to the initial trim condition before the test is repeated in the opposite direction.

5.3 MANOEUVRE STABILITY TESTING

The manoeuvre stability of a helicopter will manifest itself to the pilot as the amount of aft stick required to maintain an elevated load factor. The pilot can achieve such a load factor either in a pull-up manoeuvre or during turning flight. Therefore manoeuvre stability can be assessed in either situation. (Push-overs can be used to assess stability at reduced load factor.) The pros and cons of each method are:

(1) *Pull-ups.* The pull-up manoeuvre is easily role-relatable and does not give '*false*' data since the radius of the circular flight path is in line with the normal axis of the helicopter. The test technique which requires achievement of the elevated load factor in level flight, at zero pitch attitude, on speed and with the collective fixed at PFLF is, however, difficult to fly accurately and repeatably. It is also not possible to assess manoeuvre instability using PUPOs as it will be impossible to establish the correct balance of flight parameters.

(2) *Turning flight.* Descending turns, at constant speed with fixed collective (at PFLF), are, perhaps, less easy to role-relate. This manoeuvre is, however, easier to fly accurately and provides a method of incrementally increasing the load factor by simply increasing the angle of bank. The data obtained will be '*false*' since for a given load factor a greater pitch rate will be required than for a pull-up because the centrifugal acceleration is not aligned with the normal axis of the helicopter (except at 90° of bank). Despite the increased workload it is possible to temporarily stabilize a manoeuvre unstable aircraft during quantitative data gathering.

5.3.1 Theoretical treatment of test methods

5.3.1.1 Symmetrical pull-ups

During pull-up testing, as described earlier, the pilot endeavours to achieve the desired load factor with the aircraft at zero pitch attitude, wings level, on speed and with no yaw rate developing. The tests are conducted with fixed collective and it is also assumed that a steady pitch rate is achieved. Therefore:

$$\dot{u} = u = \dot{w} = p = \dot{p} = \dot{r} = \dot{q} = \theta_e = \delta\theta_c = 0$$

Thus the linearized form of the longitudinal equations of motion reduce to:

$$-mqU_e = w\,.\,Z_w + q\,.\,Z_q + \delta B_1\,.\,Z_{B_1}$$

$$0 = w\,.\,M_w + q\,.\,M_q + \delta B_1\,.\,M_{B_1}$$

Applying the concept of circular motion to the pull-up manoeuvre yields:

$$q_{\text{pull-up}} = \frac{g}{U_e}(n-1)$$

Substituting and solving simultaneously:

$$\frac{\delta B_1}{(n-1)} = \frac{g}{U_e}\left[\frac{M_q Z_w - mU_e M_w - Z_q M_w}{Z_{B_1} M_w - M_{B_1} Z_w}\right]$$

Assuming that $Z_q = 0$:

$$\frac{dB_1}{dn} = \frac{g}{U_e}\left[\frac{M_q Z_w - mU_e M_w}{Z_{B_1} M_w - M_{B_1} Z_w}\right]$$

5.3.1.2 Steady turns

Consider a helicopter performing a steady turn with fixed collective. Assuming that the flight path angle and bank angle are small then it is accurate to assume that the thrust vector acts parallel to the rotor shaft. For equilibrium:

$$mg = T\cos\phi$$

$$\text{CF} = T\sin\phi$$

$$\text{CF} = mR_{\text{turn}}\omega^2$$

Now:

$$n_{\text{turn}} = \frac{T}{mg} = \frac{\sqrt{mg^2 + \text{CF}^2}}{mg} = \sqrt{1 + \left(\frac{\text{CF}}{mg}\right)^2} = \frac{1}{\cos\phi}$$

The linear and angular velocities can be related by:

$$\text{CF} = mR_{\text{turn}}\omega^2 = mV\omega$$

If the flight path angle is small and w, the vertical velocity, is small then $V = U_e$. Thus:

$$n_{\text{turn}}^2 = 1 + \left(\frac{U_e\omega}{g}\right)^2 \quad \text{and} \quad \omega = \frac{g}{U_e}\sqrt{n_{\text{turn}}^2 - 1}$$

Noting that the pitch rate, q, is related to the rate of turn (in a vertical bank they would be equal) by $q_{\text{turn}} = \omega\sin\phi$, gives:

$$q_{\text{turn}} = \frac{g}{U_e}\cdot\frac{(n_{\text{turn}}^2 - 1)}{n_{\text{turn}}}$$

It is evident from the above equation that the pitch rate required in a turn is greater than a pull-up for the same load factor. Since:

$$\frac{\mathrm{d}B_1}{\mathrm{d}q} = \left[\frac{M_q \cdot Z_w - mU_e M_w}{Z_{B_1} \cdot M_w - M_{B_1} \cdot Z_w}\right] \quad \text{and} \quad \frac{\mathrm{d}B_1}{\mathrm{d}n} = \frac{\mathrm{d}B_1}{\mathrm{d}q} \cdot \frac{\mathrm{d}q}{\mathrm{d}n}$$

then:

$$\frac{\mathrm{d}B_1}{\mathrm{d}n} = \frac{g}{U_e}\left[\frac{M_q Z_w - mU_e M_w}{Z_{B_1} M_w - M_{B_1} Z_w}\right]\left(\frac{n^2+1}{n^2}\right)$$

5.3.2 Flight test methods and results

The differences in the pitch rate generated during the two tests will lead to differences in the amount of aft stick required. The main reason for the difference is the greater pitch damping moment (M_q) generated in the turn requires more aft cyclic to compensate. Since there will usually be a linear relationship between longitudinal cyclic stick position and cyclic pitch (B_1) these equations can be used to explain the difference in test results assuming all other factors are equal, see Fig. 5.7.

The figure shows that as far as the pilot is concerned the helicopter will appear more manoeuvre stable in a descending turn than during a pull-up. Note, however, that since M_{B_1} appears in both equations the actual amount of manoeuvre stability (size of M_w) cannot be quantified since the size of the control deflection is dependent on the control power.

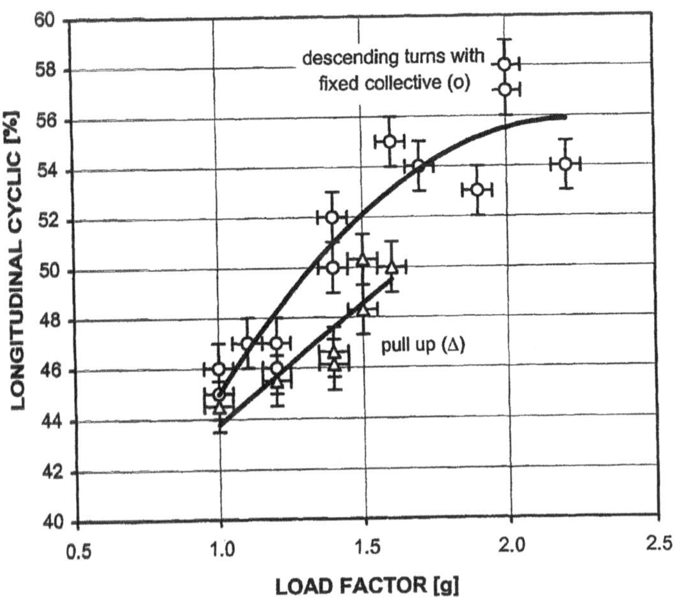

Fig. 5.7 Manoeuvre stability test data.

5.3.2.1 *Effect of changes in collective pitch*

The concept of using turning flight as a method of assessing manoeuvre stability can be extended to encompass the more role-relatable case of a level turn at constant speed. As with the descending turn, since the load factor will be related to the bank angle the severity of the manoeuvre can be incrementally increased. The complication introduced by level turns is that in order to maintain height collective pitch will be required and the pitch change with power effect will require cyclic pitch to compensate. Assuming that an increase in collective pitch causes a nose-up pitching moment it is clear that a level turn will require less aft cyclic than a descending turn at the same load factor (angle of bank). Therefore, to the pilot, the helicopter will appear to be less manoeuvre stable. In order to make the distinction between these two cases the following terminology is used: *collective fixed manoeuvre stability* (assessed either during pull-ups/push-overs or during descending turns at constant speed and fixed collective) and *apparent manoeuvre stability* (assessed during level turns at constant speed).

5.3.2.2 *Test techniques*

Although apparent static stability tests are role relatable in that most steep turns are conducted in level flight (albeit not necessarily at constant IAS), they are of limited use in determining the true characteristics of the aircraft since, as stated above, the increased collective necessary to sustain level flight at high angles of bank is usually destabilizing. Additionally the Ministry of Defence Standard 00-970 [5.1] requires manoeuvre stability to be evaluated in a turn initiated from steady straight and level flight conditions. Consequently both forms of stability testing are conducted.

Descending turns at fixed collective, or *wind-up turns*, are commenced by first establishing a trim condition in level unaccelerated, ball-centred flight at the datum altitude. The aircraft is then climbed a suitable increment above datum altitude, without re-trimming, and the trim control positions (collective and longitudinal) are re-established. A constant airspeed, ball-centred, fixed-collective descending turn is then entered aiming to be on condition in stable flight at the desired bank angle, and therefore load factor, as the datum altitude is passed. Data is normally recorded continuously in an altitude band ± 1000 ft from datum, the most stable 10 seconds or so being selected for analysis after the flight. Data typically includes airspeed, load factor, longitudinal cyclic stick position and fuel state. The test is then repeated at incrementally increasing bank angle until a limiting condition (angle of bank or load factor) is reached. At small bank angles it is often possible to complete several test points within the test altitude band before having to climb back up.

Manoeuvre stability during symmetric manoeuvres can be evaluated by studying the normal acceleration time history following a pull-up or push-over. Since a key parameter in manoeuvre stability testing is airspeed it should be maintained essentially constant. This exposes a complication with symmetric manoeuvres since the change in flight path angle results in a variation in the 'X' component of gravity causing a change in airspeed and an attitude-related change in normal acceleration as measured in the 'Z' axis of the aircraft. The corollary of these two effects is that data must be taken during the short period when forward velocity is essentially equal to the datum airspeed and when the aircraft is in the level pitch attitude. The type of manoeuvre necessary to comply with these constraints is difficult to fly accurately. The Ministry of Defence Standard 00-970 [5.1] attempts to get around these problems by requiring only that the peak increment in normal acceleration should be '*substantially proportional to the magnitude of the control input*' and that it should '*increase progressively with increasing initial airspeed*', the implication being that testing can be achieved by incremental step inputs from trimmed, level, unaccelerated flight. Nevertheless, in order to allow enough time to develop peak load factor in an agile aircraft, it is necessary to commence the manoeuvre from a nose-down attitude so that the recovery can be made before an excessive nose-up condition develops. A control fixture can be used to provide some means of incrementally increasing the severity of the manoeuvre. Push-overs are always approached with the utmost caution in helicopters with teetering rotors or with articulated rotors featuring low hinge offset. The technique is the reverse of that for the pull-ups: the aircraft is accelerated/descended from the trim condition, pulled up to a nose-high attitude and then bunted using the appropriate control input as it decelerates towards the trim speed.

5.3.3 Manoeuvring at reduced load factor

For rotors where rotation of the thrust vector is the only source of a control moment (such as teetering rotors) the ability to control the helicopter will disappear if the load factor should ever reach zero. Hingeless rotors or those with some effective hinge offset will maintain more control power at low rotor thrust, but it might be noticeable that more cyclic control is required to return to level flight from a $0.5g$ push-over than from a $1.5g$ pull-up. Not only is control power reduced at low rotor thrust, but it can set up a potentially dangerous condition leading to 'mast bumping' on teetering rotor helicopters or 'droop stop pounding' on those helicopters with articulated rotors of low hinge offset. In either case, at a low load factor the tip path plane will respond normally to cyclic inputs, but the low rotor thrust will have little effect on fuselage attitude. The ultimate effect of this critical situation is that the rotor may flap outside design limits, leading to blade/fuselage contact or rotor separation due to mast shearing. A further degradation of handling qualities at low rotor thrust results from the reduction in the rotor's contribution to rate damping (in both pitch and roll) since the damping effect of thrust tilt is reduced.

5.4 DOCUMENTING DYNAMIC STABILITY CHARACTERISTICS

5.4.1 The longitudinal long-term mode

All dynamic stability testing, but especially long-term response testing, demands that the initial trim flight condition be set up as accurately as possible. If this is not achieved then any control offset will bias the response and may cause, for example, a lightly damped oscillatory response to diverge aperiodically. If a very accurate engineering assessment of long-term response is required, then the tests are conducted in conditions of zero turbulence. In addition to quantitative tests, qualitative testing is conducted during role manoeuvres and comprises observing the pilot compensation necessary to suppress any undesirable dynamic effects such as a lightly damped long-term mode. Often it is these qualitative tests that are the most important part of the test programme as they show how the aircraft's stability characteristics will affect the operational pilot when conducting the role.

To avoid damaging the aircraft during the large excursions from straight and level flight that can occur when documenting an aggressively unstable long-term mode it is often necessary for the aircraft to be instrumented with stress parameters monitored by telemetry. Practice recoveries from unusual attitudes followed by an incremental approach with some form of in-flight prediction of the severity of the next response, similar to that described in the section on lever delay testing, should prevent any exceedence of limits.

5.4.1.1 *Methods of excitation*

Having trimmed the aircraft as accurately as possible in the desired flight condition, the long-term response is excited using one of the methods detailed below. The magnitude of any artificial excitation should be chosen to give a response which is representative of that occurring naturally. Excessive excitation may lead to an unrepresentative response because the pilot may often have to intervene before the motion has developed. As a general rule, the smallest excitation which produces a significant oscillatory response or a divergent aperiodic response is used.

(1) *Natural turbulence.* The effect of imperfect trim conditions or natural turbulence may be sufficient to excite an aperiodic or lightly damped oscillatory response. Such initiation of the long-term response is desirable in that there can be no doubt that the excitation method is representative. However, these responses are usually contaminated by subsequent atmospheric disturbance before the motion is complete and it may be impossible to extract meaningful quantitative data. Nevertheless, natural turbulence does provide the opportunity to qualitatively evaluate the difficulty of suppressing the long-term response under representative conditions.

(2) *Release to trim.* If there is a benign response to natural turbulence then verification of such a desirable aircraft characteristic is usually made using artificial excitation. Equally a pilot generated disturbance may be required if the level of turbulence is insufficient to generate a meaningful response from the aircraft. Often the best method is to accelerate or decelerate from the trim airspeed and then to smoothly release the controls back to their trim positions. The controls may then be left free (but monitored) until recovery action, or the suppression of an off-axis response, becomes necessary. A speed increment of 5–15 knots is used for this method.

(3) *Pulses.* Longitudinal cyclic or collective pulses may also be employed to verify any lightly damped or aperiodic long-term response obtained from natural turbulence. Unless of extreme magnitude and direction (that is virtually a release to trim), pulses typically provide insufficient excitation to initiate a long-term response which is well damped, especially in forward flight.

5.4.1.2 *Specification requirements*

Natural frequency and relative damping requirements for the longitudinal long-term mode vary with pilot attentiveness. As might be expected a more heavily damped mode is required if the pilot's attention is divided between the flying task and other

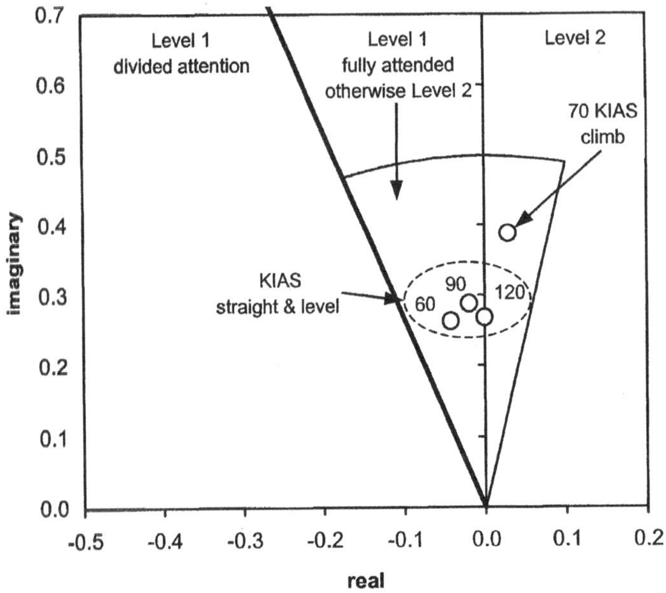

Fig. 5.8 Dynamic stability requirements (ADS-33).

duties. Figure 5.8 compares longitudinal long-term mode data from a light singled-engined helicopter with the requirements of ADS-33E [5.2]. Notice that if the pilot is fully engaged in attitude and flight path control (definition of full-attention) the aircraft has Level 1 handling qualities (as defined in Section 5.7.1). However, as soon as his attention is divided the low level of relative damping confirms Level 2 handling qualities. Ministry of Defence Standards take a different approach by assuming that the pilot will either be actively monitoring the performance of an AFCS or flying under autopilot control whilst more fully involved in ancillary duties. In each case a minimum acceptable time is specified before the pitch attitude can vary by more than 1 degree.

5.4.2 Lateral/directional oscillation

Lateral/directional oscillation (Dutch roll) characteristics are documented by first accurately trimming the rotorcraft in level flight at the desired airspeed and altitude. A lateral/directional oscillation is then excited by using any of the following methods: a release from a SHSS; a lateral cyclic, yaw pedal or collective pulse input; a lateral cyclic step input; or a lateral cyclic or yaw pedal doublet. The release from a SHSS is accomplished by returning all controls to trim simultaneously with rapid ramp control inputs. The control pulse and doublet inputs are typically conducted using a 2–3 cm displacement with a period of 1 second for cyclic inputs and 2 seconds for yaw pedal inputs. All controls remain fixed following the control input and the open loop response of the aircraft is documented.

5.4.3 Spiral stability

It has already been shown that spiral stability is dependent on the sign of the expression $(L_v N_r - L_r N_v)$, positive being stable. Now consideration of the equations of motion for TO1Cs, both pedal and cyclic, yields:

$$\theta_{tr} = \frac{-r(L_v N_r - L_r N_v)}{L_v N_{\theta_{tr}}} \qquad A_1 = \frac{r(L_v N_r - L_r N_v)}{L_{A_1} N_v}$$

For a turn to starboard, r is positive and $N_{\theta_{tr}}$ and L_v are both negative. The sign of θ_{tr} depends on the sign of $(L_v N_r - L_r N_v)$: when the bracketed expression is negative, θ_{tr} is positive. This implies that the control deflection and the yaw rate are in opposite directions since *positive* θ_{tr} implies *left* pedal. Alternatively, if $(L_v N_r - L_r N_v)$ is positive, θ_{tr} will be negative, implying that the control deflection and yaw rate will be in the same direction (to the right). Also, inspection of the equations indicates that, as N_v is positive and L_{A_1} is negative, if $(L_v N_r - L_r N_v)$ is positive, cyclic deflection and yaw rate are in the same direction Thus, the control deflection required to maintain a steady turn on one control can be used as a direct indication of the helicopter's spiral stability: if the pedal or cyclic has to be deflected into the turn (right pedal/cyclic for right turn), the aircraft is spirally stable; alternatively if the control has to be deflected out of the turn, the helicopter is spirally unstable. It is worth noticing that, although the derivatives vary with speed, altitude and configuration, L_r is usually small, so $(L_v N_r - L_r N_v)$ is often positive and helicopters are usually spirally stable at modest angles of bank.

The spiral stability test is usually conducted by flying a TO1C-C as these are generally easier to perform than TO1C-P. Longitudinal inputs are used to hold airspeed constant throughout the test. When the helicopter is stabilized at the desired bank angle, the cyclic is smoothly returned to the level flight trim value and a time history of the ensuing bank angle is recorded. During testing, inputs are made in a manner that minimizes excitation of the lateral/directional oscillation. It is normally of interest to evaluate an IFR bank angle (20°) and also to try and find the angle at which spiral stability becomes neutral or even negative.

5.4.4 Adverse yaw/proverse yaw/heading delay

Cyclic-only turns are also conducted to determine the adverse yaw or proverse yaw characteristics, whether any heading delay is apparent and whether the aircraft possesses the capability to perform co-ordinated turns with no pilot input on the pedals. The technique consists of TO1C-C initiated from a ball-centred or zero-sideslip wings-level attitude accomplished with lateral cyclic displacements of various rates consistent with the role of the aircraft. Collective remains fixed and airspeed is maintained with longitudinal cyclic. The heading, the sideslip and the yaw rate response to the control input are noted. A roll out of the turn onto a predefined heading at role-relatable roll rates is often attempted to qualitatively evaluate the ability of the pilot to select a desired heading within narrow tolerances.

5.5 TIME DOMAIN METHODS FOR CONTROL RESPONSE TESTING

The purpose of control testing in the time domain is to evaluate the dynamic modes of the aircraft using inputs that although generated in a very stylized manner represent in some way the control strategies commonly used by pilots. Typical inputs include steps and pulses. Since these inputs have been used for many years as means to determine the *control response* characteristics of rotorcraft, specific handling quality specification requirements exist which are based on these test methods. More recently as a result of the handling qualities research conducted during the LHX programme (the forerunner of the RAH-66 Comanche) the concept of manoeuvre quickness has been introduced [5.3]. Since testing to gather quickness data is an extension of the pulse input technique [5.4], it is discussed below.

5.5.1 Step response

The step input is a popular method of determining the control response characteristics of a helicopter. It involves trimming the rotorcraft in to the hover or in straight and level flight at a given airspeed and then making a pure step input using the appropriate control inceptor. In forward flight, pitch and roll inputs are made, whereas in the hover due to the differing response characteristics it is possible to make yaw inputs as well. Test data can be gathered in a rudimentary fashion using manual techniques although it is commonplace to use either on-board automatic data recording or telemetry as the primary data source. With this in mind, the test technique has evolved to one that maintains flight safety by using manually recorded data, regardless of the on-board or telemetry system adopted, to ensure an incremental increase in the severity of the aircraft manoeuvre. This method also allows factoring of pilot opinion on the difficulty of recovery into the choice of the next increment to test.

Since most unaugmented helicopters will achieve steady rates within two seconds of the input being made it is usual to hold it for the count of two and then initiate recovery. In this way the recovery action becomes part of the test and the pilot will automatically restore the aircraft to the trim condition at the end of the count unless, of course, a self-declared limit is likely to be transgressed beforehand. Control response testing using step inputs can be conducted in a low-risk, incremental manner provided a set routine is performed The routine adopted at ETPS, shown in Table 5.1, is an example of the recommended methodology. Whatever the precise routine adopted, test teams must treat changes to the direction of input with great care. There is always the possibility that having made several aft inputs the pilot will have so conditioned himself that he pulls the control inceptor aft even though the next test point requires a movement forward!

A control fixture, see Fig. 5.9, is a common means of ensuring that a step input of a pre-determined size is made. The fixture provides a rigid surface against which the control inceptor may be held thereby enabling pure step inputs. The requirements of flight safety normally require that the fixture be held in place manually by a flight test observer situated at the co-pilot station. Consequently this test method can only be used on helicopters fitted with dual controls. Alternatives to the fixture include injecting discrete control inputs via the automatic flight control system. Care is taken to avoid off-axis inputs (the introduction of lateral cyclic during longitudinal control

Table 5.1 Control response testing.

Non-handling pilot or observer	Handling pilot
	Trim aircraft to required flight condition '*On condition*'
Sets fixture for next input size '*The next input will be 1 cm aft*' Appropriately orientates fixture and shows fixture to pilot	
	Glances at fixture '*Confirmed*'
Positions fixture next to control inceptor '*In position*'	
	'*Stand by for a 1 cm aft input on the count of three, are you ready*'
Checks that fixture is still in place and correctly orientated '*Ready*'	Glances across the cockpit to ensure that fixture is correctly orientated
Notes actual input size and whether input held against fixture	'*1001, 1002, 1003 – moves inceptor – 1004, 1005, recovering*' Notes attitude at 1004 and 1005
Plots attitude at 1 and 2 seconds against input size on 'how-goes-it' chart, chooses next input size	Restores aircraft to required flight condition and reads back attitudes

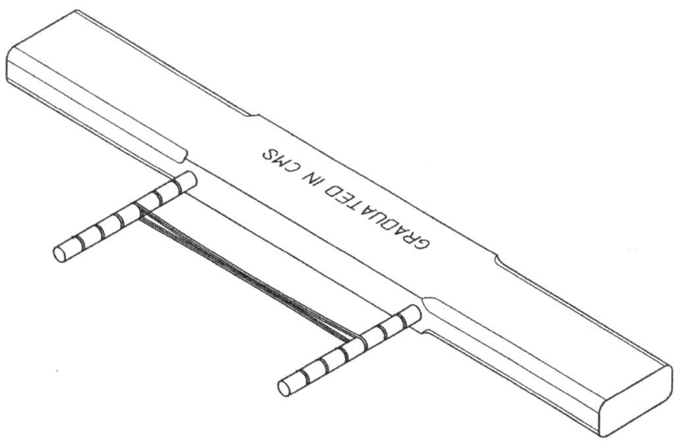

Fig. 5.9 A control fixture.

response testing for example). Using only the hand or fingers to make the input rather than moving the whole arm can facilitate this. The control is held rigidly against the fixture until the desired results have been achieved or recovery is necessary. A build-up technique is employed using progressively larger inputs so that the testing can be terminated early should any untoward trends develop. Control cross-coupling may be evaluated by observing the response in other axes. The use of telemetry or a near real-

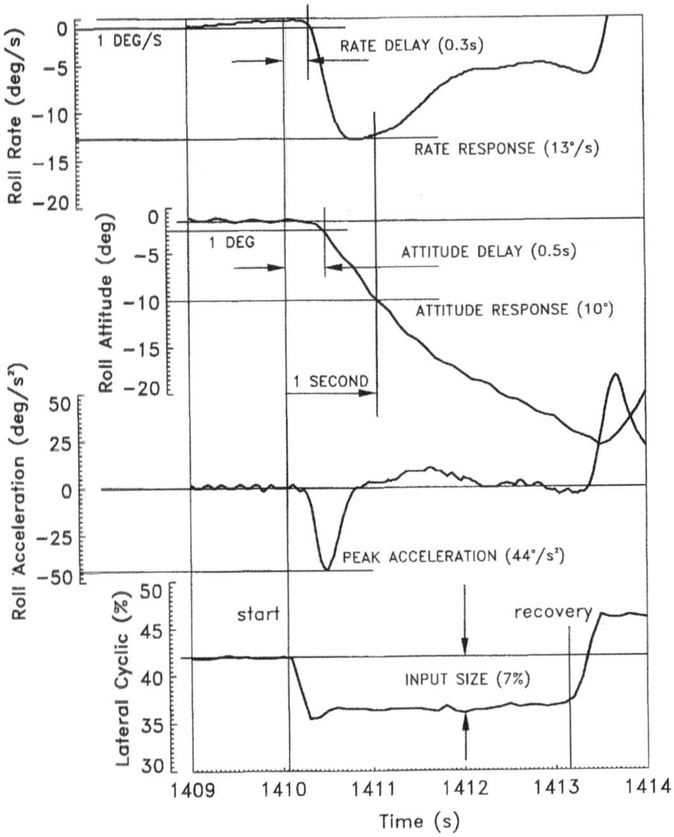

Fig. 5.10 Definition of control response parameters.

time display of input shape is often used so that a poor input shape can be recognized quickly and the test condition repeated immediately.

The control response characteristics of a rotorcraft are characterized using a standard set of *control response parameters*, see Fig. 5.10. These parameters are based on the assumption that the helicopter has a first-order rate response to control inputs. The *attitude* and *rate responses* are, respectively, the aircraft attitude and angular rate achieved one second after the input. The *attitude* and *rate delay* quantify the time taken for the helicopter to change attitude by one degree and generate an angular rate of one degree per second. The *steady-state rate* and *peak angular acceleration* are self-explanatory. A useful parameter for comparing the control response of different helicopters is the *control sensitivity*, which is the maximum angular acceleration achieved for a standard input size (1 cm or 1 inch).

Figure 5.10 highlights a typical problem in estimating steady-state angular rates. Due to the effect of sideslip it is unusual for rotorcraft to acquire a steady roll rate. Since the magnitude of the steady-state rate is only used in comparison with the rate response to judge the crispness of the control response, this phenomenum does not usually present the test team with much of a problem. In the figure it is clear that the aircraft reaches a peak rate within one second. This peak rate approximates to the steady rate that would have been acquired if the helicopter truly behaved as a first-order system. Therefore the steady rate and rate response can be stated as being equal and the control response characterized as 'crisp'.

Having gathered a set of traces and analyzed them in the manner shown in Fig. 5.10 it is possible to construct a set of summary charts to portray the control response characteristics. There are two categories of derived plots that are used to summarize control response data. One is the variation of control response parameters with airspeed: any or all of the control response parameters (rate response, attitude response, control sensitivity, steady-state rate) may be plotted against airspeed for a given input size. Alternatively the variation of control response parameters with input size for a given flight condition may be presented, see Fig. 5.11. In each case the pilot would expect some progression in response. For example, at a given speed a larger input should yield a larger, ideally proportional, response, whereas the pilot would expect a given input size to yield a larger response at a higher speed. An abundance of quantitative data can be taken from response time histories. In discussing control response it is important that the parameters chosen and presented support the pilot's qualitative opinion. Parameters that define the 'amount' of the control response are

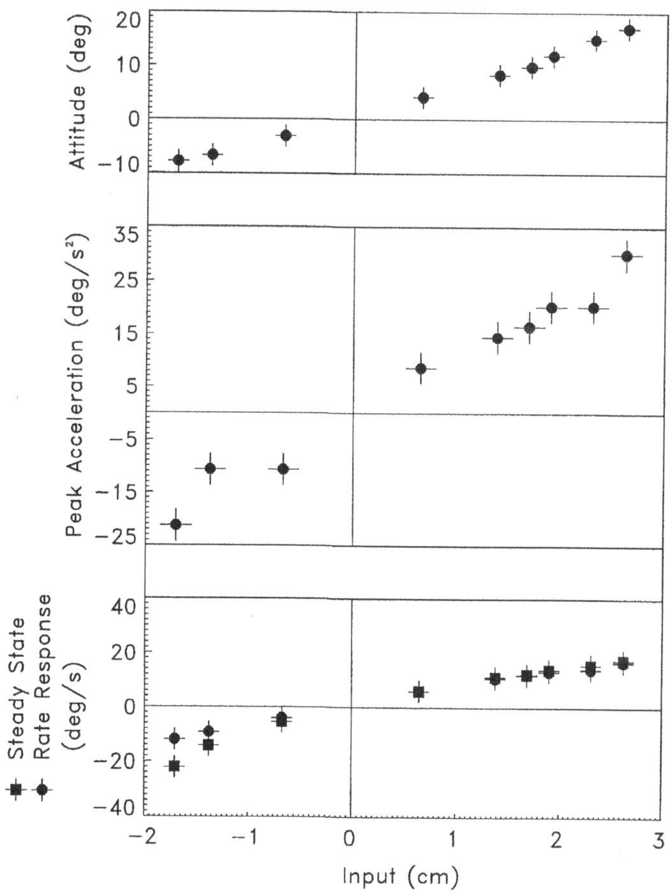

Fig. 5.11 Control response test data.

the rate response, the attitude response, and the steady-state pitch rate. Quantities which can be used in discussing the quality of the response or 'how' it got to the observed steady-state rate are the delay time, the initial angular acceleration, the time to steady state, and the inflection time.

When conducting manual step response testing there are several flight safety points to consider. Ideally visual confirmation of the input size (fixture setting) and input direction should be made before commencing the cadence count. Although this is easy to accomplish in a side-by-side cockpit, the test team will need to consider carefully how this might be achieved in a rotorcraft with a tandem cockpit configuration. Modern attack helicopters tend to have the cockpit stations optimized for pilot and co-pilot/gunner. This usually involves differences to the control inceptors in terms of location, force-feel and possibly gearing (the AH-1 Cobra is a good example). Such differences require careful consideration when establishing the procedure for in-flight testing. In addition to using an in-flight plot or 'how-goes-it' chart to track the approach to a limiting parameter, other visual indicators can be used. Large cyclic control deflections will typically reduce the clearance between the main rotor tip-path plane and the fuselage. An appropriately located set of 'chicken-sticks', see Fig. 5.12, which are designed to break if struck by a rotor blade, can warn of impending mishap. Ideally such devices will be located within the pilot's field of view so that he can confirm easily if the clearance has reduced to hazardous dimensions. Should this not be possible then the sticks are modified to include an electric circuit that will illuminate a suitable warning lamp in the cockpit if they are broken. The development of small proximity sensors of suitable accuracy has enabled a more sophisticated approach giving a greater range of information and warning.

Step response testing may require the employment of large inputs that could lead to aircraft damage if performed without due care. The following additional precautions are typical of those adopted to minimize this risk:

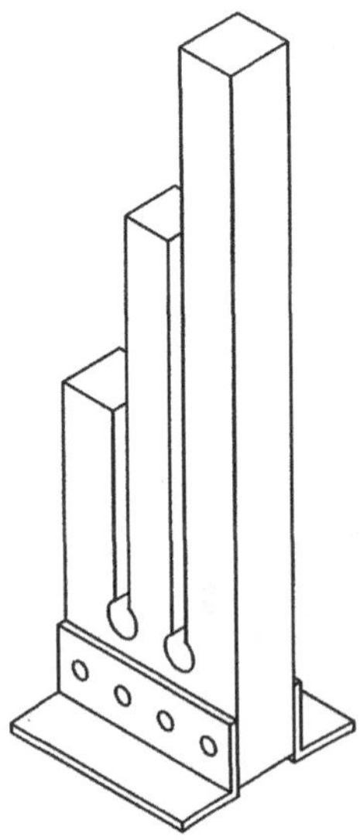

Fig. 5.12 'Chicken-sticks'.

- Awareness of the effects of manoeuvre instability during aft inputs that may lead to an unexpected 'dig-in' and possible overstress of the airframe or rotor system.
- Careful monitoring of the approach to limitations and the use of an appropriate 'how-goes-it' chart.
- Anticipation of possible series actuator saturation and a consequent sudden increase in aircraft response when testing augmented aircraft.
- Careful briefing and practise to ensure that the entire crew is aware of the control input size, direction and recovery action.

5.5.2 Pulse inputs

Unlike step response testing, the pulse input is specifically mentioned in the Ministry of Defence Standard 00-970 [5.1]. It is quoted as a suitable test technique for evaluating the transient response characteristics of rotorcraft whose handling qualities are affected by the incorporation of sophisticated stability and control augmentation systems. Although the assessment of augmented rotorcraft is the subject of a later chapter in this book it is pertinent to include pulse inputs here. Since the pulse input has the advantage of returning the control inceptor to its trimmed position any ensuing response will be dictated by both the underlying stability characteristics of the helicopter and the action of any automatic control system. In practice the input will be similar to the *release-to-trim* technique described earlier although since the magnitude of the peak response is a specification compliance parameter a control fixture is often used and the test approached in the same incremental manner as step response testing.

The desired response to a pulse input is specified using a set of quantitative data equivalent to the control response parameters introduced above. These include the *peak response*, the time required for the flight parameter to make a first pass through the datum (T_{01}), the time for a second pass in the sense of the original disturbance (T_{02}) and the time to return to datum (T_F). The rate of return to datum, the size of any overshoots and the tolerance of datum re-capture are all specified using fractions of the peak response. Thus (T_{30}) and (T_{11}) are the times taken for disturbance to reduce to 30% and 10% of the peak response. For responses that are oscillatory, unlike the dead-beat response portrayed, the magnitudes of the first and second overshoots (x_1 and x_2) are typically set at 15% and 10% respectively for Level 1 handling qualities and the tolerance for datum capture (x_F) at 10%. The initial part of the pulse, or '*boxcar*', input is also a means of assessing the control response of the helicopter. Therefore the rate of onset of the pilot-induced 'disturbance' is also specified. This is achieved by requiring that the response (y_1) exceed a certain percentage of the peak response within a given time of the input being made (T_1). Figure 5.13 shows an example time history of a pulse input and subsequent dead-beat response.

5.5.3 Manoeuvre quickness

Before discussing how manoeuvre quickness is assessed it is perhaps worth reviewing the meaning of this relatively new handling quality parameter. Padfield reported [5.4] that efforts to resolve task portrait information with pilot commentary on handling

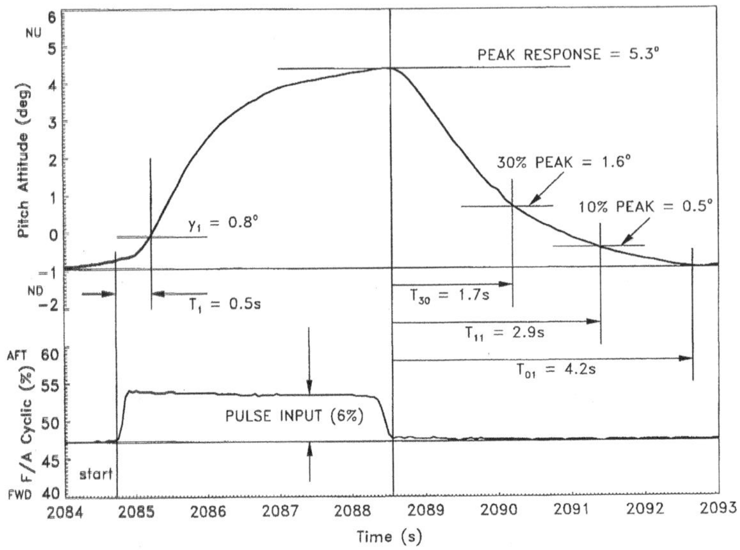

Fig. 5.13 Pulse response test data.

qualities led to a realization that there was a range of control strategies for which neither bandwidth specifications (see the section on frequency analysis, Section 5.6) nor large amplitude specifications (see Section 5.5.4) were suitable. Either the input frequencies were too low or the task did not demand the use of the maximum angular rates available.

When developing the new parameter, that eventually became known as *manoeuvre* (or *attitude*) *quickness*, it was felt that it needed to correctly reflect the varying degrees of aggression with which the pilot could perform a specified flying task. During extensive trials, in both ground based and airborne simulators, it was noted that when pilots were asked to perform a manoeuvre requiring a discrete flight path or position change, such as a lateral side-step for example, each attitude change could be associated with a particular peak angular rate, see Fig. 5.14.

When pilots were asked to fly the manoeuvres more aggressively, or with more *attack*, it was found that the attitude changes were achieved with larger angular rates by the use of larger control inputs held for shorter periods of time. Referring to Fig. 5.15, note that two of the inputs portrayed result in broadly similar attitude changes: in both cases the helicopter attitude has been changed by approximately 25°. One trace shows the result of a 3-second pulse involving full control deflection whereas the other trace results from a 6-second pulse to 50% control travel. Clearly the first input is more aggressive than the second and this is characterized by a peak angular rate that is approximately twice as great. If a *phase portrait* is constructed for the manoeuvre, see Fig. 5.16, the difference in aggression is evidenced by the area under the curve (the greater the area, the higher the task aggression).

In developing manoeuvre quickness it was decided to use a parameter akin to frequency so that it would fit with the bandwidth criteria developed for control tasks requiring much smaller control inputs. In this way a more coherent scheme of handling qualities specification was achieved. The frequency parameter was obtained by dividing

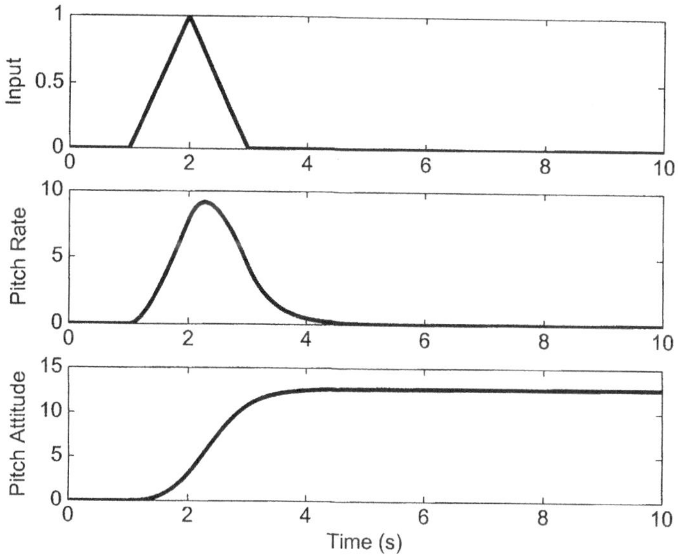

Fig. 5.14 Trace of a discrete attitude change.

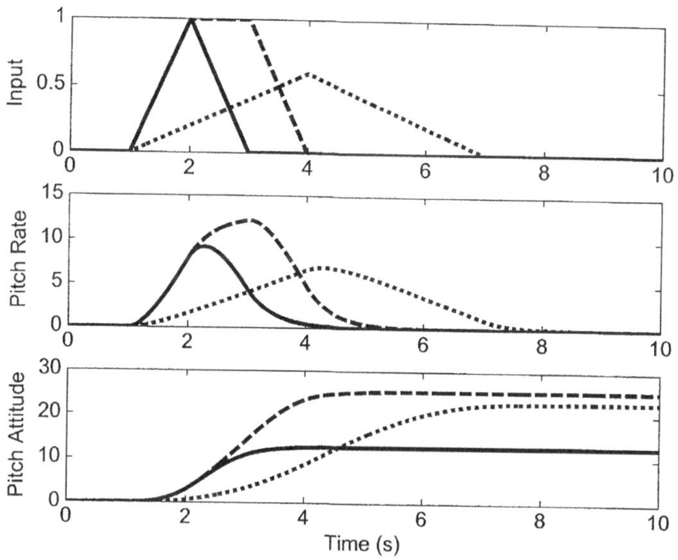

Fig. 5.15 Trace of attitudes changes with differing aggression.

the magnitude of discrete attitude changes into the peak rate achieved during those changes.

In practice the assessment of the manoeuvre quickness characteristics of a helicopter is similar to pulse response testing although it is more common to analyze the effects of a series of control inputs rather than a single one. Typically the time history is obtained whilst executing a multi-axis task that requires the pilot to make inputs of

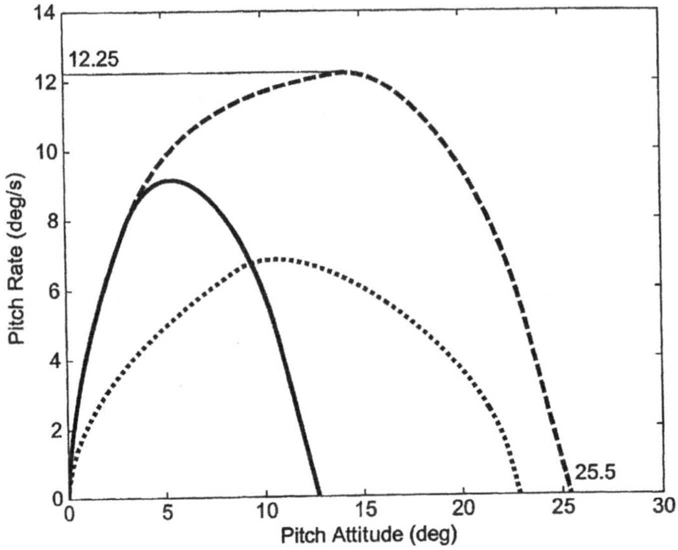

Fig. 5.16 Phase portraits of attitude changes.

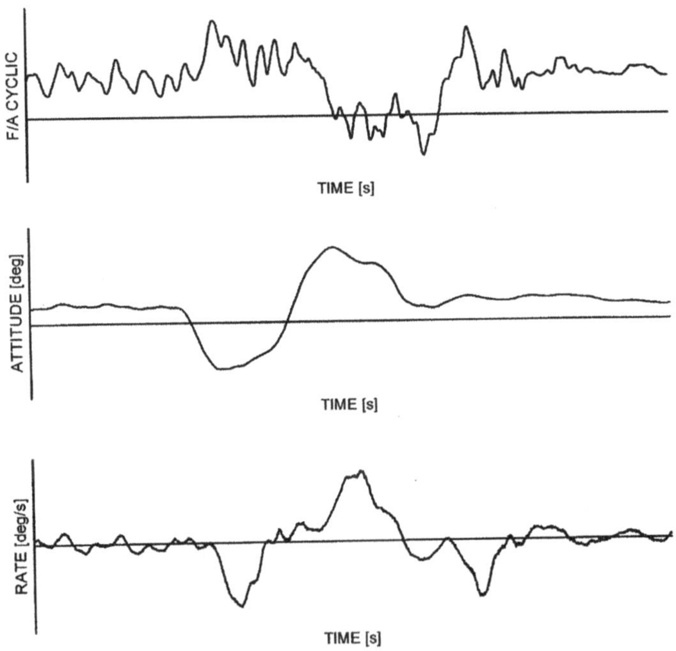

Fig. 5.17 Acceleration/deceleration test data.

varying sizes at varying rates. Examples include flying a slalom course, executing an accel-decel (see Fig. 5.17) or NOE flight. Time histories of appropriate control, attitude and angular rate are then analyzed (as in Fig. 5.18). Discrete attitude changes and the peak rate generated during such changes can be identified and processed to produce

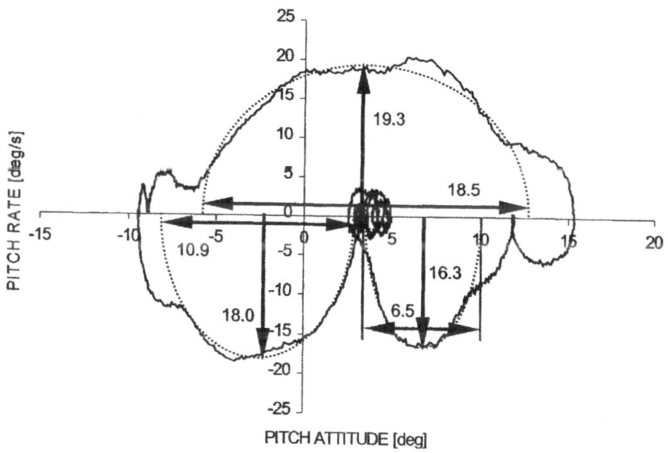

Fig. 5.18 Generating manoeuvre quickness data.

values of manoeuvre or attitude quickness. Acquiring data suitable for specification compliance will involve incremental increases in the aggression with which the manoeuvres are flown. Typical aggression metrics are time taken to fly the course or maximum attitudes used during the manoeuvres. If the aircraft is capable of generating manoeuvre quickness in excess of that specified for a particular role or range of flight tasks it will be deemed to have Level 1 handling qualities.

5.5.4 Large amplitude manoeuvres

Large amplitude manoeuvre testing is required by ADS-33E [5.2] and involves exploring the maximum capabilities of the rotorcraft. In Table 5.2, example specification requirements are listed. It is evident that to generate the required angular rates or attitude changes large inputs will need to be made by the pilot. Full control deflections, more typical of fixed wing flying, may be required, therefore it is obvious that such testing will have to be approached incrementally using an appropriately instrumented airframe.

5.6 FREQUENCY DOMAIN METHODS FOR CONTROL RESPONSE ASSESSMENT

When the handling qualities research that formed the background to the LHX programme and led to ADS-33E [5.2] was conducted the traditional methods of quantifying control response were found deficient. It became clear that the frequency response of the helicopter was of greater importance, for certain flying tasks, than its response in the time domain. Indeed it was found that pilots flying rotorcraft with vastly different step response characteristics (natural frequency and relative damping) but identical frequency response characteristics would give similar handling quality rating for tasks that required frequent small control inputs. More precisely Hoh [5.5]

Table 5.2 Large amplitude response requirements.

	Rate response types						Attitude command response types			
	Achievable angular rates (deg/s)						Achievable angle (deg)			
	Level 1			Levels 2 & 3			Level 1		Levels 2 & 3	
Agility category MTE	Pitch	Roll	Yaw	Pitch	Roll	Yaw	Pitch	Roll	Pitch	Roll
Limited agility Hover Landing Slope landing	±6	±21	±9.5	±3	±15	±5	±15	±15	±7	±10
Moderate agility Hovering turn Pirouette Vertical manoeuvre Depart/abort Lateral reposition	±13	±50	±22	±6	±21	±9.5	+20 −30	±60	±13	±30
Aggressive agility Vertical remask Acceleration-deceleration Sidestep Target acquisition and tracking Turn to target	±30	±50	±60	±13	±50	±22	±30	±60	+20 −30	±30

found, when conducting simulated deck-landing tasks, that pilot opinion remained unchanged for configurations with relative damping between 0.5 and 1.3. Later analysis showed that the common factor in all the configurations tested was the similarity in *bandwidth* or the range of input frequencies useable by the pilot.

5.6.1 Frequency response characteristics

Before describing frequency domain testing methodology, it is necessary to discuss the conceptual background to the subject. When required to control a rotorcraft using single discrete inputs experienced pilots can compensate for attitude and rate delay by making a larger input initially ('boosting') and by 'backing-off' early or *leading*. The boost over-drives the aircraft response by generating a higher acceleration (although this will be limited by the control power available) and the lead allows the aircraft to settle at the desired attitude by removing the over-large input before its full effect can be realized, see Fig. 5.19. The dotted line shows the response of a typical helicopter to a large amplitude/short duration pulse and compares it to a more gentle input of lower magnitude but lower duration. Notice that in both cases the attitude change obtained is the same but the application of 'lead' has resulted in the target attitude being obtained some 5 seconds sooner. However when performing *high-gain* flying tasks, such as target-tracking, precision hovering or deck-landings the pilot will often

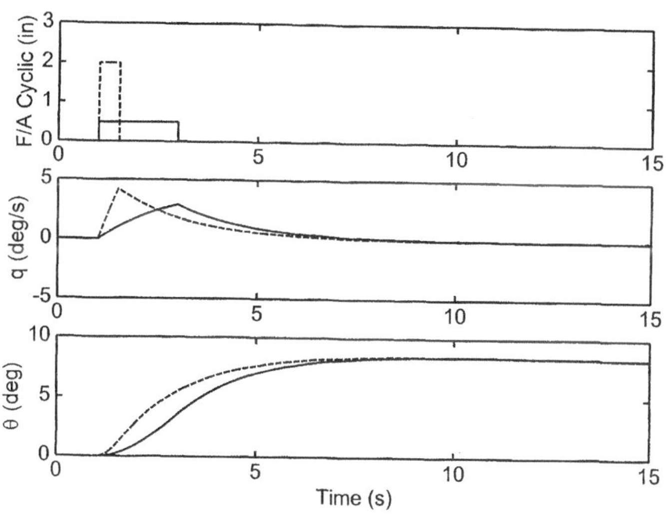

Fig. 5.19 Quickening the response by over-driving the control input.

need to make continual small inputs to keep the aircraft on the desired flight path. In these situations, he may not be able to compensate adequately for excessive time delay since in trying to boost and lead he may find himself 'out-of-phase' with the aircraft. Consider the pitch response of a typical helicopter resulting from a constant amplitude frequency sweep, Fig. 5.20. After the initial gross attitude change ($t > 10$ s) note that as the input frequency is increased two effects are seen: the output (pitch rate or pitch attitude) changes in magnitude and it lags further behind the input. The qualities of *gain* and *phase* and their variation with input frequency are fundamental in shaping pilot opinion for high-gain handling tasks. Although presenting these trends in the

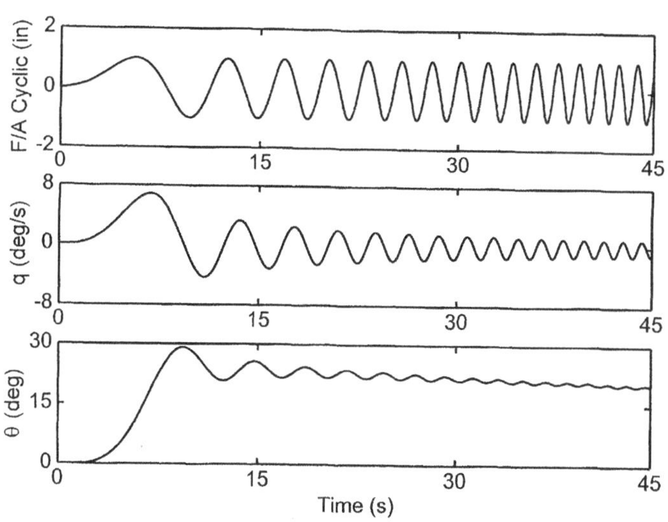

Fig. 5.20 Time history of a frequency response test.

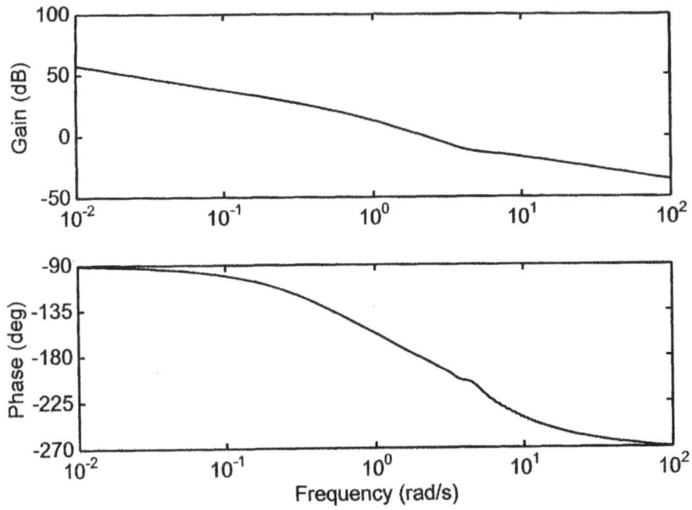

Fig. 5.21 Presenting frequency response data: Bode plot.

form of time histories makes it easier to visualize them, this form of presentation is not conducive to detailed analysis or specification compliance testing.

Two methods of presenting frequency response information are commonly used: the *Bode plot* and the *Nichols chart*. These diagrams, originally developed for use in filter and other electronic component design, have found a place in handling qualities testing as the importance of frequency response characteristics has been realized. Nowadays they are not only convenient means of documenting these characteristics but are also used for detailing handling quality specifications [5.2 and 5.6]. The Bode plot consists of two charts with input frequency as the independent variable and either gain (ratio of output to input) or phase lag as the dependent variable, see Fig. 5.21. The Nichols chart presents both gain and phase information simultaneously, Fig. 5.22. Both formats result from a spectral analysis of time histories similar to those shown earlier.

It is important to consider what constitutes the most appropriate output for both spectral analysis and requirements specification. When the pilot moves the pitch/roll inceptor he will normally be targeting an attitude change and therefore the most obvious output for analysis would seem to be attitude. Indeed as shall be seen later *attitude bandwidth* is an important specification parameter. Experience has shown, however, that for good quality results it is necessary to use attitude *rate* in the spectral analysis process since there is usually a better spread of energy across the tested frequency range. After processing, it is a simple matter to reduce the gain by 20 dB/decade and add the extra 90° phase lag associated with generating attitude from the integral of rate, see Fig. 5.23. This is usually a perfectly valid process provided the attitude excursions are not too large.

5.6.2 Test methodology

An ideal frequency sweep comprises a continuous, constant-amplitude sinusoidal control input at progressively increasing frequency. The characteristics of a good

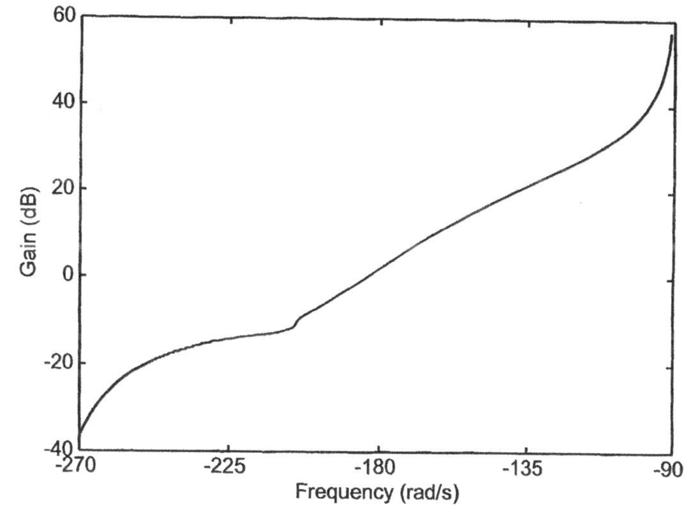

Fig. 5.22 Presenting frequency response data: Nichols chart.

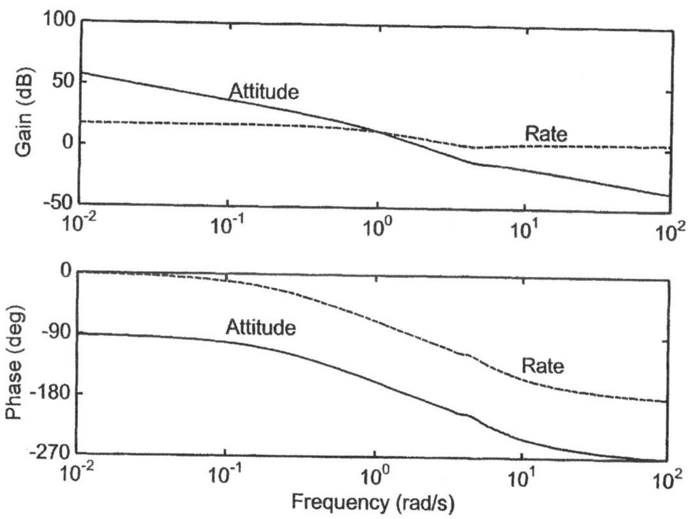

Fig. 5.23 Frequency response of attitude from angular rate test data.

frequency sweep, and the techniques that should be employed to achieve them, are detailed below.

5.6.2.1 Amplitude

Ideally the frequency sweep should be made using a constant amplitude control movement. It is very easy for the pilot to inadvertently increase input amplitude as frequency increases. This is often due to the greater forces that are required to overcome viscous damping effects in the control linkage. Increased amplitude at high frequency should be avoided as it may cause structural damage to the airframe. The use of two hands on the cyclic helps as do high bandwidth, fast-acting control position indicators. A mechanical device similar to a 'two-way control fixture' may also be used but there is a risk of squaring off the input peaks (with consequent corruption of the frequency spectrum) if the control is allowed to actually touch the fixture.

5.6.2.2 *Frequency spread*

A wide range of input frequencies (up to around 2 Hz) without 'holes' in the spectrum is essential. In other words, it is important neither to omit a particular sub-range of frequencies nor to dwell on them for too long. One of the most difficult aspects of a frequency sweep is the maintenance of a progressive increase in frequency. Moderate frequency inputs are easy to achieve and impatient pilots tend to progress to the higher frequencies too quickly. The result of this is the omission of many low-to-middle frequencies. This problem is best overcome by using a cadence count technique as well as coaching the pilot to prevent him dwelling on a particular frequency.

5.6.2.3 *Trim condition*

The maintenance of the desired trim condition (airspeed and/or attitude) is important. This is best achieved by conducting a symmetrical frequency sweep which starts and stops at an accurate trim condition. Depending on the trim condition it may be necessary to bias the central control position in order to maintain the correct trim attitude or airspeed. To prevent corruption of the power spectrum it is important that such bias movements take place at a low frequency (over several input cycles). Typically the aircraft is kept within 10 knots of the trim airspeed as it passes through the level pitch attitude. Stick trim is generally beneficial as positive centring provides useful cueing to the trim position, however, large breakout forces may introduce discontinuities in the inputs.

5.6.2.4 *Off-axis response*

It is inevitable that off-axis responses will take place and it is important to allow them to occur for parameter identification purposes and for cross-coupling specification compliance testing. If the responses become so large that the pilot must intervene then he should make corrective inputs which are not correlated with the frequency of the primary control inputs. Thus if a high frequency pitch sweep is being made and suppression of a roll response is required this should be made at low frequency and vice-versa.

5.6.2.5 *Resonance*

It is important to avoid any dominant rotor or structural modes which could cause adverse structural or aerodynamic resonance that may damage the aircraft. Such frequencies should be identified by engineering analysis prior to a sweep. A well-designed helicopter should have structural, rotor and control system natural frequencies well above the range of frequencies a pilot would be likely to generate under normal circumstances. However, the deliberate use of high frequencies during testing may erode the design safety margin in this respect.

5.6.2.6 *Flight techniques*

Frequency sweeps at low speed should only be attempted in zero wind or very light wind conditions. Similarly frequency sweeps at high airspeed should not be made if significant turbulence is present. Usually two pilots are used with one pilot performing the sweep while the other controls the remaining three axes. A 'quarter count' cadence count technique is often used for the low to medium frequencies. Using a longitudinal cyclic sweep as an example and starting with a 20-second period, the pilot would count from one to five as he moves the cyclic aft, then again as he returns the cyclic to the trim position. The process is then repeated for the forward deflection. The period is then reduced by counting to four for each quarter, and so on. Ideally the sweep is performed three times using a minimum of two pilots as each pilot will employ a slightly different technique. A 10-second 'trim-shot' is typically recorded before and after each sweep.

5.6.2.7 *Incremental approach*

There is no need to cover the entire frequency spectrum in a single sweep. In view of the independence of the results from input amplitude, it is perfectly feasible to combine the data from several test runs. From the point of view of aircraft damage, it is desirable to minimize the exposure of the airframe to this type of testing. Thus it is important that the test objectives are clearly defined as it may not be necessary to go up to potentially damaging high frequencies if adequate results can be obtained without doing so.

For ADS-33E work it is only necessary to cover the frequency range which will permit the bandwidth and phase delay calculations to be completed. It is sensible to conduct an initial frequency sweep with a low limiting frequency and analyze the results before continuing. Having determined the frequency range(s) of interest from the first 'overview' test, the optimum range(s) of inputs for subsequent, more detailed, tests is determined. Risk mitigation is achieved by minimizing the amount of testing and avoiding high frequencies wherever possible so that the potential for airframe damage is reduced. The conduct of limited range and, therefore, duration frequency sweeps will also facilitate the maintenance of an accurate trim condition throughout the test input.

5.6.2.8 *Computer generated sweeps*

Some test agencies advocate the use of automatic input devices that can generate 'pure' sinusoidal frequency sweeps by injecting inputs directly into the pitch change linkage actuators. This method generally makes analysis easier and gives a degree of repeatability. However since the input and response will be subject to spectral analysis, the exact shape and size of input is not critical. Indeed, the analysis of an 'imperfect' frequency sweep may yield information on frequencies higher than the nominal maximum due to discontinuities in the input. Additionally, computer generated sweeps do not account for the effects caused by the mechanical flying controls and the man-machine interface such as biomechanical feedback caused by vibration through the pilot's limbs. It is vital that such effects are identified so that appropriate 'notch' filters can be included in the flight control circuit if necessary. Additionally, sweep data that has not been generated by a pilot is rarely acceptable for handling qualities studies or demonstration of compliance. Finally, a pilot is able to adapt his control inputs in an intelligent way to account for drift from the trim condition.

5.6.3 Data analysis and specification compliance

Having described a suitable method for gathering frequency response data in-flight, precisely how such information is used for checking specification compliance now

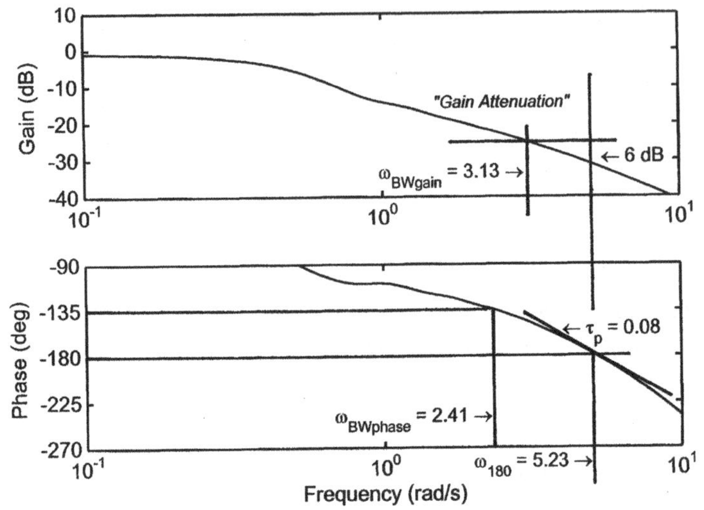

Fig. 5.24 Frequency response of a phase limited system.

needs to be considered. Since most of the requirements concern the manner in which phase lag varies with frequency, a start will be made there. Typical of all helicopters is the tendency for the output (aircraft attitude) to lag the input (control deflection) by larger amounts as the input frequency is increased. The requirement on the pilot to apply larger amounts of lead to overcome this increasing phase lag, or shift, leads to an increase in workload. Ultimately at high input frequencies, the aircraft response will reach 180° out of phase (ω_{180}) and will be *neutrally stable* [5.4] with control deflections required in the same direction as any disturbances in order to counter them. Although it is possible in such circumstances for an experienced pilot to maintain control, there will almost certainly be excessive workload that will undoubtedly prejudice mission effectiveness. For adequate handling qualities, it is necessary to avoid this situation by ensuring that the *bandwidth* of the aircraft exceeds that required for the mission. When gathering data for frequency response analysis, as indicated above, the rotorcraft is essentially operated in an *open-loop* manner that is quite different to the *closed-loop* nature of all *high-gain mission tasks.* The definition of phase-limited bandwidth considers this fact by specifying a 45° phase margin that allows for the neuro-muscular lag associated with a pilot operating with full attention but less than maximum effort [5.5]. Thus, the *phase-limited bandwidth* ($\omega_{BWphase}$) is simply defined as the frequency at which the phase lag reaches 135° (2.41 rad/s in Fig. 5.24).

In researching the use of bandwidth as a general criterion, it was discovered that pilots were also sensitive to the rate at which phase changed in the region of the bandwidth frequency and the point of neutral stability. Commensurate with the simple definition of bandwidth the phase rate, or *phase delay,* uses a two-point approximation of the phase curve between ω_{180} and $2\omega_{180}$, thereby assuming a linear roll-off in phase throughout this critical region. Thus from Fig. 5.24:

$$\tau_p = \frac{\Delta\Phi 2\omega_{180}}{57.3(2\omega_{180})} = \frac{48}{57.3 \times 10.46} = 0.08\,\text{s}$$

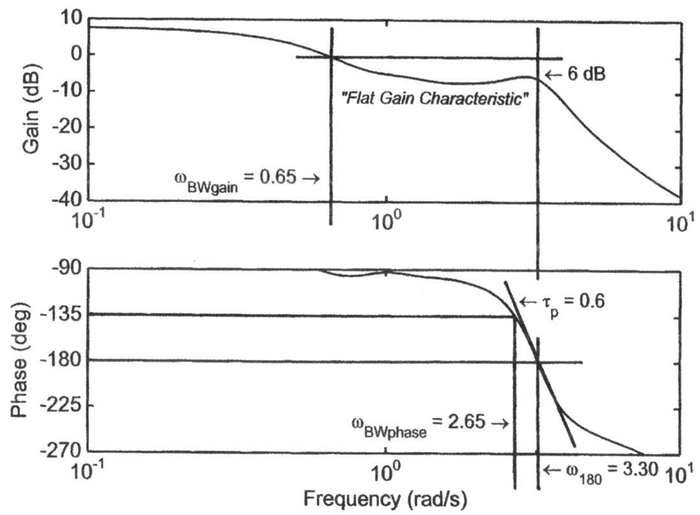

Fig. 5.25 Frequency response of a gain limited system.

When operating unaugmented helicopters or ones fitted with a rate command system, pilot inputs are required to provide attitude maintenance, especially in gusty conditions. This is different to attitude command/attitude hold (ACAH) systems, where the pilot can reduce his own personal gain and rely on the inherent attitude-keeping features of the rotorcraft. Consequently, research using rate or rate command/attitude hold (RCAH) response types highlighted the need for a further definition of bandwidth as it was found that the variation in gain with frequency had a significant impact on the handling qualities ratings awarded [5.5]. Handling problems were found when pilots attempted to operate helicopters with high phase delay and flat gain characteristics in the region of the phase limited bandwidth. When required to fly more aggressively or with greater precision, pilots have a natural tendency to increase both the size and frequency of their inputs as they attempt to 'tighten' control of the helicopter. This tendency is more evident if the phase delay is large since the helicopter will suddenly appear to be 'getting away' from the pilot as the bandwidth frequency is approached. A further complication arises if the gain characteristics are flat rather than being attenuated. With gain attenuation, the tendency of the pilot to increase input magnitude is mitigated by a reducing output/input ratio, see Fig. 5.24. However, if the gain characteristic is flat then as the pilot makes larger inputs he gets an unchanged or even a larger output which coupled with a large phase lag can result in PIO, see Fig. 5.25.

The concept of *gain-limited bandwidth* ($\omega_{BW_{gain}}$) was developed as a means of avoiding this problem. Since pilots can act as amplifiers and are capable of doubling their gain, $\omega_{BW_{gain}}$ is determined by applying a gain margin of 6 dB ($20\log_{10}(2) = 6$) to the open loop data as shown in Figs 5.24 and 5.25. Note that with flat gain characteristics and high phase delay $\omega_{BW_{gain}}$ is typically much lower than $\omega_{BW_{phase}}$ and it is a simple matter to use the lower of the two bandwidths when assessing the specification compliance of unaugmented or RCAH response types [5.2].

5.7 MISSION TASK METHODS

Test personnel have a large range of tools to document and analyze aircraft characteristics but it should always be borne in mind that it is the performance of the aircraft in the mission that is the most important part of any test programme. It is only by using a test pilot with recent experience of the role and by flying representative tasks that all deficiencies can be identified. This requires the test programme to include sufficient role tasks to be flown in an environment which is as close as possible to the conditions that will be found in service. Good test teams have always understood this concept and in recent years attempts have been made to formalize the process.

5.7.1 Handling qualities rating scales

One of the main functions of the test pilot is to report his or her opinion of the handling qualities of the aircraft after a flight. It is this opinion which will often lead to major decisions being made about a programme and, being an opinion, it is of course entirely subjective. As flight test became a more disciplined science in the period following World War Two it became apparent that some way had to be found to standardize the way in which pilots reported their qualitative results. This was needed so that comparisons could be made between the opinions of different pilots. A number of different approaches were made to try to solve this problem such as the Cornell Aeronautical Laboratory and original Cooper scales. Problems were found with these early attempts and in 1966 Messrs Cooper and Harper presented a joint paper [5.7] to the Flight Mechanics panel of AGARD in which they advocated a ten-point handling qualities rating scale. This Cooper–Harper HQR scale, as it became known, has found wide acceptance in the flight test community and is now the standard way for test pilots to report their opinions of handling qualities. Although the HQR scale is the most widely used type of scale in test flying it is not the only one. Other scales have arisen to quantify a variety of subjects such as workload (the RAE Bedford workload scale); and precise manoeuvres (the A&AEE deck landing scale).

The Cooper–Harper scale is shown in Fig. 5.26 and although it is a relatively straightforward concept it requires a considerable degree of care and skill to apply correctly. The test pilot is presented with three dichotomous questions which when answered lead to four possible categories of quality. Three of these categories are further divided into sub-categories, each of which is awarded a numerical value leading to a total of ten possible HQR values. In essence the HQR scale quantifies the amount of physical or mental *compensation* that a pilot has to expend in performing a defined task to a defined standard. In this context the term compensation is taken to mean the additional pilot workload required to complete a task, over and above that which would be required when flying an aircraft with excellent qualities. The important phrase here is *additional pilot workload*; to complete any task will require a certain level of workload even with optimum handling qualities. Cooper and Harper make this clear by stating that the total workload is comprised of the workload due to compensation for aircraft deficiencies *plus* the workload due to the task itself. The scale was not intended to be linear and therefore it is vital that the pilot enters the scale at the correct point to answer the questions and does not try to shortcut the process by simply choosing a number. This flow chart style of presentation, as shown

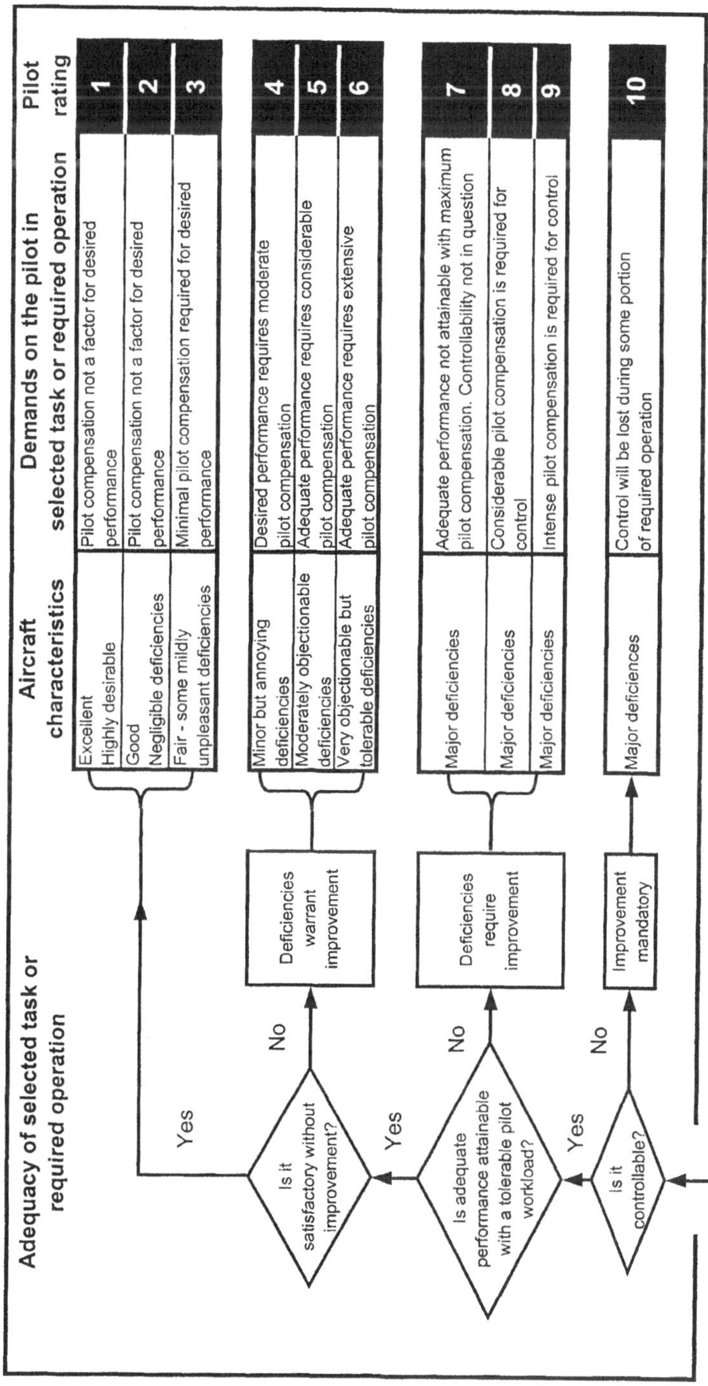

Fig. 5.26 Cooper–Harper HQR decision tree.

in Fig. 5.26, reinforces this point. Correct use of the scale depends on the evaluator understanding the precise meaning of certain key words which will be explained.

As the HQR is awarded for performing a *task* it is essential that the task is chosen with care to be relevant to the role of the aircraft. It must be a flying task, of relatively restricted scope, which the pilot must complete in pursuance of the mission. The choice of task is an important stage in the process of assigning an HQR and needs careful consideration. The task must be a distinct phase of flight that can be allocated a definable level of performance. Examples of tasks could be capturing an angle of bank or maintaining a set aircraft attitude and heading. Where a more complex task is being evaluated it may be divided into sub-tasks each of which can be allocated a rating. The highest numerical rating awarded for a sub-task then defines the rating for the task as a whole.

Particularly important terms are *workload* and *performance*. Performance can be thought of as the degree of accuracy required from the rotorcraft to execute a task, for example, maintaining the aircraft heading within 10° of the initial value. As the accuracy required increases, a corresponding increase in pilot workload will be needed. This workload can be physical, in terms of control activity and/or mental, in terms of concentration. The workload required to perform the task must be considered in the context of the mission and must take into account the amount of spare mental or physical capacity the pilot would have under operational conditions. This is an important point which is sometimes overlooked. By employing an unrealistically high workload pilots can often achieve very high standards of task performance despite poor aircraft handling qualities. It may be useful to artificially reduce the pilot's capacity for aircraft stabilization in order to simulate the demands of the mission.

Since workload will depend on the amount of precision required, it is clearly vital that not only the task itself is defined precisely but also the level of performance should be stated, for example, *maintaining airspeed during a ground controlled approach to within 5 knots*. This task performance is often termed the *tolerance* and is defined for two levels; desired performance and adequate performance. Performing the task to desired performance indicates that the pilot has been able to achieve the accuracy that he or she would wish for, whereas adequate performance means that a level of accuracy has been achieved which, although below the standard the pilot desires, is sufficient to complete the task. Deciding on the tolerances is probably the most difficult aspect of assigning HQRs and must be approached with care. The tolerance must be applicable to the role and the test pilot must be able to justify the resulting level of accuracy. For example, a desired tolerance to maintain airspeed to within 2 knots during transit flight could probably be achieved but there is unlikely to be any operational justification in setting such a high standard of performance. Setting a very wide tolerance that is unrepresentative of the role requirements would be equally incorrect.

The Cooper–Harper scale allows the performance associated with a particular task to be placed into one of four categories:

- *Desired performance*. This is the highest level of performance and is awarded when the *desired* tolerance is achieved *with a satisfactory pilot workload*. No improvement in handling qualities is required therefore (HQR 1–3).
- *Adequate performance*. In this category either the *desired* tolerance is achieved but with an *unsatisfactory* level of pilot workload (HQR 4) or the *adequate* tolerance is achieved with a *tolerable* level of workload (HQR 5 & 6). An improvement in handling qualities is warranted.

- *Inadequate performance*. This third category is defined by the failure to achieve even the *adequate* performance despite the pilot workload being increased to the maximum *tolerable* level in an effort to compensate for handling deficiencies. An improvement in handling qualities is required (HQR 6–9).
- *Loss of control*. The final category allows for the possibility of loss of control during the execution of a task. An improvement in handling qualities is obviously mandatory (HQR 10).

A term which is often used in specification documents such as 00-970 [5.1] and ADS-33E [5.2] is the *handling qualities level*. Handling qualities Level 1 encompasses HQR 1–3, Level 2 takes in HQR 4–6 and Level 3 is for HQR 7–9. There is also a direct relationship between the HQR awarded and the pilot's conclusion on the characteristic under investigation. Thus a *satisfactory* conclusion could not be supported by an HQR 4 or greater. The question of whether or not half ratings can be awarded is often debated. Cooper and Harper themselves did not disallow this in their original paper. In many test establishments, however, the practice is discouraged at least for aircraft release purposes. One common exception is the use of an HQR $4\frac{1}{2}$. This is often permitted to address the large gap between the *desired* performance with *moderate* compensation of an HQR 3 and the *adequate* performance with *considerable* compensation of an HQR 5. Of course HQR $3\frac{1}{2}$ and $6\frac{1}{2}$ are never allowed as they sit on the boundaries of the handling qualities levels.

A special case arises when considering failure modes. When applying a rating to an aircraft's handling qualities following a failure, the likely operational requirement has to be defined. Thus if the aircraft will be required to continue with its mission then an HQR should be applied in the normal way. However, if the aircraft will only be required to return to base and perform a landing then more relaxed tolerances can be applied. It may be that a significant pilot workload results following a failure, for example, landing in manual control after a loss of hydraulic pressure. In this case an HQR 5 might be awarded which would normally be associated with an unsatisfactory conclusion. However, the test pilot might conclude that the aircraft was satisfactory because the high workload might only be required for a short period when full attention can be given to completing the task. In this special circumstance it is normal practice to caveat the conclusion with the phrase, *for degraded operations*.

The handling qualities of the aircraft will clearly affect task performance but the skill of the pilot will also affect the outcome. To take into account the position of the test pilot Cooper and Harper expected that he would evaluate handling qualities with respect to his understanding of the lower degree of skill and training existent in a group of operational pilots. Thus the test pilot must analyze his workload in the context of a less skilled or less experienced pilot. As part of this process he or she must decide on how many attempts at the task will be undertaken before awarding an HQR. It would be unrealistic to award a rating on the first attempt but excessive practise runs can make it more difficult to analyze the compensation required. The compensation is usually most evident in the first few attempts.

HQRs are used in two main ways: either as data for control law research or as data to support conclusions in an evaluation report. An example of the latter is shown in Table 5.3. In research work a pool of pilots may be asked to attempt a task or set of tasks and the HQRs awarded can then be analyzed. This may involve averaging the numerical HQRs to determine the handling qualities level. Since, as already mentioned,

Table 5.3 Relationship between HQRs and other assessment criteria.

<table>
<tr><th>Cooper – Harper Task Performance</th><th>HQR</th><th>Workload</th><th>HQ Level</th><th>Conclusion</th></tr>
<tr><td rowspan="4">Desirable</td><td>1</td><td rowspan="3">not a factor to minimal</td><td rowspan="3">1</td><td rowspan="3">Satisfactory</td></tr>
<tr><td>2</td></tr>
<tr><td>3</td></tr>
<tr><td>4</td><td rowspan="3">moderate to extensive</td><td rowspan="3">2</td><td rowspan="3">Unsatisfactory</td></tr>
<tr><td rowspan="2">Adequate</td><td>5</td></tr>
<tr><td>6</td></tr>
<tr><td rowspan="3">Inadequate</td><td>7</td><td rowspan="3">extensive to intense</td><td rowspan="2">3</td><td rowspan="4">Unacceptable</td></tr>
<tr><td>8</td></tr>
<tr><td>9</td><td rowspan="2"></td></tr>
<tr><td>Loss of control</td><td>10</td><td>uncontrollable</td></tr>
</table>

the Cooper–Harper scale is not a linear measure of compensation, a certain amount of care is needed when conducting any arithmetical operations with the results. For example, an HQR 2 awarded by one pilot cannot be summed with an HQR 6 from another pilot to arrive at an HQR 4. A wide discrepancy in results like this might indicate a problem with the way the research trial had been constructed.

When using HQRs as evidence to support conclusions in evaluation reports it is vital to define the task precisely and to state the performance tolerance used. The test pilot should then use an appropriate adjective to describe the workload and should describe the amount of compensation required in terms of actual control activity and/or mental effort. Finally the rating number is stated. The following is an example of a complete HQR statement used as supporting data for a conclusion: *Maintaining heading ±5 degrees during level flight at 100 KIAS was difficult requiring constant yaw pedal inputs of up to 3 cm, once per second (HQR 5)*. It should be remembered that the pilot should reach his conclusion on the aircraft in the role; in other words the rating awarded should not drive the conclusion. Thus the test pilot should decide on the conclusion using his or her experience of the role and then check the appropriateness of the supporting HQR. If the rating is not appropriate then the task should be repeated or the tolerances reappraised. The frequency of use of HQRs in a report is a matter of judgement. They are an extremely concise and useful means of describing and quantifying the aircraft's handling characteristics, however their use is often restricted to avoid loss of impact and repetitiveness.

5.7.2 Low-speed testing

Since helicopters are typically slower, less reliable and more expensive to operate than aeroplanes, their procurement only makes sense if the mission requires operations in

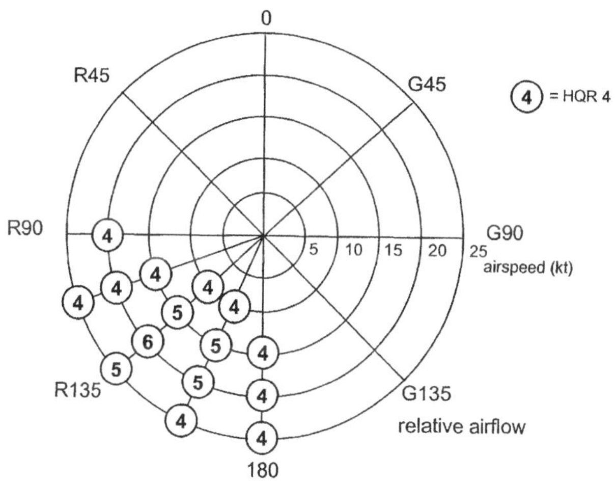

Fig. 5.27 HQRs for out-of-wing hovering. G = green winds (wind from right), R = red winds (wind from left).

the hover and at low speed. Thus, whatever the role of a helicopter, it will need to have satisfactory handling qualities in the low-speed regime. The definition of *low speed* is not consistent throughout the flight test world but it is normally taken to be from zero to 40 or 45 knots airspeed/groundspeed. Any full flight test programme will need to include qualitative tests of the manoeuvres that take place in this low-speed area of flight.

Hovering is clearly one of the most important manoeuvres to evaluate as helicopters may be required to hover for extensive periods and often with considerable accuracy due to the proximity of obstacles. Testing examines the workload to control plan position, heading and height. As well as evaluating hovering in zero wind, tests include operations in winds up to the airspeed limits from all azimuths. Although a pace vehicle can be used to generate relative winds it cannot be used to determine handling qualities in the hover due to the difference in visual cues and in the level of turbulence present. For most purposes it is only necessary to evaluate relative winds from along the longitudinal and lateral axes and the directions at 45° to these axes. In addition to the consideration of pilot workload, other factors such as the vibration level and roll angle at each wind azimuth are evaluated. Not only will these factors affect crew comfort but they may also have implications for on-board systems. A widely deployed missile system which is fitted to a number of battlefield helicopters is a good case in point. The missile will not launch at bank angles above 5° which can be below the aircraft roll angle required to maintain position in the hover with some relative winds. In addition aircraft fitted with winches often have bank angle limitations for operations in the hover. Figure 5.27 shows an example of how to present HQR data for hovering; vibration ratings (VARs) can be presented in a similar way. The stability characteristics of the aircraft and the FCMC will be the major factors that affect hovering characteristics.

Hover turns are made up to the maximum permitted rate starting in calm conditions before moving on to tests in winds up to the lateral and rearwards envelope limits. Tests investigate the ease of controlling yaw rate as well as stopping the turn on selected headings. The accuracy of height and plan position maintenance during the manoeuvre is checked. Torque spikes when initiating and stopping turns with the 'power' pedal are a common problem with responsive governor systems: transient droop may be an equivalent problem with less responsive governors. At high yaw rates some aircraft demonstrate significant cross-coupled responses. Other aircraft have suffered from loss of tail rotor effectiveness due to the tail rotor entering the vortex-ring state during high rate turns which has led to accidents or damage to the transmission when arresting the turn.

There are a number of different manoeuvres which the helicopter pilot can use to leave and return to the hover. These transitions can be longitudinal (normal approach and quickstop); lateral (sidestep); and vertical (bob-up/bob-down and towering transition). In each case the manoeuvre can be flown with different levels of aggression by varying the maximum bank or pitch angle used or varying the time to reach the maximum angle. When flown with high levels of aggression these manoeuvres will be affected by any deficiencies in the field of view; control response; cross-coupling and engine and rotor governing system. Tests of these manoeuvres are conducted in a variety of wind conditions with incrementally increasing levels of aggression up to role-relevant values.

Assessing take-offs is relatively straightforward and is mainly concerned with evaluating the control activity required to go from the stationary, on-surface condition to either the hover or forward flight. Large control inputs may not be a problem in good visual conditions but may have more serious implications at other times such as when using night vision goggles (NVG). The available tail clearance to absorb pitching motions on lift-off is one area that is investigated. For running take-offs a particular point of concern is any tendency of the aircraft to pitch nose down once airborne due to changes in the main rotor downwash on the horizontal stabilizer.

Landings are normally split into vertical and running landings. In both cases the ability of the undercarriage to absorb vertical rates that are appropriate for the role has to be proved. Ministry of Defence Standard 00-970 [5.1] defines a minimum vertical velocity of 2 m/s that the undercarriage must be capable of absorbing during landings on flat, non-moving surfaces. Higher velocities are specified for deck landings that depend on the maximum intended sea state. Vertical landings are essentially just an extension of hovering with the added complication that differing downwash effects can cause uncommanded aircraft disturbances. Assessment is also made of the behaviour of the aircraft when in partial contact with the surface to assess the likelihood of ground resonance occurring. Running landings have more considerations such as tail clearance during the deceleration; heading maintenance during the landing run; tendency of the undercarriage to dig into the surface and the effectiveness of the brakes. For skid equipped aircraft the effects of reducing collective pitch during the ground run are investigated cautiously as it can lead to the helicopter 'nosing over'.

Sloping ground operations are important for tactical rotorcraft where the choice of landing site may be very limited. Particular considerations are the control margins available and any tendency towards dynamic rollover. Dynamic rollover is a phenomenon where beyond a certain angle of bank when in contact with the ground the rotorcraft develops a rolling moment that exceeds the corrective moment that the pilot is able to generate with the cyclic and the collective. This can lead to the aircraft rolling on to its side very rapidly. It is necessary to separate workload due to the relative wind and that due to the slope by performing a landing on the same heading but on level ground. Similarly the effect of stronger winds on control margins can be predicted if the difference in displacement with increasing wind strength is known from pace vehicle tests. Ideally, the tests should be flown on surveyed slopes with a variety of surface conditions.

If required ground taxying is tested on a variety of surfaces to assess turning circles and aircraft stability on the ground. Aircraft with high CGs and relatively narrow undercarriage tracks have been known to roll over even with full lateral cyclic applied when turning through a strong wind. The ease of operation of devices such as wheel locks and steerable nose or tail wheels is also evaluated. Ground taxying tests up to the maximum permitted speed are conducted prior to attempting evaluations of running take-offs and landings.

5.7.3 Pace vehicle operations

5.7.3.1 Safety considerations

Pace vehicle operations are used extensively at test establishments to test helicopters with relative winds from all directions and a variety of strengths. The operations are relatively straightforward but there are a number of safety considerations that are addressed to ensure that all risks are reduced as far as possible. The operations start with a face-to-face brief between the aircraft crew and the vehicle crew. At this brief the area of operations is defined, including the limits of the area in the form of end markers. As the aircraft may well not be pointing in the direction of its ground track it can be difficult for the pilot to know when he or she has reached the end of the area. When the aircraft reaches an end marker the vehicle crew makes the calls, '*approaching end marker*' and '*at end marker*' indicating that the test should be halted. Usually the operations are conducted along a runway as this provides a good surface for the vehicle and an obstacle-free environment for the aircraft. As the aircraft needs to move to different positions around the vehicle when conducting testing it is necessary to reposition the vehicle to different sides of the runway to ensure that the aircraft remains over a clear area. The instructions that will be used to achieve this are briefed. The final call to be stipulated at the brief is the '*break-off*' call which the pace vehicle crew will use if for any reason they require the test to be stopped and the aircraft to move away.

Also at the brief the minimum spacing between the vehicle and the aircraft is stipulated. This is usually a minimum distance of two rotor diameters for safety reasons. In practice the actual spacing used is often greater than this to prevent the helicopter downwash from affecting the vehicle anemometer. The 'golden rule' for safety in these operations is for the pilot to keep the vehicle in clear sight at all times to prevent a collision. This means that the aircraft will not only have to turn to place the relative airflow at the required azimuth but also will have to move to a new position in relation to the vehicle to ensure that it can be seen clearly. When each test run is complete the aircraft moves well away from the operating area to allow the vehicle to reposition safely. Before starting testing, practice recoveries to the hover from high speed rearwards flight are made; this is particularly important if the aircraft shows a marked nose-down tuck on recovery due to the airflow effect on the horizontal stabilizer.

5.7.3.2 *Test technique*

Once the pace vehicle has achieved the groundspeed which gives the required test airspeed, the operator calls '*on condition*'. The pilot then positions the aircraft behind the vehicle and lines up with the roof-mounted wind vane; this gives the zero azimuth relative wind and provides the pilot with the heading from which all other relative wind azimuths are calculated. Once the data has been collected the helicopter is turned to achieve the next required wind azimuth, repositioning as necessary to keep the vehicle in sight, see Fig. 5.28. At each test point the pilot ensures that the aircraft is keeping a good formation position with the vehicle before the data is taken. At speeds at or close to the lateral or rearwards limits of the aircraft care is needed not to exceed these limits when repositioning. The technique used is to turn the aircraft into the relative airflow, reposition as necessary, then turn carefully to achieve the required wind azimuth. Making the calculation of the aircraft heading needed to achieve a particular azimuth can be difficult. To ease this process the co-pilot/FTE will often use a swivelling compass rose mounted on a board that has azimuths marked. The aircraft heading when aligned with the wind vane is set against the zero azimuth mark. The aircraft heading needed to generate any wind azimuth can then be read off directly.

5.7.3.3 *Presentation and interpretation of results*

Results from the pace vehicle tests will show the trimmed flight control positions for all the relative wind velocities tested. Particularly important results will be any indications of controls approaching limits. The apparent static stability of the aircraft can also be documented from these tests together with details of the pitch and roll attitudes adopted and vibrations experienced. The results must be used with care, however, when considering the handling qualities of the aircraft hovering with winds of different velocities. This is due mainly to the very different visual cues between a ground-referenced hover and formation with a vehicle travelling over the ground at speed. In addition stronger winds are always accompanied by some degree of turbulence which is not replicated by the pace vehicle tests. So if testing using a pace vehicle has been used to generate HQR data like that shown in Fig. 5.27 it must be considered as only a indication of a possible *critical azimuth*. This would have to be confirmed by appropriate testing in natural winds.

5.7.4 Ride characteristics

The ride characteristics of a helicopter are those qualities that influence the subjective opinion given by the occupants on comfort and pilot workload in turbulence. The characteristics are divided into two distinct areas; gust response and ride quality.

Gust response is the characteristic that governs the way an aircraft behaves in turbulence from the point of view of the pilot. The two factors of which it comprises are the tendency of the aircraft to be displaced from the set flight condition by gusts and the actions needed by the pilot to regain the condition. The type of rotor system, inertia, stability characteristics, and AFCS will all affect the magnitude of any reaction to gusts. The control response of the helicopter will determine the inputs required by the pilot to counter disturbances. In essence gust response is the change in pilot workload to maintain the aircraft attitude in turbulence.

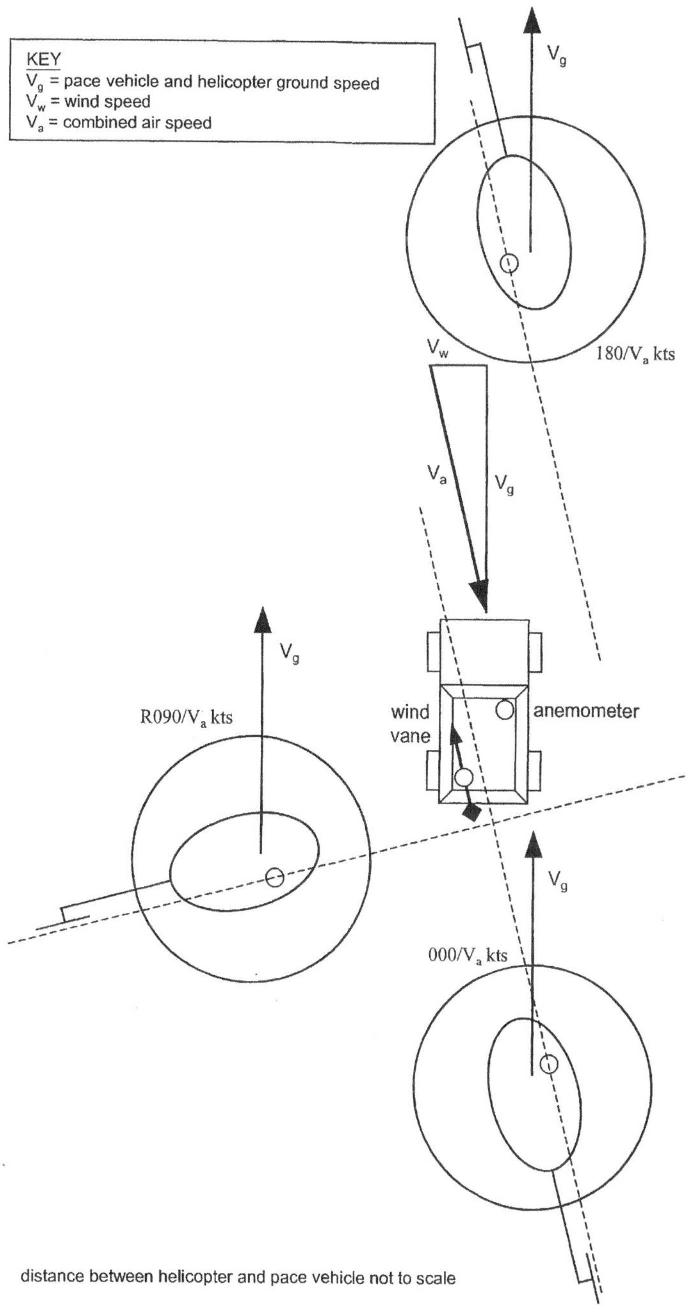

Fig. 5.28 Pace vehicle operations.

Whilst gust response can only be determined by the pilot, ride quality can be determined by any of the occupants. Ride quality is a measure of the comfort or otherwise that the crew and passengers experience in flight. It is clearly closely linked to the same factors that affect gust response (less control response), together with vibration. Ride characteristics as a whole can have a significant effect on the performance of the crew over a protracted period.

5.7.5 Mission task elements

The concept of mission task elements (MTEs) was introduced by ADS-33 as a means of proving that the designed-in characteristics of a rotorcraft actually provide Level 1 handling qualities when conducting role-relevant tasks. The definition of MTEs and of the associated tolerances is the responsibility of the procuring agency, and the choice depends on the intended role of the aircraft. ADS-33E [5.2] contains a number of definitions of MTEs which are designated as *flight test manoeuvres*. Although these all have role relevance to combat rotorcraft, they are specifically designed to facilitate testing against the quantitative (dynamic response) criteria of the Standard. ADS-33E offers precise definitions of all the MTEs together with default task tolerances.

The specification document divides the MTEs into three categories depending on the visual conditions under which the task will be performed. These categories are tasks in a Good Visual Environment (GVE); tasks in a Degraded Visual Environment (DVE); and tasks in Instrument Meteorological Conditions (IMC). Each task is further categorized by the level of agility required by a rotorcraft to accomplish it: these agility levels are Limited, Moderate, Aggressive, and Target Acquisition and Tracking. Guidance is given within the document on which tasks are appropriate for use with aircraft in the Attack, Scout, Utility, or Cargo roles. Some MTEs are also indicated as suitable for externally slung load operations. The MTEs included in ADS-33E are mainly concerned with battlefield helicopter manoeuvres but research is underway into developing MTEs for other roles. For example, a deck-landing task is being developed which requires the pilot to conduct a lateral translation and then to follow a target which moves up and down a pole. This brings out the point that an MTE does not have to be totally realistic in comparison to an operational task, it merely needs to require a similar aircraft response that will uncover any deficiencies in handling qualities.

ADS-33E gives guidance on how the MTEs should be flown and used. Each MTE should be flown by a minimum of three pilots with the HQRs awarded being averaged. Pilots are allowed to practise each manoeuvre as many times as necessary to eliminate any learning curve. The task performance can be judged by either the crew or outside observers; in both cases the test pilot should be advised of the accuracy achieved before the HQR is awarded. The division of responsibility for monitoring task performance between crew members should be defined and briefed prior to the tests. The set MTEs listed in the Standard are intended to be flown at the same level of aggressiveness relative to the capabilities of each rotorcraft. In other words the dimensions of the course will vary depending on the characteristics of the aircraft under test. It may also be necessary to change the course layout to take into account the wind on the test day.

MTEs have been employed effectively during handling qualities assessments of aircraft which predate ADS-33E. In such cases it is unrealistic to expect the aircraft to display level one qualities consistently but by using these tasks it is possible to identify deficiencies quickly. Thus knowledge of the standard MTEs and the ability to devise new ones is an essential skill of test personnel.

Chapter 6
Helicopter Systems

6.1 INTRODUCTION

Before discussing the various methods that are employed in the testing of the major systems found in rotorcraft (see Chapter 7) it is necessary to explore the underlying theory and technologies that are involved. This will not only set the testing in context but will help to indicate whether the test techniques themselves are dependent on, or affected by, the actual system under test. In this chapter three major systems are described with which the pilot has an almost continuous interface through routine flight operations: air data systems; engine control and rotor governing systems; and flight control systems (including AFCS). Although the assessment of the mechanical characteristics of powered flight control systems has already been covered technical details of these systems are covered in this chapter for completeness.

6.2 AIR DATA SYSTEMS

6.2.1 Standard atmospheres

The definition of standard atmospheres dates back to the 1920s in the United States and Europe. They were developed in response to a growing need for standardization in aircraft instrumentation and performance measurement. The US atmosphere was developed by the National Advisory Committee on Aeronautics (NACA) [6.1], while the European atmosphere was produced by the International Commission on Aerial Navigation (ICAN) [6.2]. The slight differences between these two independent standards were reconciled and international uniformity achieved by the adoption in 1952 of a new standard. This *international standard atmosphere* was sponsored by the International Civil Aviation Organization (ICAO) and was defined for altitudes up to 20 km [6.3]. Later work by the US Committee on the Extension of the Standard Atmosphere (COESA), based on data from rockets and satellites, extended this atmosphere to 700 km by 1962 [6.4]. This atmosphere was adopted by ICAO [6.5] as a new standard for altitudes up to 32 km superseding the 1952 ICAO atmosphere.

In 1975 the International Organization for Standardization (ISO) generated a standard atmosphere which covers heights up to 70 km and is based in part on the ICAO standard [6.6]. For heights below 50 km this atmosphere is referred to as the *International Standard Atmosphere* (ISA), while for heights between 50 km and 70 km it is termed the '*Interim Standard Atmosphere*'. The World Meteorological Organization (WMO) Standard Atmosphere, which is defined between −2 km and 32 km, is identical to the ISO standard.

6.2.1.1 *The atmospheric model*

The atmospheric model described below has been widely adopted and forms the background to the International Standard Atmosphere. The atmosphere is assumed to consist of a perfect gas with local values of pressure, density and temperature related by the state equation:

$$P = \rho R T$$

The atmosphere is further assumed to be in static equilibrium with respect to the earth such that the following hydrostatic analysis can be made. Consider a small element within the atmosphere of constant cross-sectional area (δA) and height (δZ) centred at some point. Assume that the pressure acting on the base of the cylinder is P and this changes to $P + \delta P$ at the top. The mass of fluid contained in the cylinder will be equal to the product of its density and volume ($\rho\, \delta A\, \delta Z$) which will generate a downwards force (its weight) equal to mass × the local value of acceleration due to gravity (g). Since the cylinder is in equilibrium:

$$(P + \delta P)\delta A + \rho g\, \delta A\, \delta Z - P\, \delta Z = 0$$

$$\delta P + \rho g\, \delta Z = 0$$

Thus:

$$\frac{\delta P}{\delta Z} = -\rho g \tag{6.1}$$

Equation (6.1) provides a relationship between pressure, density and geometric (or *tape-line*) height. When considering the pressure distribution in the atmosphere it is convenient to use *geopotential* height (H), which is based on a constant value for the acceleration due to gravity. The geopotential height is the geometric height in a uniform gravitational field that gives the same potential energy as exists at the point under consideration in the actual, variable gravity field. Now the work done in raising a body of mass (m) from one geopotential surface to a higher surface is equal to the potential energy that the body has at the higher surface. If the surfaces are separated by a distance Z then:

$$m g_{sl} H = m \int_0^z g\, dZ$$

$$H = \frac{1}{g_{sl}} \int_0^z g\, dZ$$

This equation relates the geopotential height in a uniform gravitational field, in which the acceleration due to gravity is taken as equal to the sea-level value (g_{sl}), to the actual variation of g with Z. In atmospheric modelling and aircraft performance work g_{sl} is taken as the standard value g_0 (9.70665 m/s^2) and the resulting measurement of H is called the *standard geopotential height*. The accepted value of g_0 corresponds to a geographic latitude of 45.5425° [6.7].

A relationship between standard geopotential height and pressure can be obtained by combining the above equations and by assuming that the molecular composition of air is constant over the altitudes of interest. So:

$$\frac{dP}{dZ} = -\rho g = -g\frac{P}{RT} \quad \text{and} \quad g_0 H = \int_0^Z g\,dZ$$

$$g_0\,dH = g\,dZ$$

Hence:

$$\frac{dP}{dH} = -g_0\frac{P}{RT} \tag{6.2}$$

A particular atmospheric model is defined by substituting into Equation (6.2) a given variation of temperature with height and then deducing the corresponding variation of pressure with height. Standard atmospheres consist of layers in which the temperature is either constant or varies linearly with altitude such that:

$$T = T_b + a(H - H_b)$$

where T_b is the temperature at height H_b and a is the temperature gradient, or *lapse rate*, for heights between H_b and H_{b+1}. Thus for the standard atmospheres under consideration, a will either be zero or a constant. Consider a layer where the lapse rate is zero and the temperature is therefore constant at T_b:

$$\int_{P_b}^{P}\frac{dP}{P} = -\frac{g_0}{RT_b}\int_{H_b}^{H} dH$$

$$\ln\frac{P}{P_b} = -\frac{g_0}{RT_b}(H - H_b)$$

Now consider a layer where the lapse rate is constant ($dT/dH = a$) and the pressure is given by Equation (6.2):

$$\int_{P_b}^{P}\frac{dP}{P} = -\frac{g_0}{aR}\int_{T_b}^{T}\frac{dT}{T}$$

$$\ln\frac{P}{P_b} = -\frac{g_0}{aR}\ln\frac{T}{T_b}$$

Therefore:

$$\frac{P}{P_b} = \left(\frac{T}{T_b}\right)^{-g_0/aR}$$

6.2.1.2 *The International Standard Atmosphere*

The first two layers of the International Standard Atmosphere are called the *troposphere* and the *lower stratosphere*. In the troposphere, which extends up to 11 km, the temperature falls linearly with a lapse rate of -0.0065 K/m (close to 2°/1000 ft) from a sea-level value (T_0) of 288.15 K. Above the tropopause, in the lower stratosphere, the air temperature is constant at 216.65 K. With the sea-level ambient pressure (P_0) defined at 101 325 N/m^2 it is possible to generate specific equations relating pressure to geopotential height. Since most helicopters do not routinely operate above 10 000 ft

(3000 m) we need only consider relationships for the troposphere. Since air at 288.15 K has a gas constant equal to 287.05 J/kg/K and:

$$\frac{P}{P_0} = \left(\frac{T}{T_0}\right)^{-g_0/aR}$$

$$P = P_0\left(1 + \frac{a}{T_0}H\right)^{-g_0/aR}$$

Then substituting for the defined constants into this equation gives the following for H:

$$\left.\begin{aligned} P &= 101\,325(1 - 2.2558 \times 10^{-5}H)^{5.2559} \quad H \text{ in metres} \\ P &= 101\,325(1 - 6.8756 \times 10^{-6}H)^{5.2559} \quad H \text{ in feet} \end{aligned}\right\} \qquad (6.3)$$

Note that H can be replaced by H_p (the pressure altitude) as both the sea-level ISA constants and the ISA lapse rate have been used for the formulation of the equations.

6.2.1.3 *Off-standard and design atmospheres*

Although ISA provides an internationally agreed model for the atmosphere it does not take into account actual variations in pressure and temperature due to geographical or seasonal effects. This variability is engineered by means of *off-standard* or *design atmospheres* [6.8]. Within the actual atmosphere the temperature and pressure at any given altitude is highly variable and therefore off-standard atmospheres could have been produced with non-standard variations of both pressure and temperature with geopotential height. This is not in fact done as all published off-standard and design atmospheres assume an ISA variation of pressure with height and simply define differing temperature profiles. The off-standard atmospheres also assume that the lapse rate in the troposphere is the same as ISA (−0.0065 K/m) and simply specify a different sea-level temperature. Thus for example property tables for atmospheres ranging from ISA − 15°C to ISA + 15°C are available. Another approach is the so-called design atmosphere that is intended to replicate more closely the atmospheric conditions pertaining to a particular climatic region of the earth. Four standard design atmospheres have been defined: tropical maximum; temperate and arctic maximum; tropical and temperate minimum; and arctic minimum. These extremes were drawn up on the basis of conditions unlikely to be exceeded more often than one day per year.

The tropical maximum atmosphere has a sea-level temperature of 318.15 K (45°C) and uses the ISA lapse rate. The troposphere is assumed to extend up to just above 13 km (13 077 m). In the temperate and arctic maximum atmosphere the temperature starts at 303.15 K (30°C) and reduces by the ISA lapse rate until 10 769 m is reached. Above this the lower stratosphere with its constant temperature is assumed to start. The tropical and temperate minimum atmosphere features a layer of constant temperature (253.15 K or −20°C) starting at sea-level and ending at 1219 m (4000 ft). Above this a lapse rate of −0.005 2917 K/m is assumed to apply up to 10 667 m (35 000 ft). The arctic minimum atmosphere is the most complex with three layers defined in the troposphere. Between sea-level and 1524 m (5000 ft) the temperature *increases* from 223.15 K (−50°C) at lapse rate of 0.0097425 K/m (approximately 3°/1000 ft). In the next layer ending at 3047 m (10 000 ft) the temperature is assumed to be constant at 238.15 K (−35°C). Above this and up to the tropopause, set at 10 667 m (35 000 ft), a lapse rate of −0.004 5932 K/m (1.4°/1000 ft) applies.

6.2.2 Measurement of air data

6.2.2.1 *Altitude*

As implied above the measurement of altitude requires static pressure information. This is obtained from a port located on the outer skin of an aircraft. The static pressure is converted into *pressure altitude* by the altimeter using a *calibration law* based on Equation (6.3). Altimeters feature a sub-scale marked in milli-bars (mb), or inches of mercury, which the pilot can adjust thereby altering the meaning of the height information presented on the altimeter. Since the *pressure altitude* (H_p) of a point in any atmosphere is defined as the geopotential height in the Standard Atmosphere giving the same pressure, it follows that an altimeter will give a pressure altitude reading provided it has been set to show zero at a pressure equal to P_0 (1013.25 mb). In flight testing references to height are usually in terms of pressure altitude and it is therefore very common to set the altimeter sub-scale to '1013' before commencing data gathering.

An altitude that is rarely displayed but often calculated is *density altitude* (H_D). This is the geopotential height in a standard atmosphere with the same density as that being experienced by the air vehicle. Since the forces acting on a wing, blade or body are a direct function of air density the behaviour of a helicopter will depend on its density altitude.

6.2.2.2 *Airspeed*

Determining the airspeed of an aircraft involves finding the difference between total, stagnation or *pitot* pressure and static pressure. When a helicopter moves through the atmosphere air entering the pitot probe will be brought to rest thus the pressure sensed by this probe will be its stagnation or *pitot* value. If the static pressure is subtracted from this pitot pressure (P_p) then the remaining *dynamic* pressure will be a function of the speed of flight. Thus true airspeed (V), which is a function of this dynamic pressure, is not measured directly but inferred by the action of the pitot-static system.

How the *indicated* airspeed (V_i), that is the velocity information presented on a typical airspeed indicator (ASI), relates to the free-stream velocity or *true* airspeed (V) must now be considered. Air is a compressible fluid whose characteristics can be expressed in the form of the compressible Bernoulli equation which can be manipulated to give a relationship between stagnation pressure, static pressure (assumed to equal the ambient pressure) and true Mach number:

$$P_p = P\left[1 + \frac{\gamma - 1}{2}M^2\right]^{\gamma/\gamma - 1}$$

This equation can be further modified to yield the pressure difference sensed by the ASI. Thus:

$$M^2 = \left(\frac{2}{\gamma - 1}\right)\left\{\left[1 + \left(\frac{P_p - P}{P}\right)\right]^{\gamma - 1/\gamma} - 1\right\}$$

Since $M = V/a$ and a, the local speed of sound, equals $\sqrt{\gamma P/\rho}$, then:

$$V = \left(\frac{2\gamma}{\gamma - 1}\right)^{1/2} \left(\frac{P}{\rho}\right)^{1/2} \left\{\left[1 + \left(\frac{P_p - P}{P}\right)\right]^{\gamma - 1/\gamma} - 1\right\}^{1/2} \qquad (6.4)$$

The *equivalent* airspeed (V_e) is the speed at sea-level, or more precisely the speed obtained assuming sea-level standard density, for which the dynamic pressure is the same as at the actual speed and density conditions. Thus:

$$V_e = \left(\frac{2\gamma}{\gamma - 1}\right)^{1/2} \left(\frac{P}{\rho_0}\right)^{1/2} \left\{\left[1 + \left(\frac{P_p - P}{P}\right)\right]^{\gamma - 1/\gamma} - 1\right\}^{1/2} \qquad (6.5)$$

Comparing Equations (6.4) and (6.5) gives:

$$V = \frac{V_e}{\sqrt{\sigma}}$$

It is standard practice, however, to calibrate airspeed indicators at sea-level. This is achieved by assuming a sea-level standard pressure ($P_0 = 101\,325$ N/m^2) in the calibration. Thus the *calibrated* airspeed (V_c) is given by:

$$V_c = \left(\frac{2\gamma}{\gamma - 1}\right)^{1/2} \left(\frac{P_0}{\rho_0}\right)^{1/2} \left\{\left[1 + \left(\frac{P_p - P}{P_0}\right)\right]^{\gamma - 1/\gamma} - 1\right\}^{1/2}$$

The difference between the calibrated airspeed and the equivalent airspeed is called the *scale altitude correction* (ΔV_c), thus:

$$V = \frac{V_e}{\sqrt{\sigma}} = \frac{1}{\sqrt{\sigma}}(V_c + \Delta V_c) \qquad (6.6)$$

The size of this scale altitude correction depends on the Mach number and altitude of flight. Since most rotorcraft operations take place below 10 000 ft (3000 m) and 200 kts it would be useful to determine whether a scale altitude correction need be applied during rotary wing flight data processing. At 10 000 ft the International Standard Atmosphere gives a static pressure of 69 671 N/m^2 and a temperature of 267.3 K. A true airspeed of 200 kts results in a Mach number of 0.313 and gives a dynamic pressure of 4903 N/m^2. Thus the equivalent airspeed is 171.9 kts and the calibrated airspeed is 172.5 kts. So even for high-altitude high-speed helicopters the scale altitude correction is less than 0.4% and is therefore usually ignored. Thus Equation (6.6) becomes:

$$V = \frac{V_e}{\sqrt{\sigma}} = \frac{V_c}{\sqrt{\sigma}}$$

A distinction now has to be made between the local static pressure (P_∞) of the air through which the rotorcraft is flying and the pressure (P_s) measured via the static vents located on the aircraft skin. Equally if the pitot pressure is measured incorrectly, as a result of sideslip for example, there will be a difference between the true total or pitot pressure (P_p) and that recorded by the pitot probe (P'_p). As a result of these potential error sources, the *indicated airspeed*, V_i, is given by:

$$V_i = \left(\frac{2\gamma}{\gamma - 1}\right)^{1/2} \left(\frac{P_0}{\rho_0}\right)^{1/2} \left\{\left[1 + \left(\frac{P'_p - P_s}{P_0}\right)\right]^{\gamma - 1/\gamma} - 1\right\}^{1/2}$$

6.2.2.3 *Angle of attack and sideslip*

Unlike conventional aeroplanes with wings rigidly attached to the fuselage the main rotor of a helicopter has pitch and roll degrees of freedom as well as the yaw freedom required by its rotation. As the rotor system generates most of the forces required for flight the fuselage attitude is, to a first approximation, simply governed by the balance of local aerodynamic forces and moments. Since stability axes are most often established relative to the fuselage it is evident that the velocity vector of a helicopter is rarely aligned with its longitudinal axis. The measurement of both angle of attack (α) and sideslip (β) is important for stability, control, performance and structural work. Some helicopter automatic flight control systems also use a direct measurement of sideslip as an aid to turn co-ordination.

The simplest form of instrumentation uses lightweight wooden vanes to measure the airflow direction relative to the boom attaching the vanes to the aircraft. Provided the vane is operating outside the wake from the rotor and any interference field associated with the fuselage or any external stores, the vane will align itself with the freestream airflow. The defection of the vane relative to the boom can be picked off electrically and presented to the crew as well as processed by any on-board data storage system. Depending on the orientation of the vane either angle of attack or sideslip can be sensed. Other systems determine α and β from the pitot-static system. A popular system fitted to test helicopters is the *swivelling* pitot-static probe which uses fins, like those fitted to a dart, to align the probe with the freestream flow. If correctly articulated and instrumented the angle of attack and sideslip can be inferred from the angular defection of the probe from the boom. An additional advantage of this device, provided there is no rotor wake interference, is that the true total pressure will be measured by the pitot. An alternative approach uses two fixed static pressure sources. If a static source is located on either side of the fuselage then in the presence of a sideslip the 'into-wind' source will over-read whilst the other may under-read. The subsequent difference in static pressure can be calibrated to give a measurement of sideslip.

6.2.2.4 *Low airspeed*

Low airspeed information is required for two reasons: firstly for test purposes to either ensure that the helicopter is in a true zero-wind hover or to provide accurate low airspeed so that its effect on handling and performance can be investigated; secondly for operational reasons to provide wind speed and direction information for improved weapon delivery. Thus there arises a subtle distinction between a 'hovermeter' and a true low airspeed sensor. A hovermeter is simply a device that accurately indicates the zero airspeed condition whilst giving the pilot a general indication of airspeed away from this condition so that he can bring the aircraft into the hover. A low airspeed sensor on the other hand delivers accurate airspeed information throughout the whole low speed envelope of the helicopter.

Simple hovermeter systems use a cranked boom or vane attachment that is aligned vertically to sense the direction of the downwash from the main rotor. Appropriate calibration is used to determine the vane angles, relative to the vertical, which are associated with a true zero-wind hover. Alternatively if the vanes are located sufficiently far below the rotor it may be satisfactory to assume that a true hover is indicated by each vane being aligned in a purely vertical direction.

Practical low airspeed systems generally use the rotational energy of the main rotor to boost the total pressure measured by a pitot probe. The Pacer system fitted to the AH-64A Apache consists of two pitot probes attached on top of the main rotor. When the helicopter is situated in a true zero-wind hover both probes sense the same dynamic pressure. However when the helicopter is moving or is hovering out-of-wind there will be a sinusoidal time-variation in total pressure and a phase difference between the readings from each probe. This time variation and phasing can be used along with the rotor speed and the mean pitot reading to determine the speed and direction of the airflow relative to the helicopter. If correctly calibrated a system like Pacer can also be used to measure sideslip in forward flight for test purposes and by the AFCS to enhance turn co-ordination. Another system called HADS (Helicopter Air Data System) which is fitted to the AH1 Cobra and the AH-64D also uses the dynamic pressure associated with the rotor wake but in a different manner. The HADS probe is a gimballed pitot-static probe that is located below the main rotor. It works on the principle that the main rotor wake strength and direction at any point below the rotor disk will vary with differing airspeeds and wind directions. Thus if a matrix of data points are coded into the system it is then possible for a given measurement of HADS probe angle and pitot pressure to be used as an indication of airflow relative to the aircraft. As with Pacer suitably calibrated HADS have been used to gather air data for test purposes. (See Section 3.5.6 for details of other systems.)

6.2.2.5 *Temperature*

A thermometer works by the transfer of heat energy from the medium under test to the temperature sensing element (bulb or thermo-couple). The local air temperature measured by an airborne probe will be higher than the ambient static temperature since the air will be necessarily slowed around the sensor. If the flow is completely halted and the temperature sensor is ideally screened then it can be assumed that the stagnation or *pitot* temperature will be measured by the thermometer and the compressible Bernoulli equation can be applied:

$$T_p = T\left(1 + \frac{\gamma - 1}{2} M^2\right)$$

This equation can be re-written to replace Mach number with true airspeed:

$$T_p = T + \frac{\gamma - 1}{2} \frac{V^2}{\gamma R}$$

Assuming air is a perfect gas ($\gamma = 1.4$ and $R = 277$ J/kg K):

$$T_p = T_A + \left(\frac{0.2}{1.4 \times 287}\right) V^2$$

$$T_p = T_A + 0.0005 V^2$$

Currently helicopters rarely exceed 200 kts and typical cruise speeds are closer to 100 kts (approximately 50 m/s). Thus the temperature measured by a perfect pitot-type probe would be no more than 5 K above the true ambient value and would usually be closer to 1 K greater than T_A. Also noting that practical temperature probes do not fully decelerate the flow so that the recorded temperature is less than T_p it is clear why for all practical purposes the temperatures measured by in-service probes are assumed to be equal to the ambient value.

6.3 ENGINE CONTROL AND ROTOR GOVERNING SYSTEMS

Cohen *et al.* [6.9] and Dixon [6.10] cover the basic principles and main design features of aircraft propulsion units. Here the characteristics of propulsion units suitable for use as helicopter powerplants are considered in relation to the particular requirements of this type of aircraft. Power units for shaft driven rotors of conventional configuration are discussed. Engines are usually mounted in or on the fuselage driving the rotor system via some form of gearbox. Blade mounted propulsion systems (tip drive) are not discussed although their existence should be noted. Tip drive systems have been produced in the past (Djinn, Rotodyne) but have not been considered practical for many years.

Power units for helicopters may be grouped into three main categories:

(1) *Reciprocating piston engine system.* The piston engine is now in a highly developed state and is very attractive, especially to manufacturers of small helicopters, mainly because of its cheapness. However, one of the major penalties with the piston engine is the need for a clutch in the transmission chain, to enable the engine to be started without having to turn the transmission. The need for the clutch incurs a weight penalty. There is also a growing trend away from AVGAS as a fuel due to its cost and volatility.

(2) *Fixed shaft turbine engine system.* The fixed shaft turbine engine is used in a helicopter in exactly the same manner as a piston engine. Thus, it is connected to the rotor via a clutch. It may also suffer from significant performance penalties. These are discussed briefly below.

(3) *Free shaft turbine engine system.* In a free shaft turbine engine system, a separate power turbine is included. This turbine is completely divorced from the turbine that drives the compressor. Thus, as there is no mechanical link between the power output turbine and the rest of the engine, there is no requirement for a clutch in the transmission system. The free power turbine may be held by a brake if there is a requirement to maintain the rotor blades fixed during initial start-up.

All gas turbine engines, whether of fixed or free shaft design, have several advantages over reciprocating piston engines. These advantages include:

- *Light engine weight.* To obtain a given power output, the gas turbine engine will normally be much lighter than the corresponding piston engine. This advantage is extremely important since it allows smaller turbine engined helicopters to do the same work as larger piston engined machines, or alternatively, for the same sized helicopter to allow a better performance.

- *Fuel consumption.* The specific fuel consumption of turbine engines approaches that of piston engines, and when run at full power may be even better. Since the power requirements of a turbine powered helicopter will be less, due to the lower engine weight, the total fuel used will be approximately the same. However, since gas turbine fuel is cheaper than piston engine fuel, the gas turbine engined helicopter will be cheaper to run than an equivalent piston engined machine.
- *Reduced vibration.* Since the gas turbine engine is a continuous flow machine the output from it will be uniform and hence vibration levels will be lower. This compares with the piston engined machine where the drive shaft is loaded in sequence by each cylinder of the engine.

Fixed-shaft engines are constant speed powerplants since they are directly coupled to a rotor that must operate at a substantially constant RPM. Power changes (or changes in torque) are therefore obtained only by changes in combustion temperature, the mass flow rate remaining approximately constant. Such an engine is capable of rapid changes in power, which is desirable in a helicopter especially during take-off and landing. However, as the operating speed of the turbine has to be set at a level that can be sustained continuously, there is inherent in the fixed shaft engine a power limitation which prevents the achievement of high power even for short periods. The fixed shaft engine is simpler than the free shaft engine, but the weight saving obtained as a result of this simplicity can be negated by the requirement for a clutch.

The major advantages of the free shaft engine, over one with a fixed shaft, are the elimination of the requirement for a clutch and the freedom to select a wide range of output power. The ability in the free shaft engine to vary both the combustion temperature and the air mass flow allows a wide power range to be achieved whilst the free turbine runs at a sensibly constant speed and therefore the free turbine engine is well suited to multi-engined helicopter applications. There is, however, one main disadvantage of the free turbine engine. Changes in power involve changes in the gas generator speed, including, therefore, variations in compressor speed. The inertia of the gas generator and avoidance of compressor surge will thus prevent power changes from being made as rapidly as would be possible with a fixed shaft engine.

6.3.1 Gas turbine engine and rotor characteristics

6.3.1.1 Rotor requirements

Sensibly constant RRPM is required both for mechanical reasons and for aerodynamic efficiency. If the RRPM is too high, Mach number problems are likely to be encountered on the advancing blade, whereas if the RRPM is too low, the stalled flow region on the retreating blade may become uncontrollable. If we restrict our consideration to the hover case then because we have removed the problems of asymmetry across the disk, the effect of variations in rotor speed on the power and torque required can be estimated quite easily using (see Figure 6.1):

$$C_T = \frac{W}{\rho A V_T^2} \quad \text{and} \quad C_P = \frac{P}{\rho A V_T^3} = C_Q = \frac{Q}{\rho A R V_T^2} = \frac{\kappa}{\sqrt{2}} C_T^{1.5} + \frac{s}{8} C_D$$

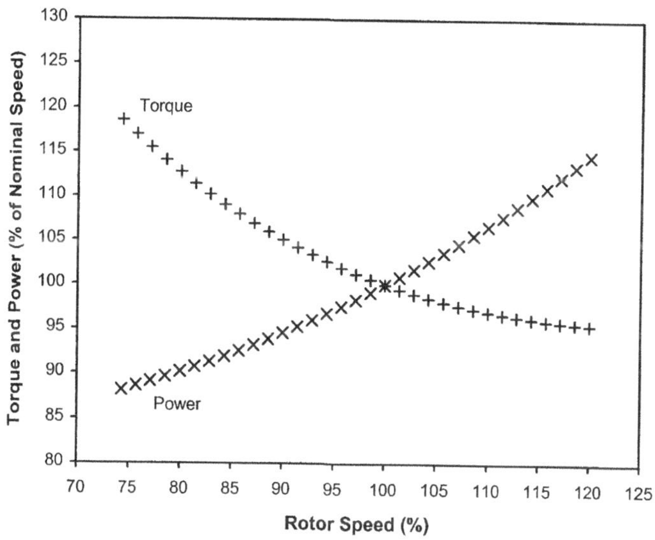

Fig. 6.1 Rotor requirements.

It can be seen from Fig. 6.1 that for a given thrust requirement a decrease in RRPM will have to be offset by an increase in the torque produced by the engine. Consequently matching the torque delivered by the powerplant to that required by the rotor will be simplified if the trend is similar. It should be noted that although a constant rotor speed is desirable, some variation in RRPM is likely particularly when the collective pitch lever is moved quickly. For optimum performance this transient should be as small as possible. However autorotative RRPM at minimum pitch should be as high as is possible to provide satisfactory engine-off landing characteristics.

6.3.1.2 Free turbine engine/rotor characteristics

The free turbine engine can be considered as two separate items: the gas generator and the power turbine. This is because in this type of engine the nozzle guide vanes at the front of the power turbine are always run in a choked condition (that is the flow Mach number equals unity at the throat of the nozzle guide vanes (NGVs)). Consequently the behaviour of the power turbine will not affect the performance of the gas generator. Thus the area of the choked NGVs will fix the position of the gas generator working line and the power turbine is supplied with a flow of high-energy gas that it can absorb at any combination of speed and torque. To understand how the overall engine characteristics are achieved it is necessary to consider the separate characteristics of the gas generator and of the power turbine.

Figure 6.2 shows a typical compressor characteristic curve for a gas generator together with the operating or working line (which as has been seen above is positioned by the area of the power turbine's NGVs). The stall margin is represented by the separation between the surge boundary and the working line. Points A and B indicate the areas where problems can be anticipated. Point A, the high $N_G/\sqrt{\theta}$ cross-over point, can usually by avoided by controlling maximum N_G; but this point can give rise to problems under high altitude, low temperature conditions (low $\sqrt{\theta}$ values). On acceleration the stall margin at point B will be reduced as the acceleration line will

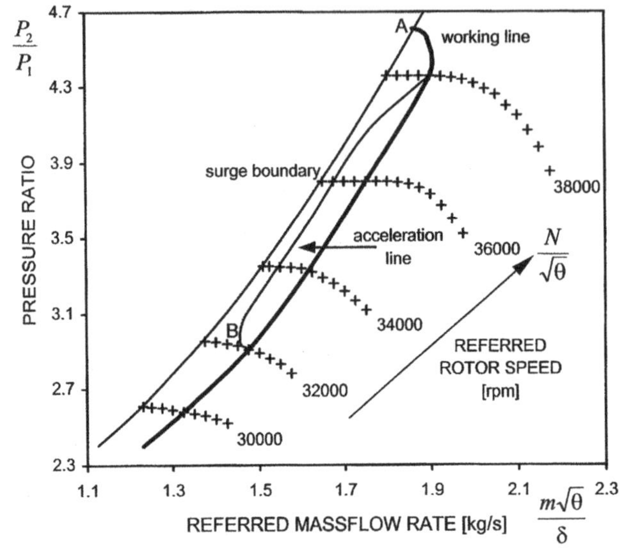

Fig. 6.2 Compressor characteristic.

always be above the steady working line. A situation like that at B usually causes the difficulties associated with rapid engine acceleration requirements. Inlet guide vanes (IGVs) and/or bleed valves are used to increase the stall margin, but there are still likely to be limits set on the maximum rate of acceleration. Rapid increases in the combustion chamber pressure due to overfuelling must be controlled otherwise downstream choking may reduce the compressor air massflow rate and hence cause surge.

In some systems a compressor delivery pressure switch is incorporated that limits the fuel flow if the rate of increase of pressure is too high. The response rate of the gas generator is a most important factor when considering recovery from autorotation and although the gas generator of a free turbine engine is inferior in this respect to a fixed-shaft engine, modern gas generators fitted with anti-stall devices usually perform satisfactorily. Attempts to improve the power-to-weight ratio and the SFC will typically reduce the stall margin further, by moving the operating line closer to the surge line; acceleration thus becomes more critical. Any distortion of the compressor, casing or intake, caused either during manufacture or in flight (erosion or corrosion, FOD ingestion, ice build-up) will also narrow the stall margin.

The torque versus speed (N_F) curves for a typical power turbine are shown Fig. 6.3. The power turbine can operate anywhere in the area bounded by the maximum torque (a transmission limit), the maximum gas generator speed, and the maximum power turbine speed limits. The power produced at a given gas generator speed (N_G) can be absorbed by the power turbine in any combination of torque and speed (N_F) – hence the *constant* N_G lines are in fact lines of constant power. The pilot has two variables under his control: fuel flow via his throttle, which varies the gas generator speed (N_G), and collective pitch that varies the torque demanded of the power turbine. Clearly if he has independent control of these two variables there is a distinct possibility that at some stage he will exceed the engine limits. For example, if he reduces throttle without

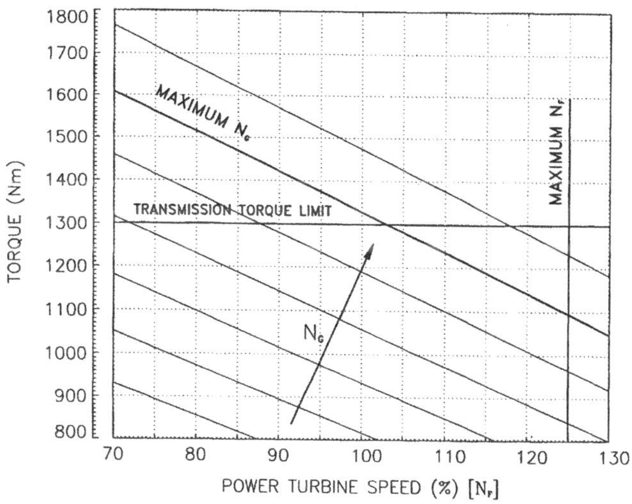

Fig. 6.3 Free power turbine characteristics.

reducing the collective pitch the engine will surge, and it is imperative therefore that the two controls are interconnected so that this cannot happen. Two extremes are possible namely a constant speed power turbine/rotor schedule or a scheme that changes N_F so that at any gas generator speed the power turbine is operating at its most efficient. The acceleration obtained using the constant speeding schedule is usually better as power turbine and rotor inertia are not involved. However, the gas generator must still accelerate and as such is subject to the normal limitations of any jet engine. In early free turbine helicopters the rate at which one could apply collective pitch (effectively the aircraft manoeuvrability) was limited by the acceleration achievable by the gas generator. Modern engines are much improved in this respect.

The characteristics of the rotor and of the free turbine engine have now been considered separately. Combining the characteristics of the two systems for a given gas generator speed (N_G) and a given flight condition gives us torque versus rotor RPM curves similar to those shown in Fig. 6.4. It can be seen from this figure that the free turbine characteristic matches the rotor requirement. If the RRPM falls then the free turbine will automatically compensate with an increase in torque for a given flight condition, such as the hover. The aircraft may well sink and therefore require a further *power* increase – this is not considered in the graph.

6.3.1.3 Fixed turbine engine/rotor characteristics

Figure 6.5 shows the torque versus RRPM characteristics for a fixed turbine engine/rotor combination. As with the free turbine engine the rotor requirement is essentially a constant power line. The engine torque, however, unlike the free turbine engine, does not match the torque requirement of the rotor in that it is of the opposite slope. As discussed above this is because a reduced rotor speed implies a reduced engine speed and consequently there is less gas power for the turbine to convert. A limiting condition, known as the *overpitch* point, occurs when the reducing engine torque output (at maximum power) just equals the increasing torque requirement of the rotor. To the left of this point the system is unstable; as N_R reduces further the rotor requires

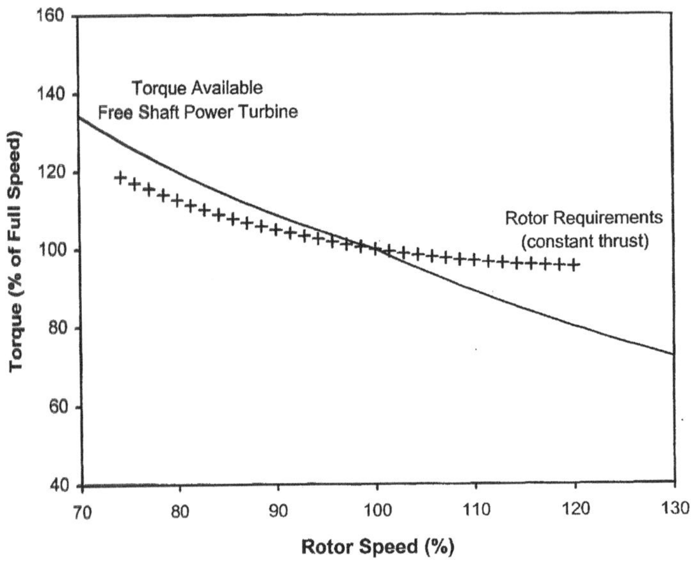

Fig. 6.4 Combined characteristics.

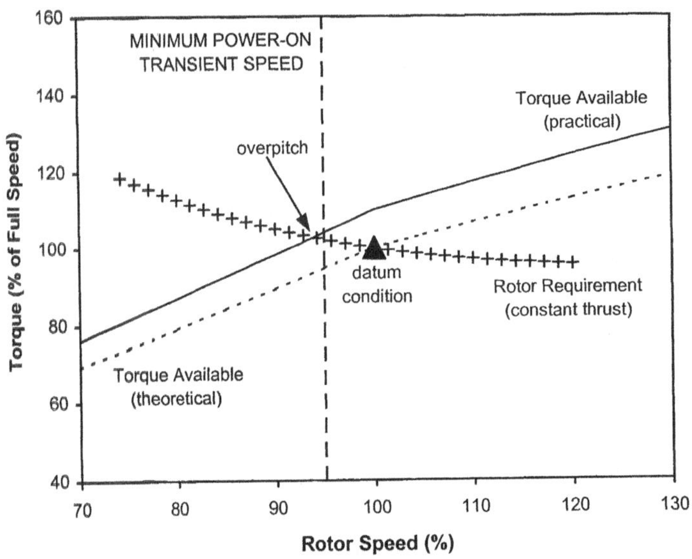

Fig. 6.5 Fixed shaft engine characteristics.

more torque but the engine produces less and less. Hence RRPM rapidly decays with a consequent loss of lift.

Overpitch can easily be induced by demanding too much power – collective pitch – at too low a RRPM. The only way to recover from overpitch is to reduce collective and increase the airflow through the rotor, thereby increasing N_R and bringing it back to the right of the cross-over point. This technique inevitably involves a large loss of height.

On the modern fixed turbine helicopter, RRPM control is achieved automatically by the engine control system. However, control systems are not perfect – if they were, an exactly constant N_R could be maintained independent of any flight manoeuvres, and one could fit an engine that just met the torque requirement at this N_R (dotted line in Fig. 6.5). In practice, allowances must be made for transient droop – there must therefore be a stable range of operating RRPMs to the right of overpitch – and so an engine must be fitted which has a torque output similar to that denoted by the solid line. When comparing this practical engine with the theoretical one it is clear, first of all, that the practical powerplant needs to be a more powerful device. At the same time all the potential power cannot be used, otherwise the transmission system would be overtorqued, and so the engine must be limited or *downrated* in some way. One other consideration is that this practical engine will always be operating at partial powers; that is at much lower temperatures and pressure ratios than originally designed. Hence the thermal efficiency will be reduced and the SFC higher. Briefly then, to achieve a satisfactory and practical RRPM range, the fixed turbine helicopter engine must be downrated to preserve torque limitations. For example, the Alouette 3 helicopter is fitted with a 649 kW (770 SHP) Artouste IIIB engine downrated to 425 kW (570 SHP). One advantage of the downrating is that a power margin exists that gives the aircraft similar performance right up to the altitude at which the engine becomes the limiting factor – normally above 15 000 ft (4500 m) – which is why rotorcraft like the Lama have such good high altitude performance.

6.3.2 Engine control system requirements

The characteristics of a gas turbine engine, particularly its characteristics during acceleration, are such that some form of engine control is essential if protection against surge is to be provided, and if the rotor RPM is to be maintained sensibly constant without creating an unacceptable workload for the pilot. In multi-engine installations it is also essential that the engine characteristics be matched. The engine control system is an integral part of the design of the whole helicopter and should, therefore, be related to its overall aerodynamic characteristics and its role. The major requirements for an engine/control system for a gas turbine engine installation are:

- The governor should maintain a sensibly constant RRPM under all power-on conditions. When required the governor should prevent the free turbine from overspeeding. At the same time the governor should be free from N_F/N_R instability at low collective pitch settings over the whole range of possible helicopter all-up weights.
- The response of the engine and/or free turbine governing system should be quick enough to maintain the RRPM (N_R) within limits during large, rapid power changes.
- The governor should be capable of providing stable control at all times notwithstanding that the engine might be ON or OFF load.
- In multi-engine installations, individual engine governors must be so configured that power matching can be achieved easily and consistently throughout the power range.

- If any part of the governor system is electronic, it must be free from electric-magnetic interference by both internal and external sources.

It may not be possible to meet all these requirements simultaneously. The emphasis placed on the importance of each requirement will depend on the operational roles of the helicopter. The above requirements may be summarized as follows:

- *Steady state function.* The system should provide closed loop rotor governing by altering the fuel flow to the engine to maintain the rotor speed constant or within allowable N_R limits.
- *Transient functions.* The system should, when required, provide protection and set limits on free turbine speed, gas generator speed, temperature or rate of change of temperature and torque. Also the system should control the acceleration of the engine during start-up and transient operation.
- *Control loop opening.* The system should provide for the control loop to be opened under certain conditions.

Part of the control system requirement is commonly achieved using limiters:

- *N_F limiter.* The free turbine speed limiter (or overspeed trip) must be of the highest order of integrity. Following a break in the transmission system under load very high free turbine accelerations are achieved. To prevent a catastrophic turbine disk failure, which could occur within one second of the transmission failure, the fuel supply to the gas generator must be cut off within a very short time (of the order of 0.05 of a second). There is a danger associated with such a limiter, in that it could lead to fuel starvation of the gas generator during a routine transient N_R overswing.
- *N_G limiter.* The gas generator maximum speed must be limited to prevent the compressor RPM exceeding the value which would produce a high $N_G/\surd\theta$ stall or cause excessive compressor blade loading. In some installations a two-position stop is provided, to permit training at lower limiting power levels and thus conserve engine life (Puma, Super Puma).
- *Temperature limiter.* A temperature limiter is required if the combustion chamber and turbine systems are not to be damaged by excessively high combustion chamber temperatures. An example is the Protection and Control Unit (PCU) fitted to the Rolls-Royce Gem engine.
- *Torque limiter.* Originally it was thought necessary to include a torque limiter to ensure that the transmission torque limits were not exceeded. However, such a limiter caused the loss of several helicopters and these limiters are no longer used. Although by mishandling it is possible to overtorque the transmission it is now considered more cost effective to sacrifice it rather than lose the aircraft. Note, however, that the design philosophy behind the transmission system varies between manufacturers. In the West, it is general policy to provide the minimum transmission power to do the task with acceptable margins, thus saving weight and cost, and leave the pilot to do the power limiting. On the other hand, the Russians have typically over-engineered the transmission such that it can cope with the maximum engine power available; thus the power limiter is the collective top stop. Collective pitch limitations are sometimes used as a compromise solution.

6.3.3 Free turbine governor characteristics

6.3.3.1 *Simple governor operation*

Most hydro-mechanical rotor governing systems are of the proportional type first developed for steam engines. When the load on the engine is increased (by the application of collective pitch) the rotor speed and consequently the power turbine speed falls. This reduction is used to signal an increased fuel flow to the gas generator. The resultant increase in power produced by the gas generator tends to restore the power turbine RPM and equilibrium is re-established when the torque produced by the power turbine equals the torque required by the rotor system.

In its crudest form the proportional governor contains bob-weights that are thrown out away from the axis of rotation thus generating a force that is *proportional* to the speed of rotation. This force is then used to close a fuel-metering valve against the action of a spring. Typically the free power turbine drives the bob-weights and their position regulates the amount of fuel fed to the gas generator. Consider a demand for more rotor thrust via an application of collective pitch. Initially the RRPM and hence the power turbine speed (N_F) will fall, reducing the force opposing the spring thereby allowing it to force the valve open and allow a greater fuel flow. This increased fuel flow accelerates the gas generator enabling it to deliver more power and arrest the decay in rotor speed. Eventually when the force from the bob-weights once again matches the spring force, N_R and N_F achieve a new equilibrium condition. At this new equilibrium state a greater fuel flow (FFR) is required to meet the higher power and therefore the fuel valve must be open further than was the case at the lower collective pitch setting. Therefore FFR is inversely proportional to N_F and the higher fuel flow rate is achieved at a reduced N_F and consequently reduced N_R. Clearly the greater the power demand, the further the fuel valve must be opened and the greater the reduction in the rotor RPM. This characteristic, termed *static droop*, leads to a steady reduction in rotor speed with collective pitch or fuel flow.

If this type of closed loop rotor governing system was functioning when the aircraft was on the ground with the rotors stopped, the system would detect a massive RRPM underspeed and demand maximum fuel flow or gas generator speed. The control loop must be broken, therefore, when the rotors are stopped. This may be achieved either electrically (as on the Rolls-Royce Gnome engine) or hydro-mechanically (as on the Rolls-Royce Gem engine). Some systems feature a single lever moving in a gate. Over the lower portion of the gate from ground idle (GI) to flight idle (FI) the lever controls the gas generator only. At FI a microswitch is made which feeds RRPM signals to the control system so closing the loop. From then on, from FI to maximum, the lever changes its function from one of controlling the gas generator to one of selecting the RRPM datum – hence its name a *speed select lever* (SSL). This system works well for a single engine installation. In the case of multi-engined helicopters the acceleration of the engine from GI to FI is often achieved via individual engine condition levers (ECLs), a separate SSL being provided for adjustment of the rotor speed datum.

In some hydro-mechanical systems the fuel flow to the engine passes through two variable area orifices in series; one controlled by the free turbine governor, the other by the gas generator governor and/or throttle. Overall system control is determined, therefore, on a *lowest wins* basis, that is the fuel flow to the gas generator will be determined by which of the orifices has the smaller opening. Alternatively a signal from the proportional governor driven by the free power turbine can be used to change the gas generator datum speed with the gas generator governor controlling the fuel flow to the engine. In a typical two-orifice system at ground idle (GI) with the rotor stopped, there is no bob-weight force on the free turbine governor and under the influence of the governor spring its orifice is wide open, allowing unrestricted fuel flow. However, the gas generator throttle is in its minimum position and the gas generator governor has the minimum spring loading (corresponding to N_G for GI) applied to it. As the ECL is moved, the gas generator throttle opens and at the same time a higher RPM setting is applied to the gas generator governor. The increased fuel flow through the throttle can now accelerate the engine up to the new higher gas generator governor setting. The free turbine governor orifice is still wide open and thus has no controlling effect. At some point in the acceleration process from GI, the rotor brake is released and a RRPM signal is applied to the free turbine governor. By the time the ECL is fully advanced, to a FI gate for example, this RRPM signal will have reached the governed range and a balance achieved with the spring force. The gas generator governor now has the maximum N_G setting applied and since in general the engine speed demanded by the rotor governor will be less than that setting, the N_G governor acts merely as a maximum gas generator speed limiter. In this way the control function is handed over from the gas generator via the ECL to the rotor via the free turbine governor.

6.3.3.2 *Static droop and governor gain*

As mentioned earlier the use of a simple hydro-mechanical proportional governor to control rotor speed leads to reductions in RRPM as the fuel flow increases and vice-versa. This trend manifests itself as static droop, that is as the torque is increased from one equilibrium point to another the rotor will stabilize at lower and lower speeds. The amount of static droop can be expressed as:

$$\text{static droop} = \frac{\text{RRPM at minimum power} - \text{RRPM at maximum power}}{\text{RRPM at minimum power}} \times 100\%$$

The change in fuel flow rate with RRPM (the static droop law) can be portrayed in a variety of ways and two typical plots are shown in Fig. 6.6. Although static droops as high as 10% have been used it should be remembered that for optimum rotor performance no change in RRPM, or *isochronous* governing, is desired. The problems associated with trying to achieve such governing by reducing the static droop are discussed below.

Governor gain is related directly to the amount of static droop. The gain is the change in fuel flow rate per unit change in RRPM. Thus a system with low governor gain will generate a high static droop and vice-versa. Consider now the effect of high governor gain, which is analogous to fitting a weak spring in a proportional governor. Although under static conditions the difference in RRPM between low and high torque settings will be low, the large changes in FFR that occur for small changes in RRPM can lead to an oscillatory and possibly unstable response. Figure 6.7 shows departures from the static droop law. Note that the oscillatory behaviour commonly associated with high gain is quite evident. (Lower chart in Figure 6.7.)

Selecting the best value for governor gain is a compromise between desired levels of stability and the need to remain within rotor limits during rapid power changes. However, the expected maximum rate of application of collective and available engine response must also be considered. Suppose the pilot moves the collective at a moderate

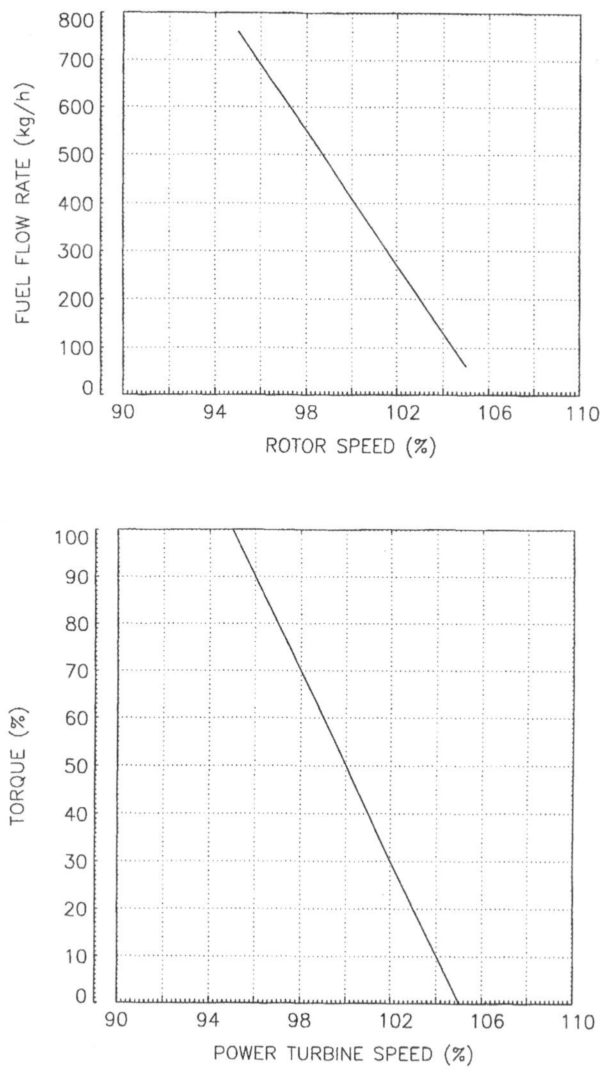

Fig. 6.6 Static droop.

rate and the engine is capable of rapid acceleration then it is possible for the control system to produce a response with no transient. If on the other hand the pilot makes a very rapid demand a transient will result if the demand exceeds the acceleration capabilities of the fuel control system.

6.3.3.3 Droop reduction or cancellation

Suppose that for a particular rotor/engine/governor combination it is impossible to satisfy the requirement for adequate stability and respect power-on rotor RPM limits. In this situation a droop reducer must be fitted, the action of which is best described with reference to Fig. 6.8.

The upper chart shows the extent of the problem, before droop reduction is

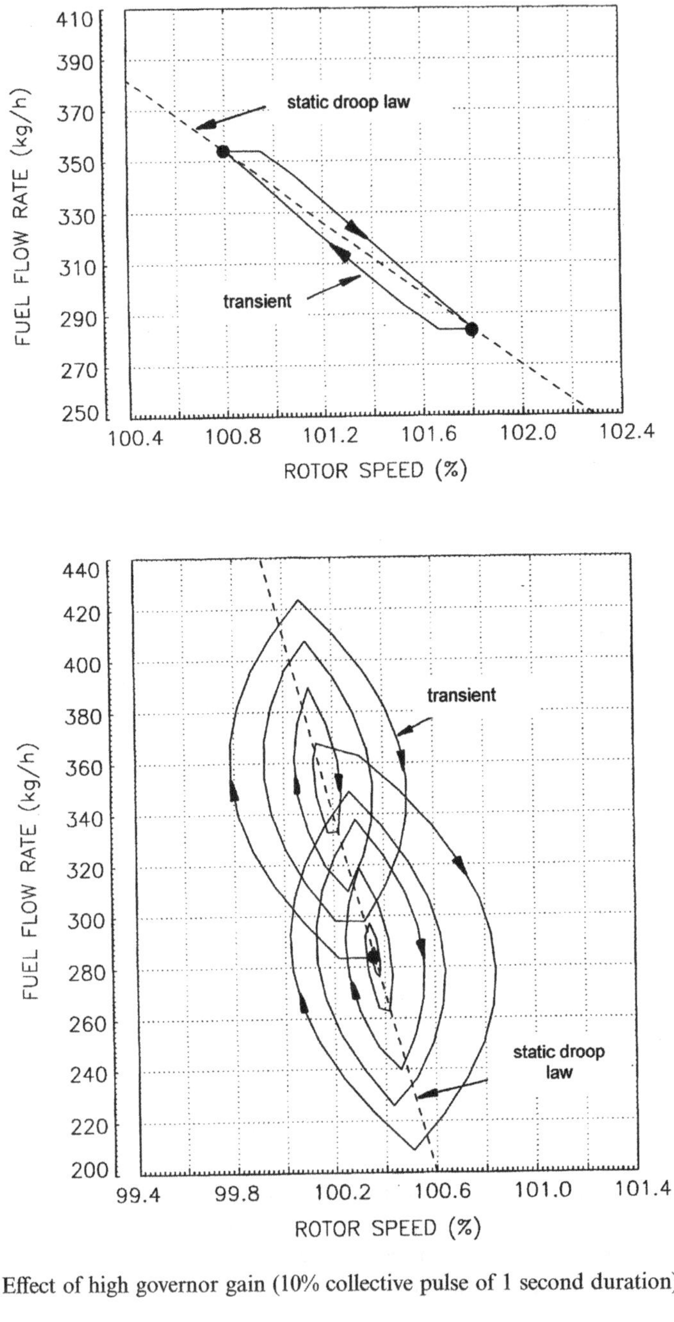

Fig. 6.7 Effect of high governor gain (10% collective pulse of 1 second duration).

employed (see the '*no compensation*' line indicated by open circles). If the amount of static droop shown (40 RPM) is required to ensure the governor has adequate stability it can be seen that at both high and low power demands (fuel flow rates) the rotor speed will pass outside the power-on limits. Since the pilot uses the collective lever to make power adjustments it is possible to change the rotor speed datum as the lever is moved. The lower graph shows a typical relationship. The combined effect of the basic

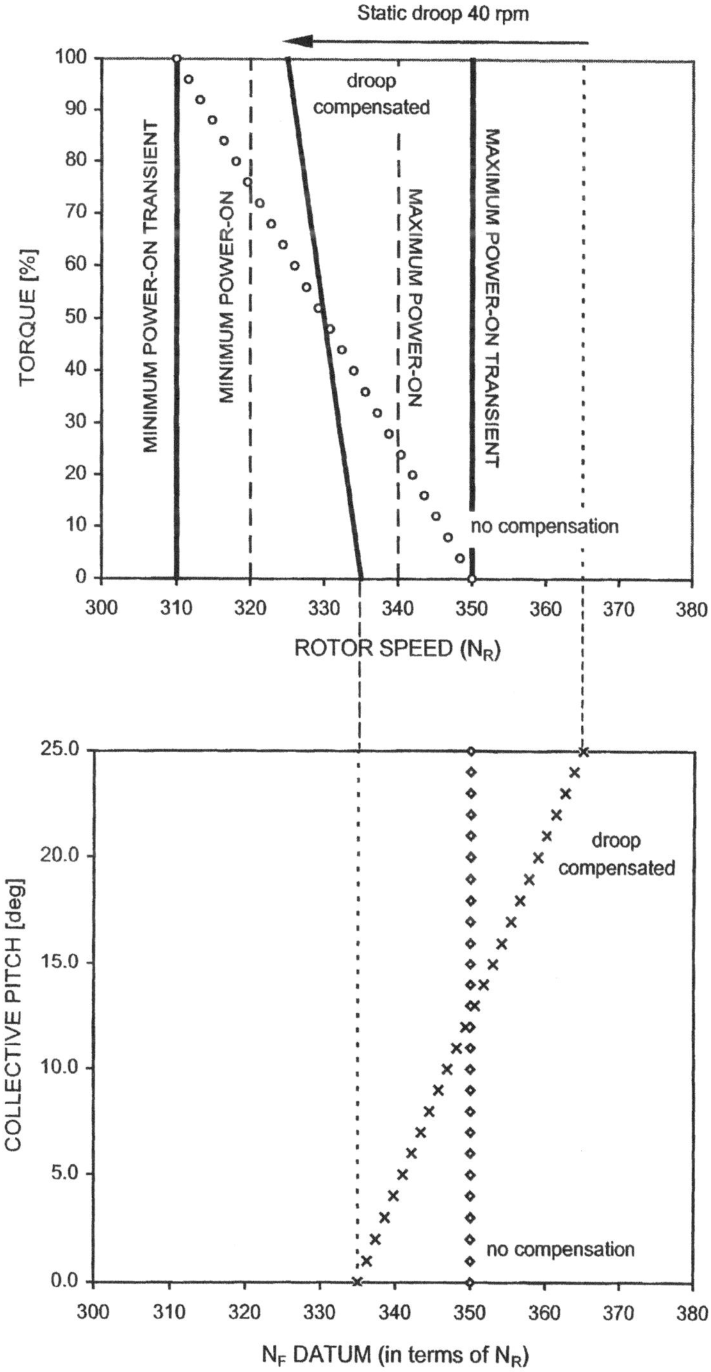

Fig. 6.8 Droop compensation.

static droop law (dependent on spring constant and governor mass) and the rotor speed datum schedule (dependent on the amount of spring pre-load) is shown in the upper chart (see the solid '*droop compensated*' line). Now the rotor RPM stays within the steady power-on limits at all times. Although it might appear that governor gain has been increased and therefore its stability reduced in fact the governor stability is dependent on the basic droop law and is unaffected.

The effect of using droop reduction, or cancellation, on the rotor response can be described relatively easily. Consider the case when the pilot makes a pulse demand or a sizeable gust strikes the rotor. The behaviour of the rotor in this situation is in broad terms a function of the basic droop law and will be little different from the case when no droop reduction was applied. The full effect of droop reduction is seen more clearly when large and rapid power demands are made. When the pilot makes a large collective pitch demand he will cause the datum N_F to increase as he raises the lever. Thus the fuel flow to the engine is increased *as* the power demand is being made rather than as a consequence of the governor sensing a reduction in rotor speed. Therefore droop cancellation will reduce the size of any transients as well as the difference between the stable rotor speed at maximum and minimum torque or collective pitch. The manner in which the transient behaviour of a rotor system is assessed and documented is described later. (Section 6.3.5 and 7.4.6.)

6.3.4 Fixed turbine control systems

Clearly to meet the RRPM requirements a fixed turbine engine should operate on a constant speeding schedule. This confers the benefit of very fast acceleration times, since there are no inertia problems and massive overfuelling can be tolerated since the low power operating point is typically far from the surge boundary. The basic requirements for the engine control system have already been discussed and, in general, these requirements can be met more simply in a fixed turbine engine. As before the pilot changes the blade pitch by means of the collective lever but in this case, however, a small closed loop system acts to keep RRPM *and* engine speed constant. The Astazou 3N2 system (as fitted in the Gazelle helicopter) is a good example of such a system. In the governed range constant pressure is maintained across a variable metering valve, so that the fuel flow to the engine depends only on the position of this metering valve. The valve position is adjusted by a servo system controlled by a pilot valve that senses rotor speed. On some systems precise control and adjustment of RRPM may be vested in a speed select lever.

6.3.4.1 Mode of operation

If it is assumed that the engine is running in equilibrium when a collective pitch increase is demanded then it is clear that N_R must fall initially. The pilot valve will therefore open and cause the metering valve to increase the fuel supply to the engine. The engine will then accelerate until the datum RRPM is restored. As it does so the pilot valve will gradually approach its null position at which point the metering valve is locked at its new more open setting.

Opening the control loop on the ground is again achieved by using a 'lowest wins' system similar in concept to that found on some free turbine engines (see Section 6.3.3.1). At ground idle the rotor speed will be very low (or zero) so the pilot valve and metering valve will be fully open. Under these conditions control of the engine is vested solely in a manually operated fuel valve (throttle). As this valve is opened to accelerate the engine to flight idle, the RRPM will increase until, with the throttle fully open, the pilot valve is nulled and the metering valve is governing the fuel flow. It is interesting to note that with such a system:

- There is no acceleration control fitted since it is not required in the flight range due to the constant speeding nature of the engine. Consequently the manual fuel valve or throttle must be handled very gently during acceleration from ground idle to flight idle.
- There is no static droop since the pilot valve always returns to the same null position. At equilibrium the rotor speed always balances the same spring force irrespective of the metering valve position.
- There will be some transient droop since a RRPM error is required before the pilot valve can move to adjust the metering valve. On some systems the transient droop may be so small that it is not discernible by the pilot. If so it is possible that torque spikes will result as the power output from the engine responds rapidly to the change in rotor speed.
- Since there is no static droop, each engine in a twin engine configuration can be delivering widely different powers and still be running at a common speed. Hence power matching is not usually feasible without the aid of some artificial stability system.

6.3.5 Transient droop and overswing

Transient droop or overswing is the variation in RRPM that occurs during changes in power initiated by a movement of the collective lever. Usually during an increasing torque demand there will be a rotor underspeed and during a power reduction the rotor will overspeed. Changes in N_R occur because, in general, engine acceleration or deceleration cannot match the rates of change of collective pitch. Therefore during, say, power increases there will be a finite delay between the power demand (changed collective position) and the required power being produced (the gas generator reaching a new speed). During this delay the RRPM will fall to a figure which is below the steady state static droop for the new power. This extra reduction in RRPM is defined as the *transient droop*, see Fig. 6.9.

The delay is aggravated by the fact that a rotor speed error must develop before the gas generator will begin to accelerate. Eventually, just as with static droop, the transient droop may be large enough to put the RRPM outside its aerodynamic or mechanical limits. Without further modification of the control system, the only way of overcoming this is to limit the rate of collective pitch movement. This means in practice that the manoeuvrability of the helicopter is limited, and indeed some early machines were affected in this way. Other factors which affect the size of the overswing or droop are:

- governor gain;
- inertia of the main rotor;
- helicopter AUM;
- density altitude;
- acceleration and deceleration characteristics of the engine.

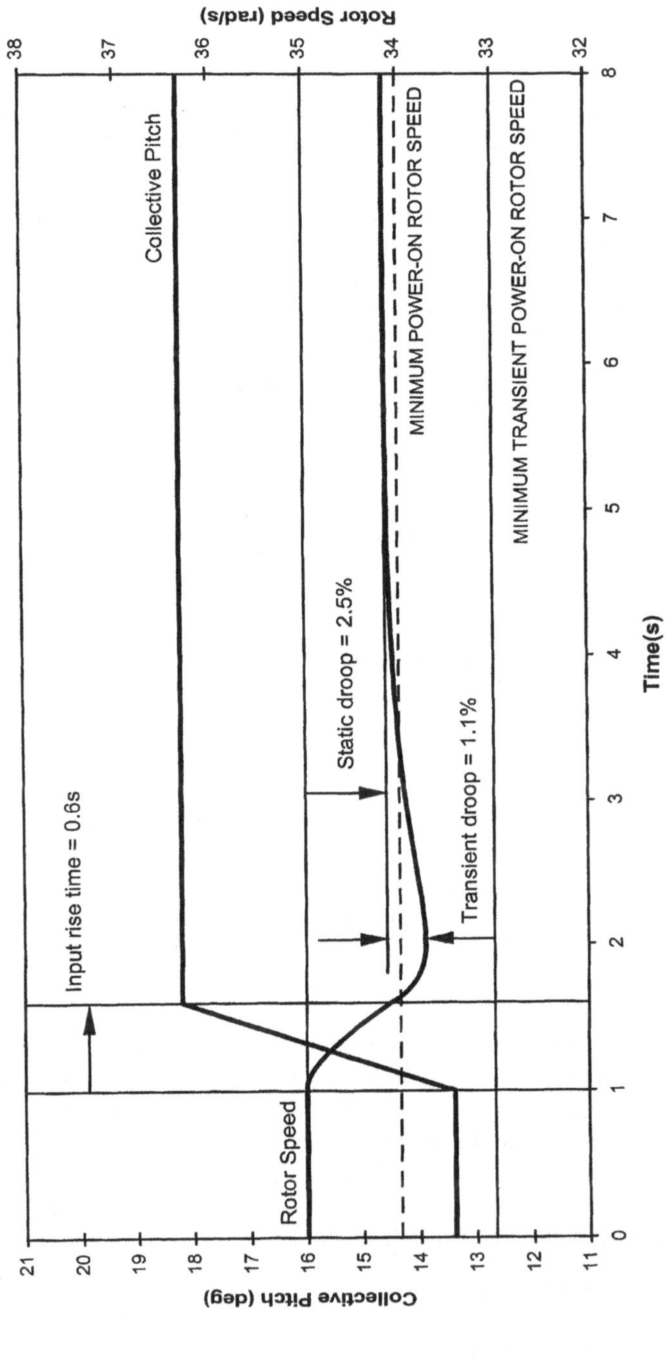

Fig. 6.9 Transient droop.

Consequently rapid acceleration is a basic design requirement of modern helicopter engines. Two factors help to make rapid acceleration possible: low rotational inertia and high idle speeds (typically flight idle = 70% of maximum N_G), which helps the engine tolerate a large degree of overfuelling. Alternatively, as discussed, a droop anticipator or compensator can be fitted which has the added benefit of reducing transient droop since it will start the gas generator accelerating without having to wait for a RRPM error. A further refinement is the generation of a signal proportional to the rate of change of collective. This signal is then used to open or close the gas generator throttle, depending on whether the collective is increasing or decreasing, so that the gas generator RPM starts to alter as soon as the collective is moved. Some fuel control systems are also configured with yaw pedal anticipator inputs.

6.3.6 Variations in the static droop law

Whilst of obvious benefit to aircraft operation the installation of collective anticipation introduces a complication in the documentation of static droop as the datum RRPM is now dependent on collective position. However, the power required by the main rotor, although similarly dependent on collective pitch, is a function of airspeed, vertical speed and yaw pedal deflection/sideslip. Therefore it is possible for the stable RRPM to change with airspeed or ROC even though the collective position is unchanged. This phenomenon is best explained by example. Consider the variation of power required with airspeed for a typical conventional helicopter (Fig. 6.10). The 'power bucket' can be clearly seen as can the fact that at high forward speed the hover

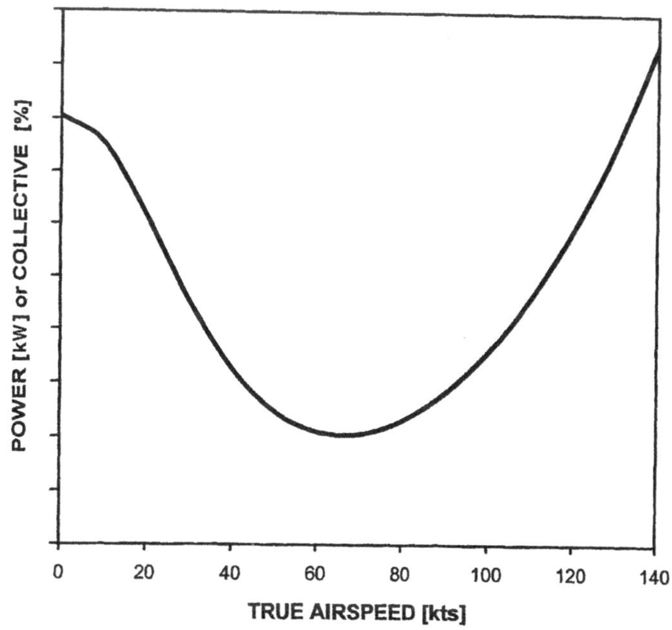

Fig. 6.10 Variation of power with airspeed.

power can be exceeded. The variation of power can be translated into the collective position required for level flight. So even in level flight it is clear that the rotor speed will vary due to changes in the N_R datum as the collective lever is raised and lowered.

6.3.6.1 *Flight at constant torque*

Flight at constant torque and in particular the collective position required to balance the forces and moments generated will give a clue to the rotor speed variation at differing airspeeds. If the helicopter is to fly at constant power, or torque, as airspeed is varied from the hover then it must adopt a rate of climb when the power for level flight at a given airspeed is less than hover power and a rate of descent when it is greater. The collective pitch required, for constant power, will depend on the thrust required to balance the weight, the vertical and horizontal drag and the degree to which compressibility and blade stall have affected the aerodynamic performance of the rotor. In general as the airspeed increases the drag penalty increases thus increased thrust is required to maintain the force balance. Note that although above minimum power speed the rate of climb required for constant power will be less; the increase in horizontal drag with airspeed more than makes up for the reduction in vertical drag and a net increase in thrust is usually required. Provided the rotor aerodynamic performance is not degraded with airspeed it can be assumed that the variation in collective lever position will match the thrust requirement. Flight test data, Fig. 6.11, can be seen to support these theoretical results.

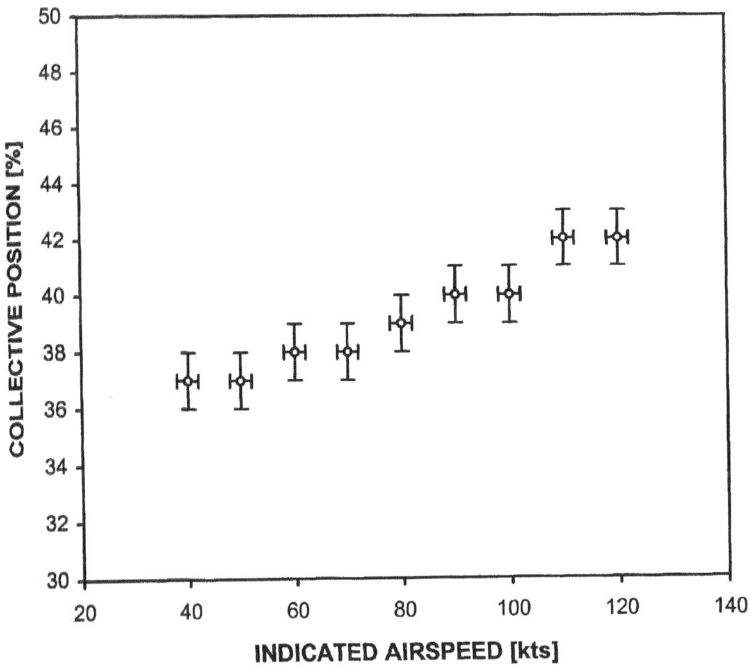

Fig. 6.11 Collective level position for constant torque – test data.

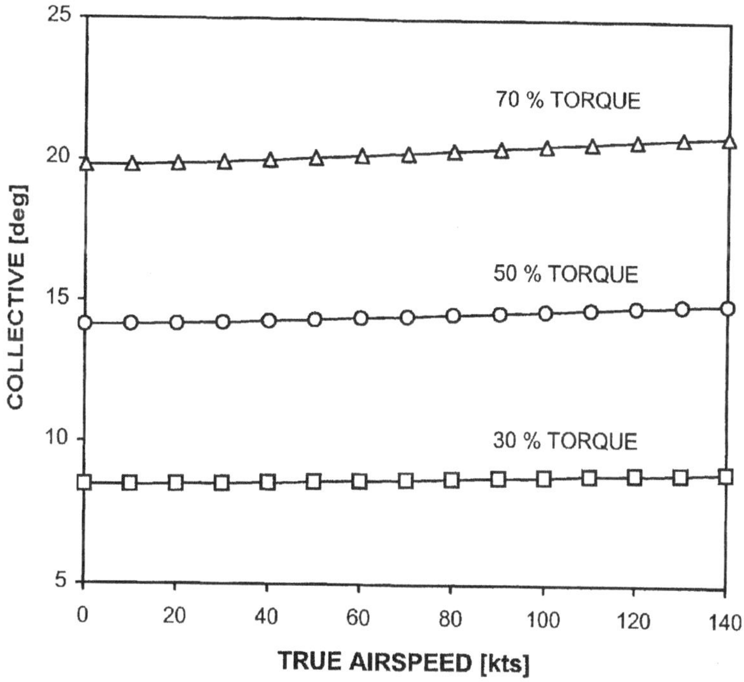

Fig. 6.12 Collective level position for constant torques.

6.3.6.2 Variation of static droop with airspeed

Having seen how the collective lever position varies with airspeed in level flight and with airspeed at constant torque it is now possible to surmise the effect of collective anticipation on the static droop recorded during climbs and descents at a range of fixed airspeeds. First the variation of collective position with airspeed, for constant torque, can be expanded for a range of torque values, see Fig. 6.12.

Now consider the basic function of collective anticipation; as the collective lever is raised the N_F datum is increased in order to compensate for static droop. Suppose that for the example helicopter the standard rotor speed is 330 RPM and that with the lever fully lowered (0%) 0° of collective pitch is applied. Now assume that with the lever fully raised (100%) a collective pitch of 30° is set at the rotor head and that the basic static droop law causes a reduction of 20 RPM between 0% torque and 100% torque. Likewise assume that the engine control system designer intended to eliminate static droop by arranging that the N_F datum be set to a rotor speed of 330 RPM with the collective lever fully lowered. As the lever is raised the N_F datum is increased linearly to a maximum value equivalent to a rotor speed of 350 RPM. With these simple assumptions in mind it is now possible to determine the actual effects of the collective anticipator in flight. In the hover, 30% torque requires a collective pitch of 7.5° (a lever position of 27.2%). This will signal a N_F datum equivalent to 335.7 RPM. The basic static droop law will result in a reduction of 6 RPM at 30% torque and so the rotor speed with collective anticipation will be 329.7 RPM. Similar calculations for 50% and 70% torque can be made; these are summarized in Table 6.1.

If the process is repeated for 140 KTAS different results arise due to the slightly

Table 6.1 Variation of rotor speed with torque – hover.

Torque (%)	Collective pitch (°)	Collective position (%)	N_F datum (RPM)	Basic static droop (RPM)	Resulting rotor speed (RPM)
30	8.5	28.3	335.7	6	329.7
50	14.1	47.1	339.4	10	329.4
70	19.8	66.0	343.2	14	329.2

Table 6.2 Variation of rotor speed with torque – 140 KTAS.

Torque (%)	Collective pitch (°)	Collective position (%)	N_F datum (RPM)	Basic static droop (RPM)	Resulting rotor speed (RPM)
30	9.0	29.9	336.0	6	330.0
50	14.9	49.8	340.0	10	330.0
70	20.9	69.7	343.9	14	329.9

different collective lever positions required for the same three torque values, see Table 6.2. If, therefore, a static droop assessment is conducted on this aircraft in the hover and at high forward speed the results will not be exactly the same due to the influence of the collective anticipator.

6.3.7 Flight idle glide

6.3.7.1 Flight idle setting and rotor rigging

The speed of the engine/gas generator at flight idle (the minimum speed to which it will fall with the collective lever fully lowered) is a compromise between two requirements. It must be high enough to permit good initial acceleration without risk of surge and low enough to avoid a large power contribution (and therefore rotor overspeeding) when minimum collective pitch is selected. Most governors feature an adjustable minimum flow by-pass to prevent the governor cutting off all the fuel to the engine/gas generator in the event of a transient overspeed. The by-pass adjuster is used to vary the shape (and hence gain) of the governor's droop law especially at the low end of the law. This enables the minimum N_G in FIG to be increased to achieve stall-free engine acceleration at the expense of increasing the likelihood of a power contribution.

So far in our study of governor characteristics only the upper part of the droop law has been considered. Attention must now be turned to the lower part of the droop law that represents FIG conditions. Although stability within the control system is still important a further factor must also be taken into account: the ease with which the pilot can set a desired rotor speed at low torque. Some droop laws feature a 'knee', that provides a lower gain at low torque in an attempt to meet this requirement.

6.3.7.2 Power contribution in flight idle glide

In many rotorcraft it is not practical to retard the throttles when practising autorotations or 'forced landings'. Therefore, for realistic training flight idle glides should

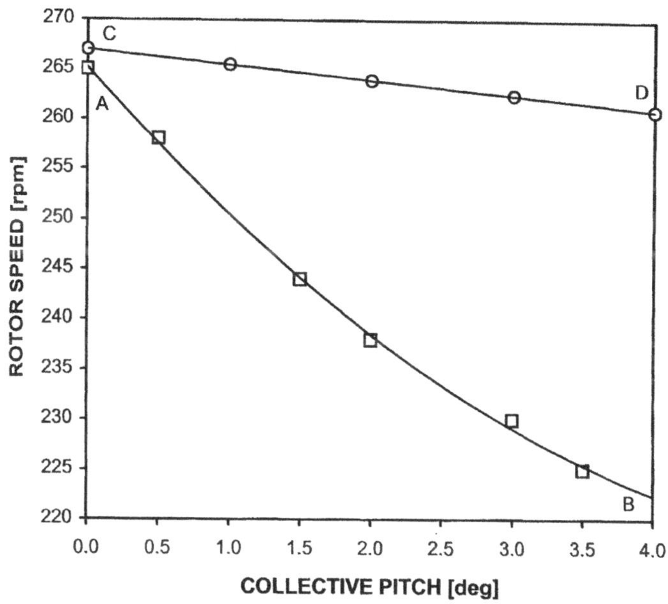

Fig. 6.13 Static droop/autorotation characteristics – FIG power contribution.

be representative of a true autorotation from the perspective of N_R, rotor speed control and ROD. To achieve this the power contribution from the engine(s) must be zero. The characteristics of the rotor system, the engine and the engine control system may prevent this condition from being achieved. For the purposes of this discussion a FIG is a descent with engine condition levers ECL(s) forward and an autorotation is a descent with the ECL(s) retarded or with the engine(s) shut down.

During a stabilized FIG, or powered descent, the engine and a component of the ROD will maintain the rotor speed. On the other hand in an autorotation at a given AUM, airspeed and density altitude there will be a fixed relationship between the collective position and N_R. At a constant airspeed in autorotation the rotor speed achieved at a given collective position will the maximum that can be generated by ROD alone, see line *AB* in Fig. 6.13. At the same conditions in a FIG there will also be a fixed relationship between collective position and N_R. This will be the same as the static droop plot for the same flight condition (line CD). In a FIG if the rotor speed demanded by the governing system is greater than the N_R that can be generated by the ROD, the engine will, as shown in Fig. 6.13, drive the rotor to the required speed. Consequently there will always be a power contribution in FIG even with the collective lever fully lowered.

The magnitude of any power contribution in FIG is not fixed as the autorotative relationship will vary with AUM, density altitude, airspeed and rotor rigging. The by-pass adjuster and other elements within the engine control system may be used to alter the static droop law. If the relative position of the autorotative N_R line and the static droop law are as shown in Fig. 6.14, the response of the rotor to changes of collective pitch in a FIG will be different. At point E the ROD is capable of driving the rotor to the required N_R and, therefore, there will be no power contribution from the engine. Note that at point E torque will be zero and N_R will be at the value defined

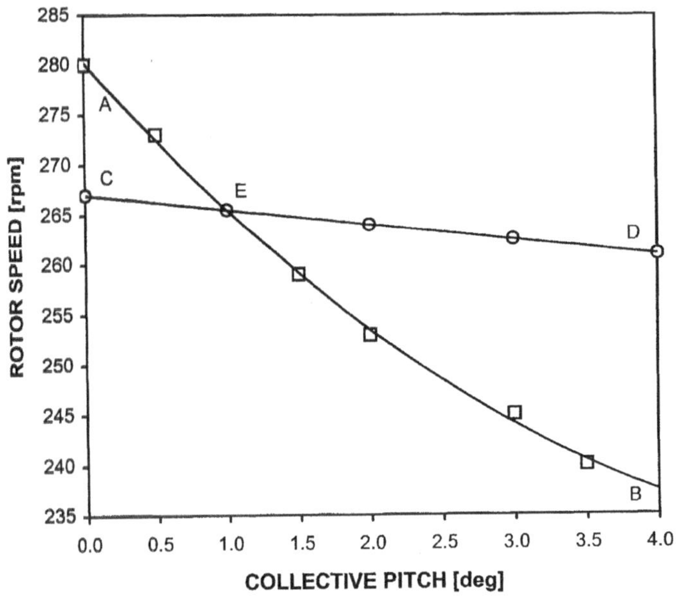

Fig. 6.14 Static droop/autorotation characteristics – no power contribution in FIG.

by the low power end of the static droop law. For the configuration in Fig. 6.14 as the collective pitch is lowered below 1°, rotor speed will follow the line EA whilst N_F will follow the line EC. Thus in this flight regime N_R and ROD will be the same regardless of the position of the ECL(s) and there will be a discernible N_F/N_R split. Therefore a FIG descent will have the same flight characteristics as a true autorotation.

6.3.7.3 *FIG power contribution at autorotative rotor speed*

Following a power failure the manufacturer may recommend a suitable value of rotor speed at which the subsequent autorotation should be flown. The value of N_R will be chosen such that it is unlikely to exceed limits, the autorotative performance is good and the rotor contains sufficient energy to perform a safe landing. Despite some manufacturers only specifying a range of permitted rotor speeds it will still be necessary to target a particular N_R or collective pitch for the purposes of quantifying any power contribution in FIG.

Figure 6.15 shows three possible positions of the autorotative N_R versus collective pitch trend relative to the static droop law. In each case a target autorotative N_R of 265 RPM has been assumed.

Curve 1. The value of rotor speed generated by an autorotative descent is always below the value demanded by the engine governor. In a power-off descent the pilot will have to lower the collective lever fully to set a N_R of 265 RPM. In a FIG at the same collective position the rotor speed demanded by the governor is higher at 267 RPM. Consequently there will be a power contribution in FIG, as the rotor will be driven up to this value by the engine. Alternatively the pilot could raise the collective to around 1° CP and set the target N_R of 265 RPM thereby increasing the power

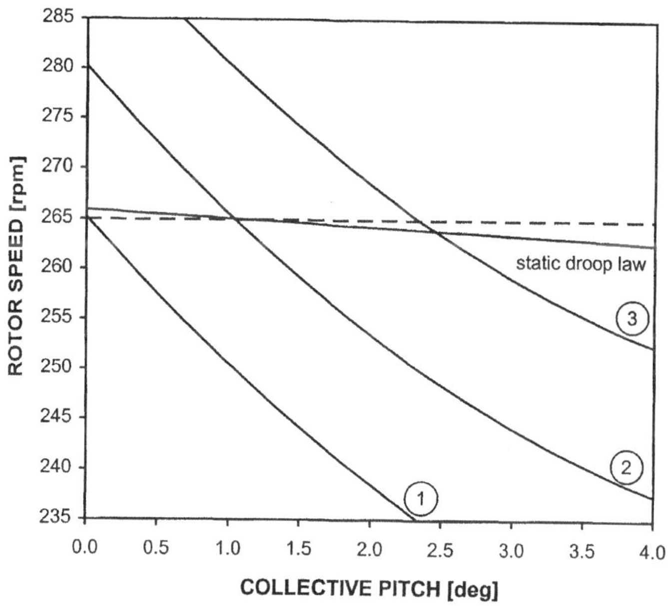

Fig. 6.15 Power contributions at autorotative N_R.

contribution. In either case there will be a noticeable difference between the rate of descent power-off and in a FIG.

Curve 2. At a higher AUM or density altitude the N_R that can be generated by the ROD is higher and in this situation the governed rotor speed and power-off NR are identical at the CP required to set 265 RPM in autorotation. Thus there will be no power contribution in a FIG and the rate of descent will be the same as in the autorotative case. Since the rotor speed in each case is identical if the pilots sets 265 RPM in a FIG then although the torque will be zero there will be no N_F/N_R split.

Curve 3. At still higher values of AUM or density altitude the situation becomes clearer. In setting 265 RPM, even with the engines operating at flight idle, the rotor speed can be fully sustained by ROD alone and therefore no power contribution is required. The pilot will need approximately 2.3° CP to set 265 RPM and if the engines were at flight idle he would notice that as well as zero torque being indicated, the free turbine would be governed at approximately 263.5 RPM so a clear N_F/N_R split would be discernible.

It is perhaps worth reiterating that despite the variations of autorotative N_R with AUM and density altitude if during a FIG there is zero torque indicated *and* a clear N_F/N_R split then there can be no power contribution and the FIG is fully representative of a true autorotation.

6.4 FLIGHT CONTROL SYSTEMS

Early helicopters had fairly simple control systems, consisting of cables and rods connecting the pilot's controls to the pitch change mechanism at the rotor head. With these early manual systems, the pilot had to apply the necessary force to overcome the aerodynamic loads on the rotor blades, with only some mechanical advantage to aid him. For this reason blade design had to be optimized to maintain nearly constant blade pitching moments over a wide range of collective pitch angles and rotor RPM. These blades were produced in matched sets and were usually of tapered profile and symmetrical blade section. These types of blades presented problems in quantity production and an attempt was made to use metal blades of constant chord which were easier to mass produce, but the control forces were generally unacceptably high. These problems and the growth in AUM led to the widespread adoption of powered controls for helicopters.

6.4.1 Types of helicopter flying control systems

6.4.1.1 Manual flying controls

This system uses a direct mechanical link, in the form of rods, bell cranks, and cables, between the pilot's controls and the pitch change linkage. Exceptions are some of the Kaman helicopters where the pilot's cyclic and collective controls are connected to servo tabs on the main rotor blades. Aerodynamic forces acting on the servo surface cause flap pitch changes on the rotor blades. Control systems that move a pitch change linkage have been found to be rather inflexible with some poor control characteristics especially at high AUM. They have been found to be unacceptable on helicopters with AUM in excess of about 10 000 lb (4500 kg). These systems, however, are still fitted to light helicopters and to some heavier machines as a reversionary mode in the event of failure of the powered control system. Refinements can be fitted to improve the characteristics:

- Spring bias units to reduce steady state cyclic control forces.
- Spring arranged to provide a pre-load on the collective lever or a more complex arrangement of levers and weights attached to the rotor head to reduce collective pitch loads.
- A one-way hydraulic lock unit or damper to reduce fluctuating loads fed back from the rotor head to the pilot's controls.

6.4.1.2 Powered flight controls

Most modern military helicopters are fitted with either simplex or duplex power control systems of varying complexity to suit the characteristics of the aircraft and its roles. The system is a remote position, closed loop servomechanism and to be satisfactory must meet the basic requirements of such a system. It must have a satisfactory performance, with good response and stability characteristics and it must be safe and reliable.

It is important that the performance of the system is such that the servos are capable of producing the necessary thrust to overcome the blade pitching moments under all conditions of flight otherwise it is possible that control of the rotor will be lost during some critical manoeuvre. Usually the required servo performance can be obtained by suitable design, since the thrust produced by a hydraulically operated servo is proportional to the cross-sectional area of the piston and the effective pressure of the hydraulic fluid. For precise control the response characteristics of the servo must be appropriate for all conditions of flight. The rate of movement required from a servo will depend to some degree on its function. Servos required to follow pilot's inputs and stabilize the helicopter in the cyclic channels will be rapid; slower-acting servos can usually be tolerated in the yaw control and collective circuits. To obtain satisfactory control, the lag between a demanded input and the resulting movement must be small otherwise the pilot will complain of a lag in the response of the rotorcraft.

Since powered flying controls are basically high gain servo mechanisms, it is necessary to ensure that the system is a stable one. This is usually achieved by designing the system with a high natural frequency, the inertia of the moving parts being made small in relation to the thrust produced, and by connecting the servo unit output direct to the pitch change mechanism of the rotor blades. The aerodynamic loads from the rotor that have to be overcome by the servos are oscillatory in nature. Great care is therefore taken to ensure that these oscillatory loads are not fed back into the control system by making it as stiff (irreversible) as possible. For obvious reasons, powered flying control systems must be made as safe and as reliable as possible. A single system may be used with a manual reversion capability but with the increased AUM and higher speeds of modern aircraft, the trend has been to fit multiplex control systems for safety.

6.4.2 Basic power flying control systems

A basic power control system consists of a pilot valve/main servo arrangement. The pilot's input is transmitted by mechanical means to the pilot valve. Hydraulic fluid pressure controlled by this valve causes the body of the servo to move in the required direction and this movement acts on the rotor blade through the normal pitch change mechanism. As with all servo mechanisms, there must be some feedback to stop the movement at the required position so that the input at the rotor head is proportional to the pilot's input in the cockpit. Typically this is achieved by arranging for movement of the servo body to cancel the pilot valve displacement. The force required to move the pilot valve is small with the servo providing the power to move the rotor blades against aerodynamic and inertia loads. Thus this basic system provides a large power amplification.

The basic powered flying control system reduces the pilot's workload but has no force feel, apart from inherent friction in the control runs. Thus the pilot is unable to release the controls to carry out other tasks, as they may move under gravitational or vibratory forces. In addition, there is no cue to tell the pilot how far the control has been moved. There is, therefore, a requirement to provide control retention and control centring force cues, which ideally can be trimmed over the full range of control movements required for flight. Below is a list of additional devices that may be fitted to tailor the control characteristics to the pilot's needs, to the aircraft and its roles.

(1) *Friction control device.* A friction control device is the simplest addition to a powered flight control circuit. It normally employs an adjustable sliding friction device acting on the control or control rods. The major drawback of this device is that the force required to overcome static friction (stiction) is usually greater than that required to overcome the sliding friction with the control in motion. This tends to lead to jerky control movements and possible overcontrolling,

although the device does provide control retention. With wear and the ingress of dirt and oil, this system is prone to binding and the generation of non-linearities in the force required to move the control over its full range, all of which is not conducive to the smooth and precise control of a helicopter. Friction devices are common on the collective pitch control where small and rapid movements are not normally required.

(2) *Spring feel and clutch systems.* The spring feel and clutch system provides a synthetic force gradient about the trim position, control centring, and an instantaneous trimming capability. The clutch is usually disengaged by a *trim release* button on the pilot's control. Fail-safe operation tends to vary from aircraft to aircraft, some leaving the controls free or at a set trim position. One problem with this type of system is the potential for *stick jump*. If, while holding a control force against the spring, the clutch release is operated, the force resisting the pilot drops to zero faster than the pilot can relax his applied force so the stick will jump. Very light forces do not present a problem in this respect but experience shows that forces in excess of 1 daN tend to produce a rapid jerk of the control when the clutch is released.

(3) *Cyclic 'beeper trim' system.* This system meets the requirements for positive control retention, force cues and the ability to trim the control over its full range of movement. In most aircraft the system can be disengaged by the use of the *trim release button* which releases an electromagnetic clutch between the motor and the spring box. The trim rate has to be tailored to the control gearing, the aircraft's stability characteristics, control force gradients and the mission of the aircraft. For example, for a given aircraft and gearing, a slow trim rate would be ideal for instrument flight, giving almost vernier adjustments, but this rate would be unsuitable for more aggressive manoeuvring particularly if high force gradients were used. In this latter case, aggressive manoeuvres would require the pilot to make large control displacements leading to high control forces unless the trim rate was fast enough to relieve them quickly.

(4) *Viscous damper.* Rate damping, in the form of a viscous damper is used in a number of control systems to limit the rate of application of a control for stress reasons or to prevent a pilot induced oscillation in some mode. The characteristics produced by a viscous damper can enhance or detract from controllability depending on the combined effects of the damper, friction, breakout force and force gradient. From a pilot's point of view a rate damper gives him a sense of how fast the control is being moved and tends to remove some of the jerkiness from control movements. A damper may also be used to prevent stick jump in a spring feel and clutch system.

6.5 AUTOMATIC FLIGHT CONTROL SYSTEMS

Helicopter automatic flight control systems (AFCS) are many and varied. In some cases they provide slight compensation for the 'raw' aircraft characteristics whereas in others these characteristics are completely masked (so-called *superaugmentation*) and the pilot's perception of the aircraft is solely based on the nature and performance of the AFCS. This section is intended to give the reader an understanding of rotorcraft AFCS without describing in great detail all the various systems currently available or envisaged for the future. After a brief review of the important stability characteristics of helicopters, and a discussion of the levels of augmentation available, the major system components are described. The section concludes with descriptions of generic systems.

6.5.1 Helicopter stability deficiencies

Before discussing AFCS in detail it is instructive to review the important stability characteristics of helicopters. The inherent stability problems associated with generating lift and thrust from an 'edgewise' rotor mean that only the lightest and cheapest helicopters on the market are unaugmented.

6.5.1.1 Longitudinal long-term mode

The longitudinal long-term mode contributes to the multi-axis oscillation, the 'falling leaf mode', found in the hover. This mode is usually unstable and will only be satisfactory if the time to double amplitude (T_2) and period are sufficiently long that mission workload is tolerable. In forward flight the long-term mode is similar to the classic phugoid found in conventional fixed wing aircraft, although with greater changes in pitch attitude. Depending on the characteristics of the horizontal stabilizer and, therefore, the degree of speed stability, the mode can be either convergent or divergent. Once again the rate of divergence, the period of the motion and the intended role of the helicopter will determine if this mode can be tolerated without augmentation. In some extreme cases, such as a semi-rigid rotor at high speed, this mode may become aperiodically divergent, in which case augmentation will be imperative.

6.5.1.2 Manoeuvre stability

The manoeuvre stability of all helicopters will degrade to instability at high speed and high load factor due to the progressively greater destabilizing effect of the main rotor. Therefore, for helicopters required to operate at high load factors, some form of augmentation will normally be fitted. The greater pitch damping provided by rate stabilization is a possible solution, as is the use of a programmable stabilizer.

6.5.1.3 Lateral/directional oscillation

The Lateral/Directional Oscillation (LDO) of a helicopter can be regarded as the lateral/directional equivalent of the longitudinal long-term mode. In most rotorcraft the LDO is very similar to the Dutch roll mode found in conventional fixed wing aircraft. Depending on the relative magnitudes of lateral static stability and directional static stability the Dutch roll will be either convergent or divergent, and highly oscillatory or deadbeat. The characteristics of the mode may also change markedly with flight condition as fin effectiveness can be strongly influenced by the skew angle of the main rotor wake. The ease of excitation, ease of suppression, the frequency, the roll-to-yaw ratio and the intended role of the helicopter will determine if the LDO can be tolerated without augmentation.

6.5.1.4 *Spiral stability*

For rotorcraft, the aerodynamic factors contributing to good LDO characteristics will often degrade the spiral stability. At high angles of bank the weight of the vehicle will produce a large into-turn moment which may overcome the existing aerodynamic forces thus degrading stability still further. It may be impossible to endow the helicopter with acceptable LDO and spiral mode characteristics throughout its flight envelope without resorting to artificial aids.

6.5.1.5 *Directional static stability*

The directional static stability characteristics of a helicopter can be poor, especially at low values of forward speed and low angles of sideslip. The main cause of this phenomenon is the lack of adequate fin effectiveness at low values of lateral velocity or total velocity. Blanking of the fin by the fuselage or inadequate tailoring of the main rotor wake are typical reasons. The resulting low values of N_v may lead to poor natural sideslip control that may be unacceptable in certain missions, such as ground attack. Directional stability may be further degraded if there is a large amount of side area towards the front of the fuselage, such as when external stores are carried or floats fitted. Aerodynamic fixes may be difficult to engineer and are usually expensive to install. Often the most suitable solution is stability augmentation in the form of a yaw damper or the provision of a heading hold.

6.5.1.6 *Lateral static stability*

The lateral static stability can also be degraded by the addition of external stores or floats. In this case the increase in side area below the mid-point of the fuselage may produce a large destabilizing contribution to the overall lateral stability. Weak positive to neutral static stability may result with consequently poor sideforce cues. If this characteristic occurs with poor directional static stability the pilot may find it very difficult to control sideslip. In addition weak lateral static stability will increase the workload required to maintain wings-level flight. Once again augmentation, this time in the form of roll rate feedback is likely to be the most cost-effective solution, although a bank angle hold may be necessary.

6.5.1.7 *Control response*

The control response of the helicopter will be of great concern to the pilot. If the response is poorly tuned to the role the aircraft may still have unacceptable handling qualities regardless of how well the stability deficiencies have been overcome. Generally the nature of the control response is a function of the rotor system fitted to the helicopter and only minor changes can be made via modifications to the mechanical characteristics of the cockpit controls.

6.5.2 Levels of augmentation

The term automatic flight control system covers a multitude of systems that are used to improve, or remove, the stability deficiencies described above. In addition the more sophisticated systems can improve, or tailor, the handling qualities and permit certain role manoeuvres to be flown automatically. The term AFCS will be taken to apply to all forms of stability augmentation and automatic flight path control.

Stability augmentation (or *inner-loop*) systems can be further sub-divided into systems that use angular rate information as their primary data source and those that rely on direct measurement of aircraft attitude. Typically rate-based systems are cheaper and simpler to design and are, therefore, found on small-to-medium sized helicopters. Attitude–based systems are often enhanced by a range of *outer-loop* modes, see below, and are found on larger helicopters or those aimed at single pilot IFR operation.

The autopilot, or outer-loop system, is a higher level of augmentation. As the name suggests it can be considered as an outer loop around a stability augmentation system that allows automatic control of flight path through the action of parallel actuators. Holds other than pitch and roll attitude will often be included such as heading, speed, height (radar and barometric) and vertical speed. Special modes may also be included: coupled approach, auto transition, 'go-around' and hover position hold.

Fly-by-wire (FBW) technology, for long the sole preserve of flight control systems for fixed wing aircraft, has flown successfully on several research rotorcraft and is being implemented on two helicopters (RAH-66 and NH-90) destined for full-scale production. As in fixed-wing systems FBW affords the helicopter FCS designer greater flexibility of design. This has been necessary to reduce pilot workload by providing a set of control laws and holds that are role suited, selectable and blended so that the most appropriate form of command and stability augmentation is available for each phase of the mission. This will enable the crew to devote more time to operating on-board sensors or weapons systems rather than piloting the aircraft. With FBW the flying controls need not be placed in the conventional positions as all piloting commands are passed through a flight control computer (FCC) by means of electrical signals. A sidestick and a small collective lever are alternatives to the conventional controls.

6.5.3 System components

The constituent parts of most AFCS can be grouped into four classes of components, namely:

- *Inceptors.* Devices that the pilot uses to communicate with the AFCS and to make control inputs.
- *Sensors.* These are used to measure the relevant reference parameter and transmit the necessary information to the computer.
- *Computers.* These convert the sensor information and pilot demands into signals to drive the output devices.
- *Output devices.* These convert the computer signals into a form that will result in the required helicopter movement. The output may be direct, in the form of actuators which move the pitch change links, independently or via the cockpit controls, or indirect in the form of a cockpit display, a 'flight director', which will direct the pilot to make the necessary control movements himself.

6.5.3.1 Inceptors

Inceptors refer to the devices used by the pilot to make inputs to the FCS. The simplest and most conventional are the traditional cyclic stick, collective lever and pedals which the pilot uses to command changes to the aircraft attitude and power setting. The selection and adjustment of holds, such as heading, height or airspeed are often made by means of switches and knobs, either located on the primary inceptors or close to the associated flight instrument. It is worth noting that the use of electronic flight information systems (EFIS) have allowed some consolidation of these controls.

The absence of a mechanical link between the cockpit control and the rotor head, which has been made possible with the advent of FBW, has given the cockpit designer greater freedom. Although mechanically coupled side-stick arrangements have be used in some non-FBW attack helicopters, the mechanical tailoring has been necessarily limited. Consequently there has been a great deal of research into the best style and location for active versions of these inceptors. It has now been generally agreed that a small stick, located centrally or to the right, and some form of collective lever, or second side-stick controller, will be used. Such an installation can improve pilot comfort by allowing the optimization of inceptor position, FCMC and adjustment, and improve cockpit layout by allowing more appropriate positioning of displays, instruments and switches, due to the smaller envelope occupied by the side-stick controllers (SSCs) and the possible removal of the yaw pedals. At present little agreement exists concerning the controls that will be assigned to these novel inceptors. Three configurations have been evaluated, these are:

- A two-axis SSC located to the right (2R), a collective (1C) and pedals (1P) or (2R + 1C + 1P).
- A three-axis SSC located to the right and a collective or (3R + 1C).
- A four-axis SSC located to the right or (4R).

Research conducted in Europe as part of the EuroACT programme concentrated on a passive (3R + 1C) system and an active (2R + 1C + 1P) system. The passive system provided the pilot with a fixed set of spring feel characteristics, whereas the active system allow the tailoring of the feel characteristic to suit the current handling task. The more conventional configuration was selected for the active system in order to simplify the basic stick design so that incorporation of actively variable FCMC was possible. Recent research has highlighted the following:

(1) SSC design characteristics, such as grip shape, arm support and the length of the pivot arm, can have a profound effect on the handling qualities due to anatomically induced cross-coupling.
(2) Isometric (or rigid) and low compliance (or movement) SSCs are undesirable because of their lack of control magnitude and control position cues.
(3) Four-axis SSCs have the advantage of freeing a hand for other tasks. But significant handling deficiencies are experienced due to anatomical coupling. It is very difficult for the pilot to make single axis inputs, particularly in heave, thereby complicating certain piloting tasks, such as slope landings. Dynamic multi-axis tasks such as quickstops are also difficult to perform with this type of inceptor.
(4) Surprisingly, three-axis SSCs with twist for yaw control are well accepted by pilots provided it is coupled with an accurate heading hold incorporating an integral trim that obviates the need to hold-off a force during out-of-wind hovering for example.

Control of autopilot modes is generally by means of appropriate knobs and switches. These may be duplicated so that a range of the more important functions can be selected or deselected from the primary flying inceptors. The selection and control of certain modes, such as airspeed hold and vertical speed hold may be via controls incorporated into the appropriate flight instruments. Alternatively, a repeater screen may be fitted that displays the current hold setting. The displayed data changes depending on the hold selected and the datum is adjusted by means of switches on the primary flight inceptors.

6.5.3.2 Sensors

Sensors are required in order that the flight control computer is aware of the state of the aircraft and position of the primary flight controls. The simpler stability augmentation systems require information from rate gyros whilst attitude stabilization requires a signal from a vertical gyro, or from integrating the rate signal. A gyrocompass will be used for heading hold. More complex autopilot systems will require more sensors, many of which may also be used to provide the pilot with flight information. Sensors include:

- a pitot/static system for speed and height holds (for safety reasons this system is usually separate from that which feeds the first pilot's instruments);
- the navigation suite for waypoint steering;
- lateral accelerometers for turn co-ordination;
- ILS receivers for coupled approaches;
- Doppler or GPS receivers for automatic hover position hold.

A fuller list of the sensors currently used in helicopter automatic flight control systems is given in Table 6.3.

Table 6.3 Automatic flight control system – parameters and sensing systems.

Parameter	Sensors
Pitch rate, roll rate and yaw rate	Rate gyroscope or differentiated attitude
Pitch, attitude and roll attitude	Vertical gyroscope or integrated rate
Yaw attitude (heading)	Gyrocompass
Airspeed and vertical speed	Pitot-static system
Low airspeed	HADS or other low airspeed sensor
Sideslip	Yaw vane or differential static pressure
Sideforce	Lateral accelerometer
Normal acceleration	Accelerometer
Altitude	Static system or separate aneroid
Height	Radar altimeter
Along and across track velocities	Doppler radar, IN or GPS
Hover plan position	Doppler radar, GPS or cable angle
Flight path data	ILS, VOR, DME, GPS or IN
Control positions	Position pick-off or force sensor
Control position (discrete)	Microswitch

6.5.3.3 *Computers*

The computation performed by the data processing part of an AFCS may be simple or complex depending on the type of system and degree of augmentation provided. Many older systems based around an analogue computer are still in operation, some of which provide limited outer loop modes such as height hold and automatic transitions. More modern and sophisticated systems such as fully capable autopilots suited for single pilot IFR operations and FBW installations require the flexibility and ease of control law implementation offered by digital computers. Some of the more common functions of the computer are:

- *Amplification*. To increase the signal strength from the sensor to a level high enough to be effective at the output stage.
- *Integration*. To derive synthetic attitude information from a rate gyro signal or translation rate from acceleration.
- *Differentiation*. To derive synthetic rate information from a vertical gyro signal.
- *Summation*. To generate an error signal by comparison of the pilot demand with the output from the appropriate sensor or to blend angular rate and attitude feedback signals.
- *Limiting*. To contain the effect of certain parameter changes within predetermined limits or to contain the effect of sub-system failures.
- *Shaping*. To adapt the control strategy and handling qualities to suit the particular mission phase.
- *Programming*. To produce a precise manoeuvre, such as an automatic transition, in the absence of positional data.
- *Blending*. The use of a particular parameter, such as airspeed, to modify the gain or functioning of an AFCS feature.

6.5.3.4 *Output devices*

Clearly the AFCS needs to move the controls of the helicopter in order to achieve the aims of the control law design, therefore, some form of actuator connected to the controls must be used for this purpose. The actuator may be either electro-mechanical or electro-hydraulic and it may operate in series with, or parallel to, the control run.

Series actuators can be considered as an extendable portion of the control run whose length is under the control of the AFCS (see Fig. 6.16). Usually the breakout force at the cockpit control is greater than that at the pilot valve of the main control servo and so, as the AFCS actuates the pitch change linkages, no movement of the primary inceptors occurs. In this way the AFCS designer can augment the stability of the host aircraft without causing unwanted movement of the cockpit controls. Transients on engagement/disengagement of the system are possible, as the actuator must re-centre and become a rigid link when the AFCS is inoperative so that no lost motion is introduced. Most control laws involve some enhancement of the damping of the host aircraft by responding to undemanded angular rates. Consequently the series actuators need to be rapid acting and therefore must have limited authority so that the aircraft response to an active failure, such as a hardover or runaway, is not catastrophic and that the available control range following such a failure remains adequate. About $\pm 10\%$ is typical, which means that series actuation alone is unlikely to have the control authority to be suitable for all autopilot functions, especially those

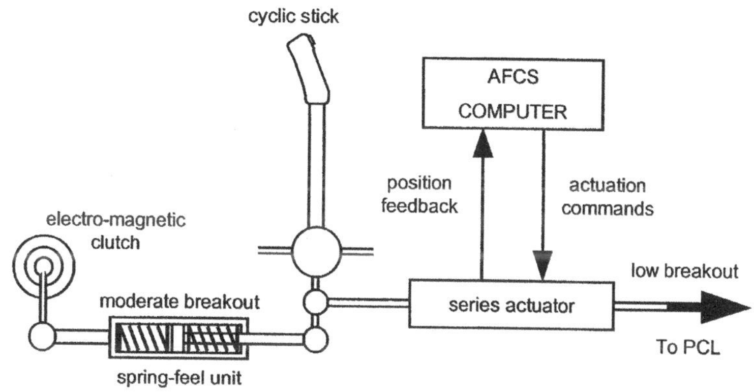

Fig. 6.16 Series actuation.

involving large changes in flight path such as automatic transitions and turn co-ordination.

The parallel actuator is coupled to the control run in such a way that the cockpit control is moved as the actuator operates (see Fig. 6.17). Consequently the actuator has full authority and so the actuation rate must be a compromise between the requirement to move the control run rapidly, so that the associated series actuator is prevented from saturating, and the active failure characteristics. Unlike with series actuation there is no AFCS reason for breakout at the cockpit control. The actuator is often coupled to the control run by means of an electro-magnetic clutch and, therefore, disengagement can be made at any time without disturbance to the flight path. It is common practice to provide parallel actuation in the collective and yaw channels because outer loop modes, such as heading steer, often suffer from series actuator saturation. Trim motors acting on the cyclic stick can also serve as parallel actuators in providing a *follow-up trim* mode that uses the spring-feel unit to move the complete control run, thus keeping the series actuator nulled and able to exert approximately equal authority in both directions at all times. Parallel actuators are usually electro-mechanical as high operating speeds are neither required nor desired.

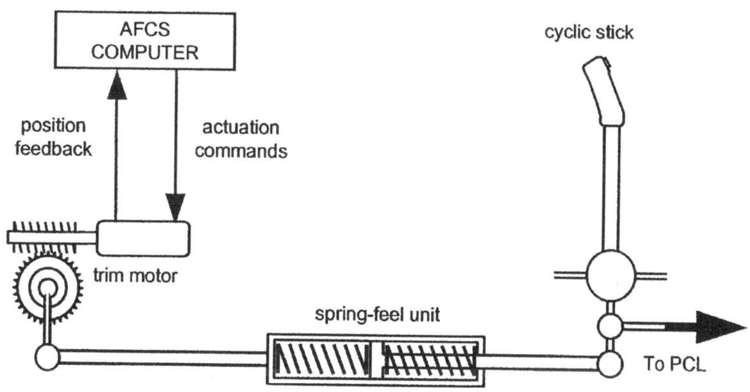

Fig. 6.17 Parallel actuation.

6.5.3.5 *Transparency and override facilities*

Automatic flight control systems that feature outer loop modes need to be carefully designed so that the pilot is able to make control demands without having to overcome an opposing input from the AFCS. Ideally movement of the controls in the cockpit should cause the relevant hold to be disabled, either permanently or temporarily, depending on the circumstances. If well designed the functioning of the basic holds (attitude, heading and height) can be *transparent* to the pilot.

Of particular importance is the blend between heading hold and turn co-ordination or more simply heading hold on and off. In forward flight the initiation of a turning manoeuvre is signalled by the application of lateral cyclic and this is used by the AFCS designer to unlock the heading hold and use the yaw series actuator (and parallel if fitted) to help generate the necessary yaw rate for a balanced turn. Thus a natural blend is achieved and, as far as the pilot is concerned, the helicopter changes transparently from maintaining a heading in straight and level flight to performing a Rate 1 turn for example. In the low speed regime the situation is different since during a lateral transition or side-step the pilot will expect the heading to be maintained. Therefore some form of airspeed switch is required to prevent heading hold unlock at low forward speed.

Rotorcraft AFCS without automatic turn co-ordination or an airspeed switch present the designer with more of a challenge in that he can no longer use lateral cyclic to unlock the heading hold. Instead a sensor is placed in the yaw control run so that if the pilot attempts to initiate a yaw rate by application of pedal the heading hold is disengaged. Unless this is carefully engineered the AFCS may suffer a lack of transparency that *in extremis* makes the helicopter harder to turn with the AFCS on than with it off.

Some AFCS provide the pilot with a range of programmed manoeuvres such as automatic transitions, auto-ILS and waypoint steerage, that often require large movements of the controls within the cockpit under the action of a parallel actuator. It is usual, therefore, for the pilot to be provided with an override or cut-out button so that in the event of a runaway, some other failure or a change of mind, he can disengage part or all of the mode either permanently or temporarily as the situation dictates. So for example the Westland Sea King ASW helicopter is provided with both cyclic and collective cut-outs to disengage parts of the automatic transition programme. Similarly helicopters fitted with height holds have cut-outs, usually located on the collective, to disengage the hold and some have a manoeuvre button to allow temporary disengagement of the hold as a new datum altitude is set.

6.5.4 Reliability and availability

When an aircraft is evaluated for a particular role the reliability of the complete vehicle will be almost as important as the handling qualities that it possesses. Thus the vehicle specification will often include some statement of the minimum acceptable reliability. When a helicopter is fitted with an AFCS, on which it depends for normal operation, the reliability of this system has increased importance. It is usual for the system availability to be specified when serious degradation of the handling qualities results from total AFCS failure. Thus some minimum number of major failures is specified before any degradation in AFCS performance is allowed.

The reliability of an AFCS is usually defined as the probability of total system failure, but can also be expressed as a mean time between failure (MTBF). In the case of an ACT helicopter such a failure would be catastrophic, most likely resulting in the loss of the aircraft, and so the specified reliability of the overall flight control system would be very high. Typical figures are 10^{-9} per flight hour for civil applications and 10^{-7} for military systems. Associated with reliability requirements is the availability of a system. This is defined as the requirement that it should continue to operate after a specified number of major failures. For example, a system may be specified such that it shall function correctly after one or two failures. Clearly, this implies duplication, or *redundancy*, in the system. Redundancy is often required to ensure that the AFCS meets the necessary reliability since individual component reliability cannot be guaranteed for the extreme MTBFs specified.

The degree of multiplexing, or redundancy, incorporated into an AFCS will depend on the handling qualities of the unaugmented aircraft and whether successful completion of the mission is dependent on AFCS integrity. Degrees of multiplexing can be ranked in order of increasing fault tolerance. To gain some idea of the theoretical improvement in reliability, an individual component MTBF of 1000 hours has been assumed in the following examples.

6.5.4.1 *Simplex*

A simplex system has no built-in fault tolerance and will cease to function following a single failure. Thus a simplex AFCS would only be acceptable in a helicopter with acceptable unaugmented handling qualities and when the majority of the mission could be completed satisfactorily with the 'raw' aircraft. A simplex system may be reduced to the main components (sensors/computer/actuator), all of which are required for satisfactory operation. The probability of failure of the example system would be, typically, 3×10^{-3} per hour, giving a MTBF of approximately 330 hours.

6.5.4.2 *Duplex*

The duplex system consists of two completely separate AFCS systems, from sensors through to actuators although the actuation will be combined at the pilot valve of the main servo jack. Without any form of system monitoring it is impossible to arrange for automatic deselection of a malfunctioning lane and therefore the system would have a probability of first failure similar to the simplex AFCS. In fact it might be worse since there are twice as many components in this system. The advantage it has over a simplex system is that a runaway in one lane will be sensed and countered by the other, assuming that a simultaneous failure has not occurred, therefore the ensuing departure from controlled flight will be more benign. Unfortunately fault diagnosis will be more difficult since the pilot has no way of determining which lane has failed and which operated correctly to counter the disturbance. Therefore the handling qualities of the raw aircraft must still be acceptable to ensure flight safety during the fault diagnosis and deselection process. More modern duplex systems feature digital computation that enables self-monitoring at the expense of increased computation time. This is often achieved by engineering a pseudo third lane that monitors sensor information and determines the appropriate actuator response. Using this approach it is possible to indicate a failed lane or to arrange for automatic deselection, in which case the probability of system failure resulting in visibility of the raw handling qualities reduces to 9×10^{-6} per hour giving an MTBF of over 110 000 hours as two failures are now required before augmentation is lost.

6.5.4.3 *Triplex*

Three separate lanes are necessary in aircraft where the handling qualities dictate the need for stability augmentation following two failures or where their impact on manual fault diagnosis and manual deselection of a failed lane are unacceptable. As noted above this latter requirement can now be satisfied using a self-monitored duplex (or pseudo triplex) system. In all cases however automatic monitoring and deselection of a single failed lane requires the presence of three signals so that the two good lanes can 'vote' out the bad one. Such a system can survive a single failure with no change in handling qualities although it provides no protection against a common mode failure that causes two signals to go bad at the same time. Following a first failure, however, in the event of a disparity between the two surviving channels the voting logic will fail and the aircraft will suffer a total loss of augmentation just like that caused by a single failure in the duplex system. It is at this point that the greater the *availability* provided by triplex systems becomes evident in that the crew can regain augmentation after the second failure provided the surviving lane can be correctly identified. Once again, through the use of self-monitoring, modern digital systems can improve the situation by providing automatic deselection following a second failure, thereby reducing the probability of total failure to 2.7×10^{-7} per hour, a MTBF of 37 000 000 hours.

6.5.4.4 *Quadruplex*

Increasing the level of redundancy to four separate lanes of AFCS would be necessary if voting logic was used and continued augmentation after two failures was required. The system reliability may be less than expected, however, due to mechanical complexity involved in this level of redundancy particularly as rotorcraft do not have multiple means of generating control forces and moments about a given axis. The advent of digital computation and self-monitoring, as mentioned above, have enabled a triplex system to perform at the level of fault tolerance formerly possible only with a quadruplex system. It is probable, therefore, that in the future, a quadruplex, digital AFCS with self-monitoring would only be fitted on an aircraft with catastrophic failure modes. Such a system, based on a component MTBF of 1000 hours, would have a theoretical probability of total failure of 7.1×10^{-11} per hour or an MTBF of over 1 200 000 000 hours.

6.5.4.5 *Signal consolidation*

All the systems described above have consisted of separated lanes of AFCS with consolidation only at the main servo jack. Alternative architectures arrange for all the sensor signals to go to all of the computers and for all of the computers to drive all of the actuators. This serves to improve the availability of the system because a lane containing a failed component can continue to operate using signals from the surviving components in the other lane(s). Although theoretically the reliability is not changed, actually there will be an improvement because common failures or simultaneous dissimilar failures in the different lanes will be required before the AFCS integrity is in question.

6.5.4.6 *Failure modes, effects and criticality analysis*

The systems described above have been necessarily simplistic and have assumed independent sensor packages and electrical/hydraulic supplies to each lane. Actual systems are generally more complicated with the precise system architecture varying with aircraft type or AFCS manufacturer. Typically a triplex, or quadruplex, system may still offer augmentation following a range of failures but at a degraded level and within a reduced flight envelope. In addition, optional sensors may be used to provide a synthetic signal following failure of the primary sensor. For example, a rate gyro signal may be integrated and used to replace the signal from a failed vertical reference system (VRS).

It should be clear that to assess an AFCS requires a detailed knowledge of the system architecture and its intended modes of operation, including operation following failures. An evaluation called a Failure Modes, Effects and Criticality Analysis (FMECA) is usually conducted in order to determine the likely effects of all conceivable failures to ensure the validity of any degraded modes assessment conducted as part of the flight test programme. Due to the limited time available only the more probable failure cases will be subject to an in-flight assessment. So, for example, although there may be a variety of failures that could lead to the loss of a primary sensor, such as a vertical gyro, or the loss of a lane in a multiplexed system, flight testing would simply involve assessing the permissible envelope with a single lane disengaged or with the feed from the gyro disabled. It is, however, worth remembering the number of occasions that the FMECA has proved, through bitter flight experience, to be incomplete. For example, supposedly dissimilar components in individual lanes that are all susceptible to the same common mode failure due to a similarity in design.

6.5.5 Characteristics of typical stability augmentation systems

Before describing the assessment of AFCS equipped aircraft in detail (see Section 7.5) it is worth discussing the characteristics of typical systems that arise from their hardware implementation. In our experience we have found such a discussion to be very useful as the precise design of an AFCS not only affects the conduct of stability and control flight testing but often affects the interpretation of the results obtained from standard test techniques and manoeuvres.

Stability augmentation systems are designed to suppress the longitudinal long-term mode and the LDO. The SAS, in its basic form, consists of a device sensitive to the rate of change of attitude, a rate gyro, which feeds a series actuator placed in the control run. Corrective cyclic pitch inputs are thus made by the SAS in opposition to and proportional to, the rate of change of pitch or roll (see Fig. 6.18). A similar system operates in the yaw channel by feeding corrective inputs to the tail rotor control. An airspeed switch may, however, disable it as the fin tends to provide an increased contribution to yaw damping at high airspeed.

As discussed briefly in a previous chapter increased rate damping (angular rate feedback) can be used by the control law designer to enhance the control response characteristics of the helicopter by making the time to steady rate shorter (improved predictability) at the expense of a reduced output. Thus there is a strong argument for retaining the rate feedback path whilst the pilot is manoeuvring the helicopter.

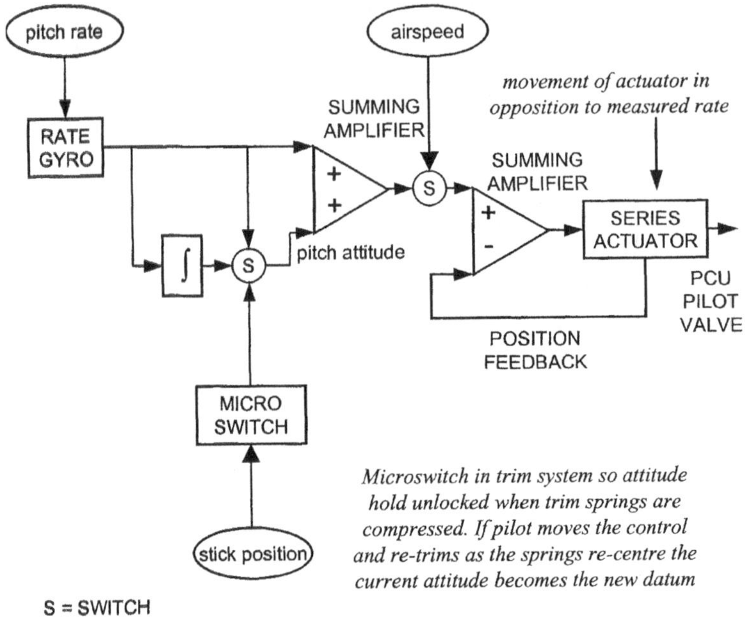

Fig. 6.18 Block diagram of a typical rate-based stability augmentation system.

Unfortunately it may be difficult to find an optimal gain that provides adequate gust rejection as well as appropriate control response shaping.

In order to improve the handling qualities of the helicopter, beyond the application of rate feedback, it is usually necessary to apply corrective cyclic inputs in response to attitude changes (integral feedback). This can be achieved either by integrating the signal from the rate gyro (*pseudo-attitude*) or by direct measurement of the aircraft's true attitude. In each case the signal is fed back to the series actuator to provide a command that opposes deviations from some attitude datum. Note that integrated rate is not incorporated into the yaw channel, as it does not relate well to aircraft heading. The key issue arising from this control law improvement concerns how the attitude signal is disabled during manoeuvres and what response type results: Rate Command/Attitude Hold (RCAH) or Attitude Command/Attitude Hold (ACAH).

6.5.5.1 Rate command and short-term attitude hold

If integrated rate is used then enhanced gust rejection can be expected, as even small angular rates will, over a period of time, generate sizeable attitude errors. However if the pilot moves the control to a new position then, within the authority of the series actuator and all the time an angular rate is present, the AFCS will attempt to drive the attitude back to that occurring before the input was made. One solution to this problem is to arrange for the attitude error signal to 'leak away' with time (Fig. 6.19). Thus if the pilot makes a deliberate control input the AFCS will, after a short period of time, take the current attitude as the one about which to hold. Applying an aft pulse (Fig. 6.20), shows a leaky integrator's effect on the handling qualities. Application of the pulse indicates that the SAS is capable of holding an attitude, selectable by the pilot, for a short period.

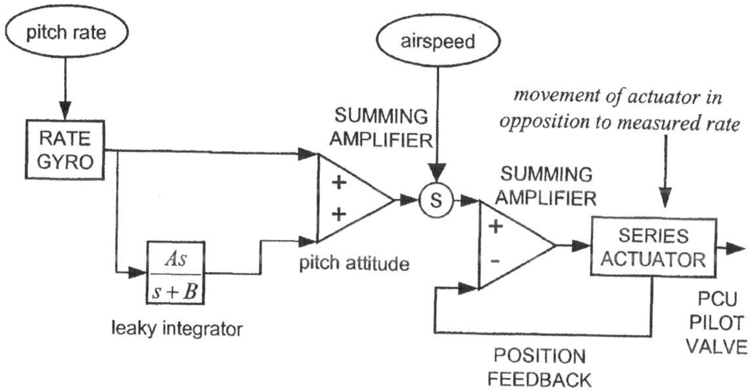

Fig. 6.19 Block diagram of a stability augmentation system incorporating a leaky integrator.

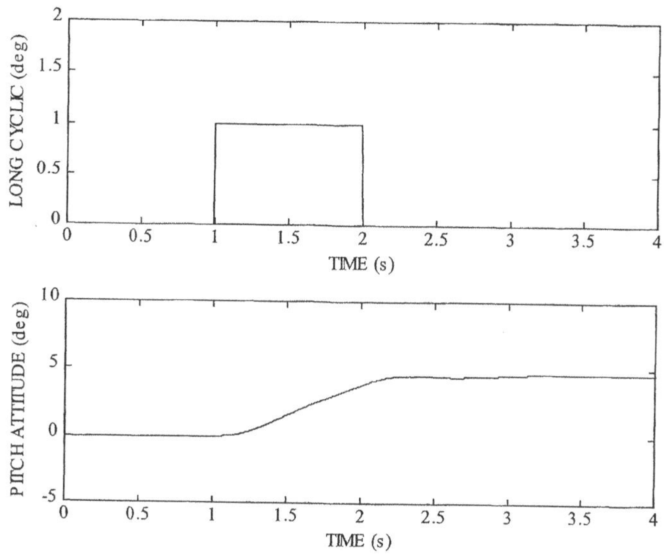

Fig. 6.20 Pulse response of SAS with leaky integrator.

Another solution to unwanted retention of pseudo-attitude hold is simply to switch it off when it is not required. This can be achieved by installing microswitches that disconnect the integrated rate signal when the stick is in motion. When stick movement has ceased the hold can be re-engaged and divergence from the attitude prevailing at that instant will cause movement of the series actuator. Re-engagement of the signal is often dependent on a low rate signal being sensed, 2 degree/second, for example, as well as the lack of any stick movement. The gust response of such a SAS is shown in Fig. 6.21. From this figure it would appear that a satisfactory solution can be achieved as the long-term attitude hold seems reasonable.

It must be remembered, however, that if the attitude signal has been derived from the rate gyro any drift or noise occurring within the rate gyro will compromise the attitude signal thus reducing the effectiveness of the attitude hold. Figure 6.22 shows

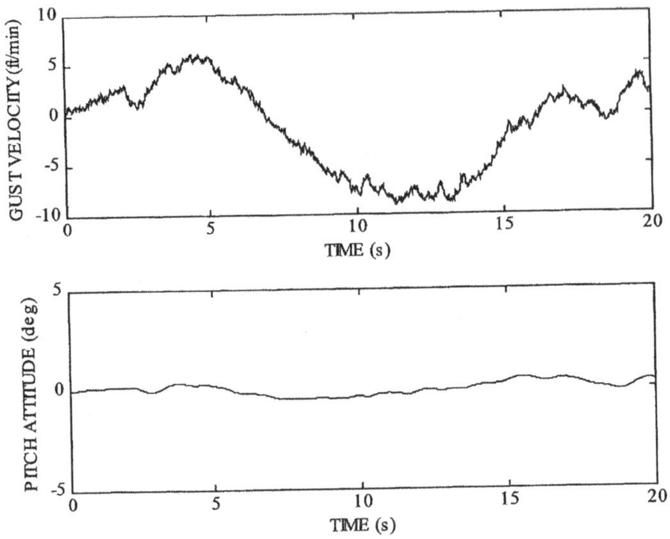

Fig. 6.21 Gust response of SAS with stick cancellation.

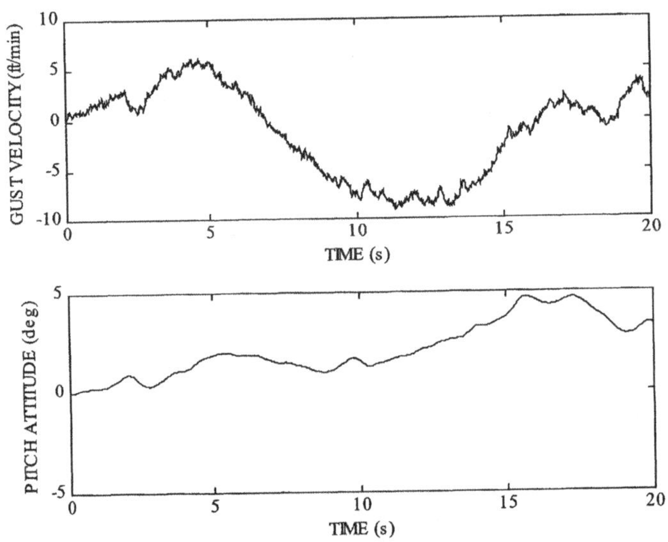

Fig. 6.22 Gust response of SAS with noise and drift on rate gyro signal.

the result of including realistic levels of noise into the rate signal causing a deterioration of the attitude hold as expected.

In summary, we can characterize stability augmentation via rate and integrated rate feedback, with the pseudo-attitude signal either allowed to leak or being disabled during manoeuvres, as a system that provides rate command and short-term attitude hold.

6.5.5.2 *Rate command or attitude command and attitude hold*

Complete elimination of all long-term 'nuisance' modes requires accurate attitude information and so a device sensitive to aircraft attitude, a vertical gyro for example, is used to feed the series actuator (see Fig. 6.23). The power of accurate attitude feedback can be demonstrated by reference to the longitudinal dynamics of the example helicopter when excited by gusts in the form of vertical turbulence. Without attitude stabilization the unstable long-term mode is revealed (see Fig. 6.24). If unity attitude feedback is used this mode is easily stabilized (Fig. 6.25).

As suggested earlier the challenge for the AFCS designer is in obtaining the right

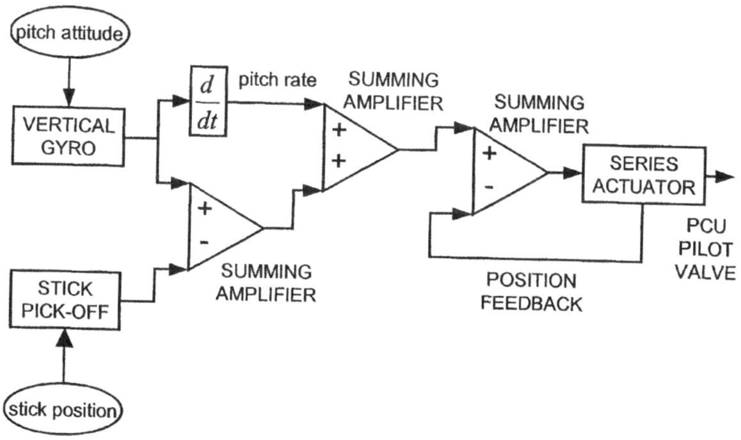

Fig. 6.23 Block diagram of a typical attitude-based stability augmentation system.

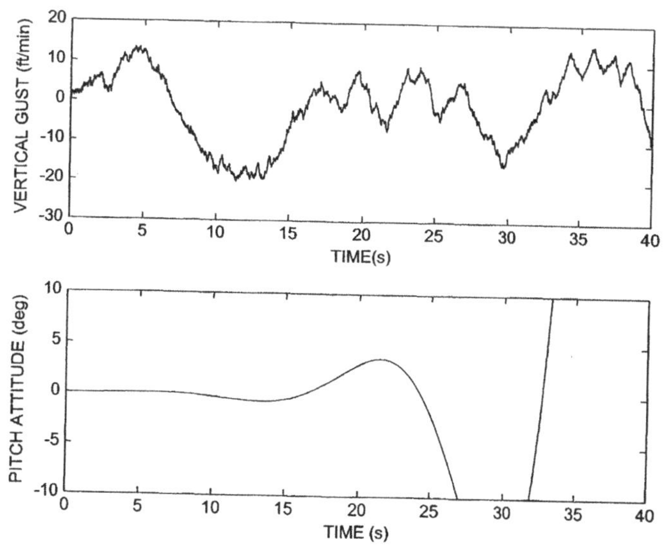

Fig. 6.24 Unaugmented gust response of example helicopter (note the y-axis scaling set for comparative purposes).

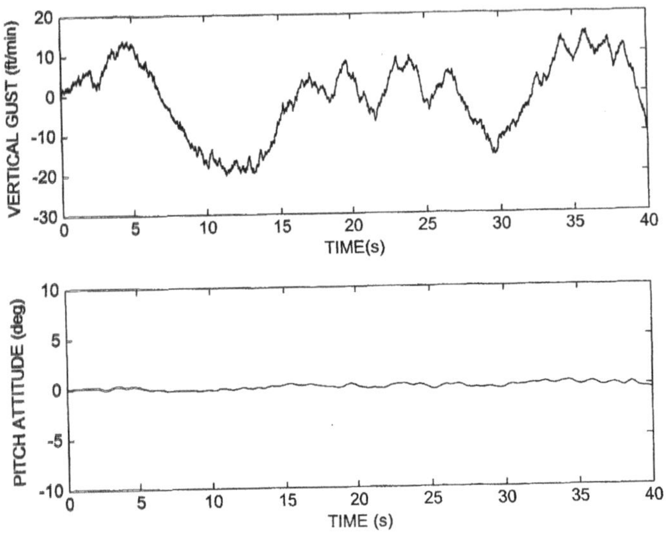

Fig. 6.25 Gust response of AFCS with unity attitude feedback.

compromise between gust rejection and control response. In its current form the modelled AFCS would completely oppose the pilot since it would treat any pilot induced attitude change as an unwanted deviation from datum. Indeed if the series actuator has sufficient authority it will return the cyclic pitch at the rotor head to the value extant before the pilot made the input. It has already been noted that angular rate feedback can improve the control response by reducing the time constant and enhance gust rejection by generating an error signal before the aircraft attitude has deviated significantly from the datum. Thus a satisfactory situation usually results if pitch or roll attitude is differentiated, or the angular rate is measured directly, and used as an additional error signal provided the attitude feedback signal is inoperative during pilot inputs. The manner in which the attitude signal is disabled will dictate whether the AFCS provides rate command or attitude command.

Two basic methods exist: one uses a signal from the stick to switch out the attitude feedback loop whilst the other uses the signal to update the attitude datum. A microswitch placed in the spring-feel unit that only allows attitude feedback when the aircraft is trimmed is an example of the former whereas a device that generates a signal in opposition to the attitude error coming from the gyro (a *stick-canceller*) is an example of the latter. The microswitch approach would provide the pilot with RCAH provided he is happy to manoeuvre the aircraft without re-trimming and then use a trim-release switch to rapidly re-centre the spring-feel unit. Such a mechanization would also require storage of the datum attitude in some form of memory with the stored value being updated as the cyclic stick is moved against the trim springs.

The stick-canceller approach operates by converting stick position into an attitude command that is used to oppose, or cancel, the signal from the attitude sensor. With this type of AFCS, therefore, a step input results in the aircraft capturing a new attitude and thus provides the pilot with the ACAH response type. Another consequence of using a signal from the stick to oppose a steady error signal from the vertical gyro manifests itself during the initial phase on a control input as 'control quickening'. In

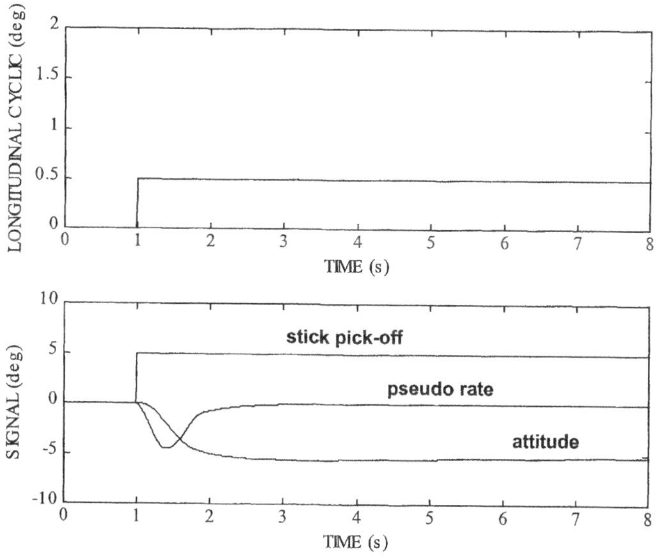

Fig. 6.26 Effect of rate feedback and stick cancellation on actuator drive signals.

order to function correctly the position signal from the pick-off must oppose that coming from the gyro and the series actuator must oppose any angular disturbances sensed by the gyro. The net effect of subtracting the stick signal from the gyro and feeding the result to the series actuator is that the contribution from the stick pick-off becomes *additive* whereas that from the gyro remains subtractive. Therefore, if the pilot moves the stick rapidly from trim the series actuator will move initially in the same direction. It will only begin to oppose the input when sufficient angular rate and attitude changes have developed (see Fig. 6.26). Note that the series actuator drive signal will be the sum of the three signals presented.

6.5.5.3 Trim follow-up systems

Some systems (Westland Sea King and Lynx) use only a combination of stick cancelling and pseudo-rate (differentiated attitude) feedback to provide a satisfactory control response and gust rejection. Whereas others are more complex, employing both attitude and rate gyros with force disconnects to alert the AFCS to an intended deviation from the current attitude datum. Ultimately, possibly even without the requirement to provide outer-loop modes such as airspeed hold or auto-ILS, the limited authority of series actuation will lead to some degradation of the long-term attitude-keeping performance. Whilst simply extending the travel of the series actuator would overcome this deficiency it is not a favoured solution since it would impact the safety case for the system (actuator runaways would have more severe consequences) and may degrade the control response characteristics due to excess damping. Usually a better alternative is to provide a mechanism that attempts to centralize the series actuators as they approach their end-stops thus ensuring that they operate with approximately full authority at all times. The key to a trim follow-up system is a device that can move the cockpit control thereby shifting the cyclic, collective or tail rotor pitch datum about which the series actuator operates. Usually the trim motor, which

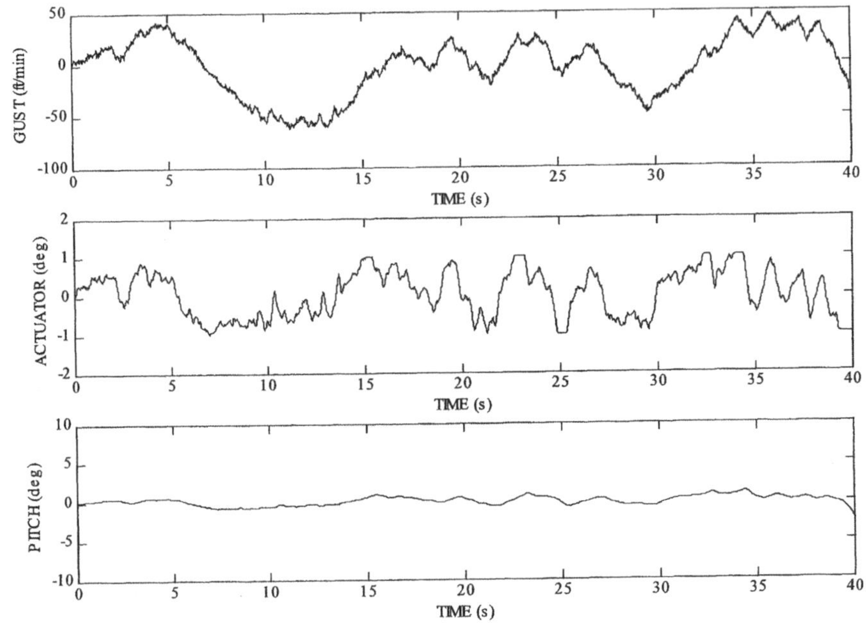

Fig. 6.27 Effect of trim follow-up on gust response.

doubles as the parallel actuator in the cyclic channels, provides this trim follow-up capability. Activating a trim follow-up system also requires a logic circuit that can determine when to send a drive signal to this actuator. Since the series actuator position is often measured in order to provide a feedback signal to the actuation circuit, and possibly drive some form of cockpit indication, it is a simple matter to use this data to trigger movement of the parallel actuator/trim motor. As expected the application of trim follow-up improves gust rejection by simply increasing the effectiveness of the series actuator (compare Fig. 6.27 with Fig. 6.25).

Figure 6.28 confirms that trim follow-up extends the authority of the system by showing that the activity at the main servo now exceeds $\pm 1°$. Figure 6.28 also highlights one of the potential disadvantages of trim follow-up systems for some pilots: unnatural and unexpected control activity in the cockpit. Although the pilot has selected a trim position for the desired airspeed the follow-up system causes the cyclic stick to migrate fore and aft depending on the degree of extra authority required by the series actuators to counter the atmospheric turbulence. Since the pilot has not selected a true autopilot mode he may be operating 'hands-on' and could find the uncommanded stick movements disconcerting.

Inappropriate stick forces during rapid control movements are perhaps of greater concern than stick activity during trimmed flight in turbulence. Suppose the pilot makes a small aft step input. Initially the series actuator will move in the same direction. This may cause the trim motor to activate driving the stick further aft. A short time later the aircraft starts to pitch-up and the rate feedback signal, augmented eventually by the attitude feedback signal if it is still retained, causes the series actuator to oppose the initial input. If this reversal is sufficiently large the trim motor may again activate but this time the stick will be driven in a direction opposite to the

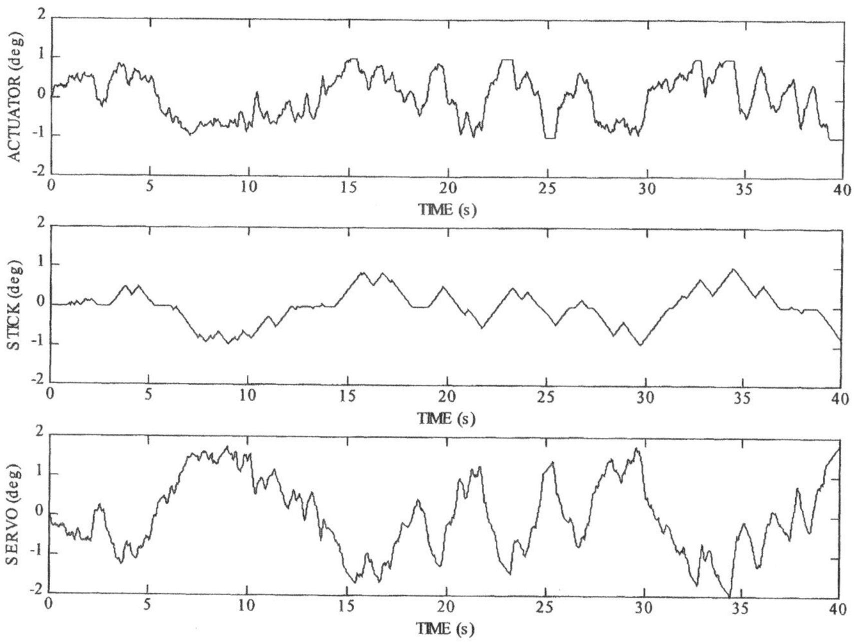

Fig. 6.28 Effect of trim follow-up on main servo travel.

original input. Eventually the series actuator will approach null and the trim motor will be signalled to drive the stick back to the position selected by the pilot at the start of the input. If the pilot grips the control firmly during this process he may notice a disconcerting change in the feel of the control as the trim system is moved relative to the control. Problems with control response can be alleviated by reducing the trim rate. A slower acting trim motor is less likely to give perceptible stick migration during rapid controlling and will reduce the risk of catastrophic damage following a hard-over of the trim system. Rapid removal of unwanted stick forces following a gross change in aircraft attitude, a possible failing of a slow trim system, may be achieved by a trim release switch. Figure 6.29 shows the step response of an AFCS with a slow-acting trim follow-up system and Fig. 6.30 confirms that the gust rejection is still satisfactory.

Note that the '*unrestricted stick migration*' portrayed shows the hypothetical situation that would arise if the pilot made a step input and then allowed the stick to move in his hand as a function of the trim follow-up requirements. In reality the pilot would maintain a firm grip on the stick commensurate with the type of input made, in which case he would feel a change in the stick forces instead.

6.5.5.4 Typical AFCS systems for IFR flight

The foregoing discussion has served to detail the qualities of typical attitude-based stability augmentation systems. For IFR flight and the more demanding operational roles the pilot often requires more than pitch and roll attitude stabilization and enhanced yaw damping in the low speed regime. Thus these systems usually provide heading hold/turn co-ordination and possibly height hold. Consequently they will

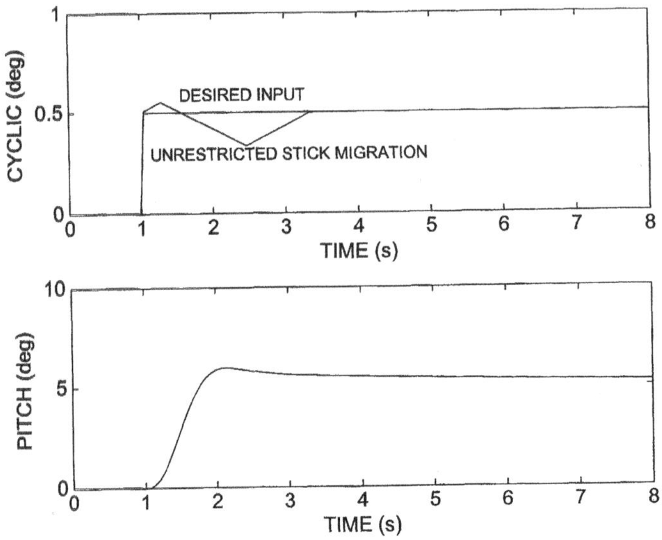

Fig. 6.29 Controls response of an AFCS with trim follow-up.

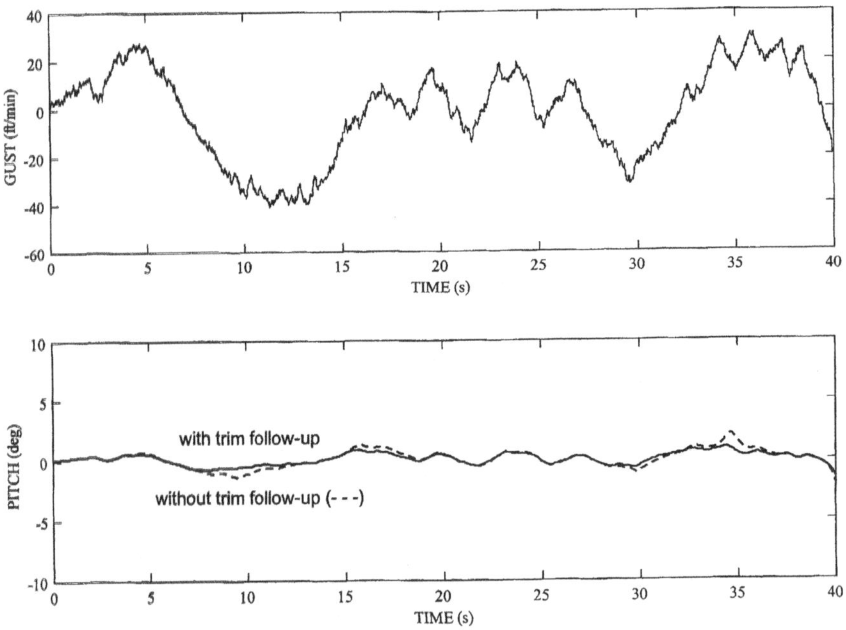

Fig. 6.30 Gust response of optimized trim follow-up AFCS.

usually consist of a vertical gyro and at least a yaw rate gyro that feed signals to series actuators for stability with a gyro-compass feed for heading hold and an altitude sensor (barometric or radar) for height hold. Although four-axis follow-up trim is becoming increasingly popular, many in-service AFCS have parallel actuation only in the yaw and collective axes with conventional trim provided in the pitch and roll channels. Where autopilot modes, such as automatic transition to/from the hover, are provided the trim motors serve as parallel actuators. Schematics for each channel of a typical AFCS of this class are shown in Figs 6.31 to 6.34.

Differentiated attitude (or true angular rate) can be used to generate rate feedback in the pitch and roll channels but is inappropriate in the yaw channel due to the large voltage change associated with crossing magnetic north. Thus it is usual for AFCS manufacturers to use direct measurement of yaw rate (see Fig. 6.33). Heading hold is usually achieved by comparing the actual heading from the gyro-compass, with the value stored in memory at the instant of engagement of the hold. It is common for

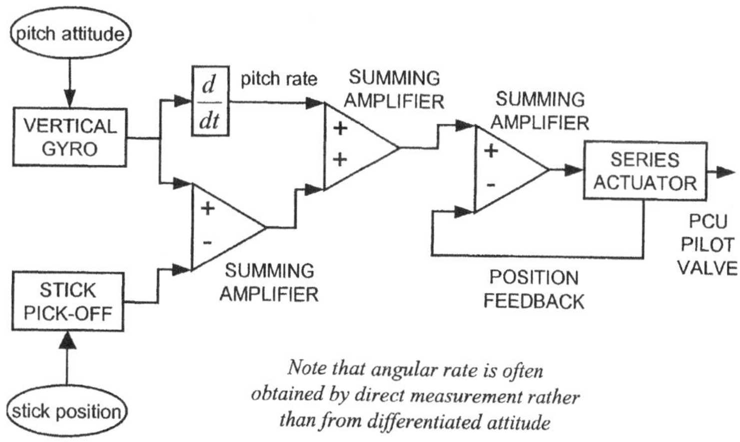

Fig. 6.31 Generic AFCS pitch channel.

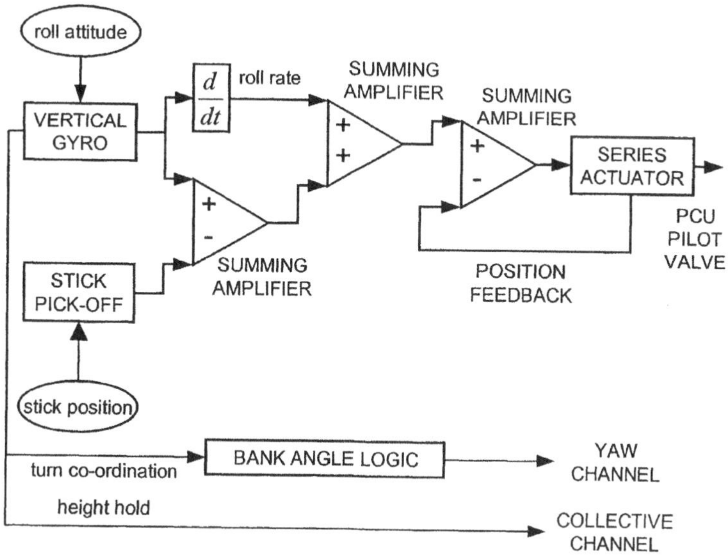

Fig. 6.32 Generic AFCS roll channel.

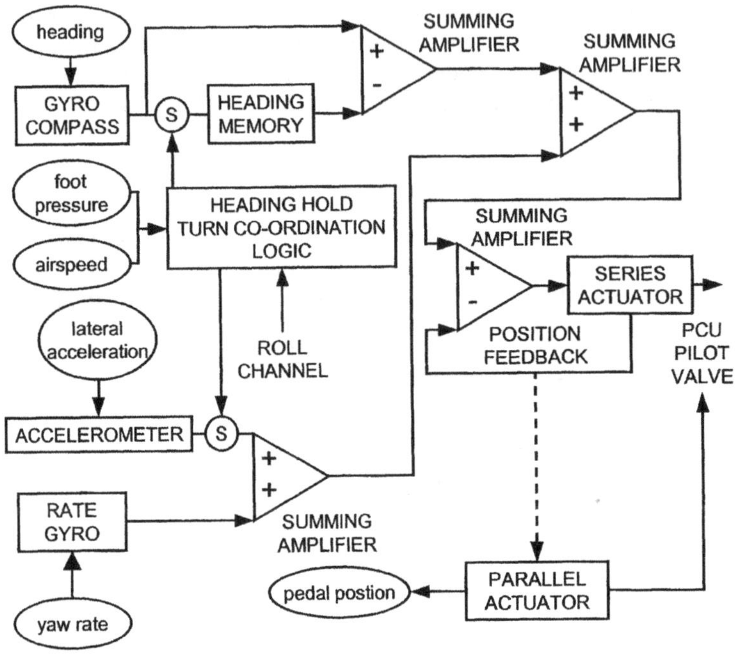

Fig. 6.33 Generic AFCS yaw channel.

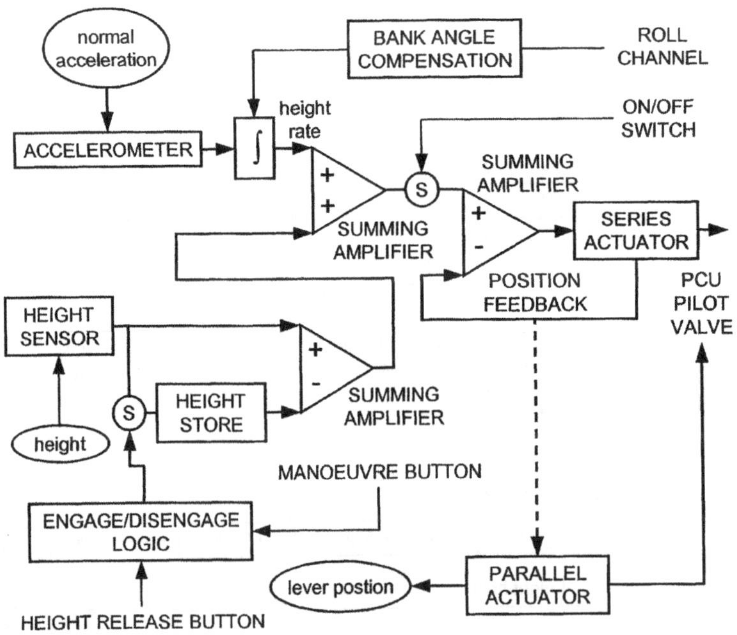

Fig. 6.34 Generic AFCS collective channel.

turn co-ordination to be provided, as well as ensuring adequate heading hold up to the lateral velocity limits of the helicopter. Therefore, in addition to featuring parallel actuation to extend the authority of the series actuator, it is usual to find a sensor measuring lateral acceleration or sideslip. Some manufacturers choose to mechanize heading hold directly through a parallel actuator whereas others use the series actuator and follow-up trim.

The response of a rotor to a change in vertical velocity means that the heave axis of a helicopter is inherently stable and therefore it is not always necessary to have series actuation in this axis. A slow-acting parallel actuator is often sufficient to ensure adequate barometric altitude hold. A tighter control of altitude, often associated with a radar altitude hold, will require both series and parallel actuation in the collective channel. Altitude hold is analogous to heading hold in that the actual altitude is compared with a value in memory. Improved altitude keeping can be obtained by using integrated normal acceleration to generate a pseudo altitude rate. This pseudo rate will require bank angle compensation in turning flight since the accelerometer is fixed relative to the body axes whereas the vertical velocity relative to the earth is needed for altitude keeping. Engagement and disengagement of an altitude hold is often less complex than a heading hold system. Typically there will be an altitude hold release switch on the collective lever along with an on/off switch located within the AFCS control head. Some systems feature a 'manoeuvre button', also located on the collective, to allow rapid selection of a new altitude datum by the pilot. Whilst this button is depressed the altitude memory is continually updated with the actual measured altitude.

The requirement to provide turn co-ordination complicates the basic stability augmentation system. Heading hold and sideslip control are functions of the yaw channel but it is most natural for the pilot to initiate a turn by means of lateral cyclic stick deflection. It is therefore necessary for the AFCS to incorporate some linkage between the roll and yaw channels so that heading hold can be unlocked when the pilot moves the cyclic laterally or a bank angle above a certain threshold is sensed. Unlocking the heading hold is simply achieved by allowing the heading store to be updated by the current reading from the gyro-compass. In the low speed regime the pilot will need to re-position the helicopter laterally without initiating a turn and thus an airspeed switch is often needed. Simpler systems use foot pressure to make a switch on the yaw pedals as an indicator that the pilot wishes to commence a turn thereby obviating the need for any bank angle logic or airspeed switching. However even sophisticated turn co-ordination logic will require a sensor in the yaw circuit to enable the pilot to perform spot turns.

Although stick cancelling can be engineered to provide a convenient and satisfactory method of informing the AFCS that the pilot wishes to change the aircraft attitude it is not the only possible solution. Some systems (see Fig. 6.35), store the datum attitude in memory and arrange for the pilot to modify this memorized value by various means. For small attitude changes the pilot can use a beeper-trim to increase or decrease the stored attitude as well as changing the stick position. Alternatively, when making large attitude changes the pilot can depress a trim release switch, which in addition to removing the trim forces causes the attitude memory to be updated with the current measurement from the vertical gyro. It should be noted that the application of trim release will also temporarily remove rate stabilization since the open electromagnetic clutch will prevent the series actuator from effecting control. Stick movement

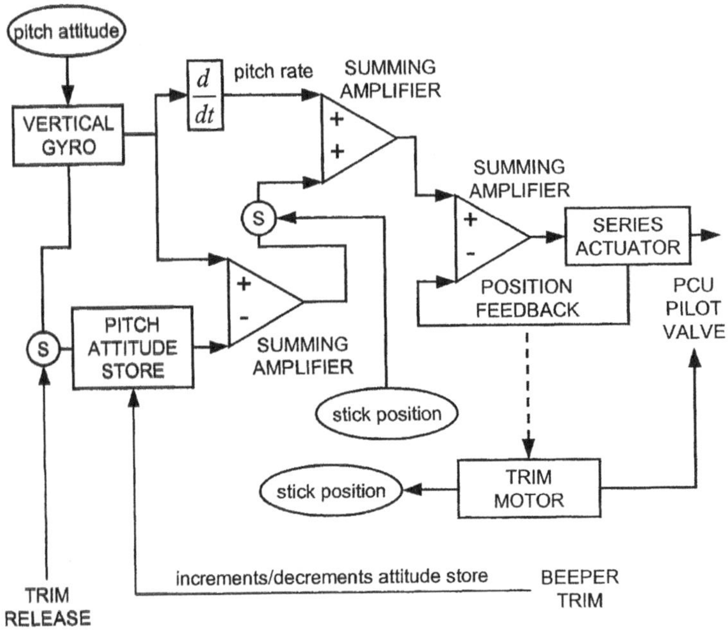

Fig. 6.35 Alternatives to stick cancellation – pitch channel.

without retrimming (against the spring feel loads) is used to open the attitude feedback loop allowing the pilot to manoeuvre around the datum attitude without updating this datum.

6.5.6 Movable horizontal stabilizers

As noted in Chapter 4 helicopters have a horizontal tailplane to improve both manoeuvre stability and dynamic longitudinal stability. Sometimes these surfaces need to be large in order to provide the necessary influence. However, a large tailplane can itself engender several problems such as excessive TCWP, excessive download (and associated poor performance), pitch-up during low-speed flight and high nose-up pitch attitudes in the hover. At the expense of weight, complexity and reduced reliability, a number of these problems can be overcome by making the tailplane incidence adjustable under the command of the AFCS. Such devices are called programmable stabilizers or, in American parlance, 'stabilators'. Movable horizontal stabilizers will be discussed separately from the host AFCS as only a few helicopters are fitted with these devices and they offer a range of novel AFCS enhancements and some unique failure characteristics.

There are several reasons for incorporating stabilators: some associated with enhancing performance whilst others overcome the handling problems often associated with large tailplane surfaces. The key reasons are listed below:

- To align the tailplane with the rotor downwash in low speed flight in order to improve performance, reduce the nose-up hover attitude and to reduce the tendency to pitch-up during transitions to forward flight.

- To improve 'cockpit' longitudinal static stability (both collective fixed and apparent) by programming the stabilator trailing edge up (TEUP) as airspeed increases thus causing the pilot to have to apply more forward cyclic to stabilize at an increased IAS and vice-versa.
- To improve longitudinal dynamic and manoeuvre stability by programming in response to a pitch rate signal and, possibly, a pitch attitude signal.
- To reduce TCWP by programming trailing edge down (TEDN) with increased collective and vice-versa. As this function is ineffective in low speed flight it is often phased in with airspeed.
- To oppose the effects of a pitching moment due to sideslip (M_v) on aircraft with a canted tail rotor by programming in response to a sideslip or sideforce signal.
- To improve FOV by programming, or being commanded, TEDN at NOE airspeeds, typically 30 to 70 kts, thereby reducing nose-up attitude.

Increasing the stick migration with speed can be matched to a reduction in the pitch attitude change with airspeed that may result in improved crew and passenger comfort due to a level fuselage deck in high-speed cruise flight. Equally performance may be improved since the relatively level fuselage will typically produce less drag although the extra trim drag from the stabilator will tend to reduce this effect. However, reducing the variation of pitch attitude with airspeed is not always beneficial. The pitch attitude hold will be less effective at maintaining airspeed leading to a requirement for an actual airspeed hold function which, unfortunately, is not always satisfied.

As stabilators are normally large and in the hover they may deflect up to 45° trailing edge down any failure that could cause it to runaway TEDN at high forward speed could be potentially catastrophic. The large nose-down pitching moment generated would probably exceed the available cyclic control power even if the pilot were able to react in time. A similar, although potentially less serious, trailing edge up failure case also exists. Consequently, even though stringent reliability requirements are applied, the stabilator slew rate is usually a compromise between normal operation when a fast rate may be necessary and the failure case when a slow runaway is desirable. Most stabilator systems are thus duplex and feature extensive safety and monitoring devices. These normally take the form of a comparator that checks the positions of the duplex actuators and the signals that drive them. A mismatch typically provokes a complete stabilator freeze and an aural warning. Once the automatic functioning has frozen the stabilator can usually be controlled manually provided there has been no mechanical seizure.

Although normally controlled by the AFCS, most stabilators have a manual mode which may be used in flight to give a measure of control over pitch attitude and, possibly, vibration. As controls for this type of operation need to be accessible from the primary inceptor they are normally placed on the collective where they may be operated by the left thumb. A safety device may be fitted to prevent manual operation above certain airspeeds to prevent the inadvertent overpowering of the available cyclic control. Some installations feature a semi-automatic NOE or approach mode which, when preselected by the pilot, will cause the stabilator to slew TEDN according to a revised schedule designed to cause a nose-down pitch attitude and so improve the FOV.

In an emergency, following a system freeze for example, it may be necessary to slew the stabilator manually. In this case, the stabilator is set to a fixed position that corresponds to a particular airspeed schedule as indicated in the flight reference cards or pilot's notes. This is normally a simple selection of stabilator level above a certain airspeed or maximum TEDN below about 40 KIAS. With a frozen stabilator dynamic functions such as rate damping and sideslip correction are no longer available and so a minor degradation in aircraft handling qualities usually results. Prompt pilot action may be required to prevent a disastrous nose-down tuck if the stabilator should fail to programme correctly during a rapid transition from the hover. As the pilot will have little time to identify and operate the slew-up switch one is usually incorporated into the cyclic grip. The switch is usually designed so that pulling it aft causes an immediate slew up. This is the natural sense in terms of its action on the pitch attitude of the aircraft.

The stabilator is normally programmed by an airspeed signal derived from the normal pitot-static system; its movement will thus be subject to any PEs that may be present. This is particularly important at low IAS where pressure errors may adversely affect stabilator programming during transitions, which is exactly when the fastest and most accurate programming is required.

6.5.7 Autopilots

In order for an autopilot to work successfully it requires authority over the relevant flying controls. This is usually exercised through an existing, and in most cases separate, AFCS that acts as a SAS and/or ASE when the autopilot is not operating (see Section 7.5). Although it was stated above that some ASE installations include heading hold and altitude hold, these functions will often be provided by the autopilot circuitry when one is fitted, simply because other, more complex, modes will make use of heading and altitude control. Before progressing it is necessary to review the terms used by certain AFCS manufacturers. The term 'autopilot' has been defined herein to mean an AFCS offering automatic flight path control, navigation and mission-related modes. Unfortunately the term 'autopilot' is used by certain manufacturers to refer to the SAS/ASE since it is this system that flies the helicopter when under automatic control. Others refer to this device as a 'helipilot' for much the same reason. The device that drives the SAS/ASE is often called a 'coupler', or 'flight path computer', since it takes navigation information and couples it to the flying controls. Most couplers offer a lesser mode whereby the flight path information is fed to command bars on the ADI and HSI in order to guide the pilot.

There are two basic types of autopilot: 3-channel or 4-channel. A 3-channel system controls the helicopter in pitch, roll and yaw with the pilot providing the necessary compensation in the collective channel. A 4-channel system provides automatic control of all flying controls and no compensation is required. The typical system will have series actuators operating in pitch, roll and yaw (for rate damping and initial attitude stabilization) and parallel actuators (or AFCS operated trim motors) operating in all channels for autopilot control and/or trim follow-up. Vertical gyros, and possibly rate gyros, will be fitted to provide signals for the SAS/ASE. The autopilot computer will require signals from the gyro-compass, pitot-static system, lateral accelerometer or sideslip ports, radar-altimeter and navigational radio/satellite receivers. Other equipment may be required depending on the mission modes available, these include Doppler receivers and inertial/GPS navigation systems. The modes provided by a typical autopilot can be divided into three groups: forward flight holds, navigation modes and mission modes. The navigation modes are often provided to enable single pilot IFR operation and are therefore usually civil orientated. Mission modes will consist of modes peculiar to the role of the helicopter, such as autotransition, hover position hold and programmed search patterns.

6.5.7.1 *Forward flight holds*

Altitude hold. If the autopilot is a 4-channel system then the altitude hold will operate with the collective parallel actuator making corrective inputs in response to altitude deviations sensed by either the static system, a separate barometric capsule or a radio altimeter. A 3-channel system, however, operates somewhat differently with altitude corrections being made through the pitch channel. Since pitch attitude is being used to control altitude it is not possible to retain control of airspeed. Compensation for the actions of the altitude hold on the airspeed is left to the pilot. Indeed if the pilot were to raise the collective the airspeed would increase but it would be difficult to estimate by exactly how much. Obviously, the altitude hold mode of a 3-channel system will only work satisfactorily above V_{imp} and therefore the hold is often disabled below a certain speed, typically 60 KIAS.

Airspeed hold. In all autopilots airspeed hold is achieved using the pitch channel. Once engaged the pitch parallel actuator will make corrective inputs in response to airspeed deviations, sensed by the pitot-static system, from the value set at the instant of engagement. Once again if a 3-channel system is installed the pilot will have to compensate for the effect of the autopilot by making appropriate collective inputs, this time to maintain altitude, and as before the hold will only operate satisfactorily above V_{imp}. It is clear that a 4-channel system is required if simultaneous operation of altitude hold and airspeed hold is desired.

Vertical speed hold. The operation of the vertical speed hold is very similar to the altitude hold. A 4-channel system will use collective and a 3-channel system will use the pitch channel. In a 3-channel system the pilot will have to apply collective to maintain airspeed as the autopilot controls the rate of climb or descent. Static pressure signals will be used by the autopilot to generate the appropriate error signal.

Heading hold. Naturally heading *hold* is achieved through the yaw channel, although most autopilots use the roll parallel actuator for large heading *corrections*, greater than 2° for example, with the yaw actuator maintaining the helicopter in balance in response to signals from a lateral accelerometer or sideslip ports. The yaw channel is used primarily because it is assumed that any deviations from the datum heading will be small and therefore sufficient control can be exercised through this channel without any changes to the position of the roll parallel actuator. Some AFCS make a more positive distinction between heading hold, which is retained as the yaw channel ASE/ATT mode, and heading *selection*, or steerage, which is an autopilot mode achieved primarily through the roll axis. Most systems alter the control law at some value of forward speed to take account of the increasing effectiveness of the fin.

6.5.8 Navigation modes

Now that the helicopter is capable of airspeed, altitude, heading and vertical speed control it is possible to use these modes in combination to provide automatic navigation and approach to a landing site. Navigation modes include heading selection, VOR, ILS, Back Course, Go-around and On-route Navigation.

Heading selection. The heading selection (HDG) mode differs from heading hold in that large heading changes may be required and, therefore, the mode is achieved through the roll channel. As mentioned above, heading selection may be a mode performed by the autopilot separate from the ASE heading hold function. Alternatively both modes can be performed by the autopilot. The heading error signal may be scheduled as a function of TAS in order to ensure the helicopter executes 'Rate 1' turns. In addition a bank angle limit may be included to prevent excessive roll attitude changes at high speed.

VOR tracking. VOR tracking is a mode common to most autopilots and is designed to provide automatic intercept, capture and tracking of a selected VOR radial. Typically the pilot will tune the navigation receiver to the desired VOR frequency, select a VOR radial and using the HDG mode set a heading that will intercept the desired course. As the helicopter approaches the beacon the VOR signal is monitored for beam deviation, beam rate and validity. At the appropriate time the helicopter is steered towards the beacon and eventually the radial will be captured, with the helicopter flying in the desired direction. Steerage on to the radial is usually achieved in stages either by using a 45° 'cut' or successive cuts of up to 30°.

This feature is designed to avoid overshooting the beam when large intercepts are used at high speeds. Some autopilots automatically reduce the bank angle and roll rate limits to avoid over-controlling as the beam centre is approached. On entering the cone of confusion located overhead the beacon an Over Station Sensor (OSS) warning will be given to the pilot. The link between the radio receiver and the autopilot is severed and the system will usually revert to heading hold. Once the signal becomes usable again the link is restored and radial tracking is resumed. Some systems will increase the bank angle and roll rate limits on re-engagement to ensure rapid re-acquirement of the beam centreline in gusty conditions. If the pilot selects a new course, or radial, whilst over-flying the beacon the autopilot will usually steer the helicopter on to this heading using the HDG mode. On leaving the cone of confusion when the VOR signal is reacquired the system will adjust the heading to track the centreline of the new outbound radial. Consequently the VOR tracking logic may cut the corner avoiding the need to overfly the actual VOR beacon. If this logic is also applied to GPS or IN based waypoint tracking problems can occur if the pilot wishes to use the automatic navigation function to overfly a particular point of interest and then execute a turn immediately afterwards.

ILS mode. Initially the ILS mode operates in a similar manner to the VOR mode described above. The navigation receiver is tuned to the appropriate localizer frequency and a beam intercept course acquired using the HDG mode. As the helicopter closes on the localizer beam the radio signal is monitored and at the appropriate time a steerage signal is sent to the roll channel. The helicopter will thus capture and track the beam. Once again the helicopter may be steered on to the correct heading through a series of 'cuts'. Additionally the heading error signal may be modified as a function of TAS to avoid an overshoot at high speed.

Typically overcontrolling is avoided by reducing the bank angle and roll rate limits as the beam centre is approached. The helicopter will continue along the localizer beam until the glide slope beam is acquired. Following acquisition of this signal existing holds are disconnected. If the autopilot operates in only three channels any longitudinal mode will need to be decoupled as the pitch channel is used for glide slope maintenance. Such decoupling usually takes place automatically leaving the pilot to monitor the airspeed and adjust the collective lever position as appropriate. If, however, a 4-channel system is fitted then airspeed hold can be retained since the glide slope command will be fed to the collective channel.

It is usual for the gain applied to the glide slope signal to be varied as a function of altitude as sensed by the radio altimeter and perhaps passage over the middle marker. The reduction in gain is designed to avoid overcontrolling as the glide slope beam narrows towards the landing site. At a certain radio altitude the helicopter will 'autolevel', through the pitch and/or collective channel, to avoid ground contact. The helicopter will then continue to fly along the runway under localizer guidance awaiting pilot action to either disengage the hold, and commit to a landing, or abort the landing by engaging a Go-Around mode.

Go-around mode. Although not a navigation mode, Go-Around (GA) is often associated with an automatic ILS mode. GA is activated, in most cases, by means of a switch placed on the collective lever. The GA mode enables the pilot to abort an automatic approach with the autopilot causing the helicopter to adopt a positive rate of climb. In 3-channel systems the pilot will have to raise the collective to maintain the airspeed. With a 4-channel system the airspeed can also be controlled and the AFCS would usually program a speed change to set and hold the airspeed at the speed for best rate of climb, V_Y.

Back-course mode. The Back-Course (BC) mode operates in a similar manner to the initial phase of the VOR mode. The BC mode provides for automatic intercept, capture and tracking of the back course localizer signal. The control law gains may be adjusted since the helicopter will be closer to the localizer by the length of the runway. Capture of the glide slope is usually inhibited to prevent any possibility of a rate of descent being commanded.

On-route navigation modes. The coupling of a navigation computer to the autopilot provides on-route navigation modes. The navigation computer will use signals from internal or external sources to determine the helicopter's position in relation to a pre-programmed track. On receipt of an error signal the autopilot will use the HDG logic to maintain the desired ground track. Sources of ground position data include, Doppler, GPS, or an INS. Modern systems feature combinations of these systems, the signals of which are mixed and filtered to generate a very accurate fix of the present position of the helicopter. Navigation systems intended for ASW or SAR usually allow the pilot to program combinations of waypoints, or leg-lengths, in order to execute set search patterns over a target area. Navigation modes such as these help the pilot maintain a good lookout whilst ensuring a precise ground track.

6.5.9 Mission modes

Mission modes are included in autopilots intended for specific roles. As such they are not usually included in AFCS fitted to civilian helicopters unless they are expected to have a SAR capability as well as being suitable for single pilot IFR.

Auto-transition mode. Helicopters intended for ASW or SAR operations will usually have an autotransition mode provided by the autopilot. From a start point at a radio altitude of about 200 ft and airspeed close to that for minimum power, the AFCS will execute a programmed manoeuvre using both the pitch and the collective channel. Thus under that action of the autopilot the helicopter will decelerate and descend to hover at pre-selected radio altitude. It is usual to arrange that appropriate movements of the longitudinal cyclic or collective inceptor by the pilot will cause the programming of either altitude or airspeed to cease thereby enabling him to fly a modified profile or abort the manoeuvre. At the end of the program the helicopter will usually enter a hover with plan position held using signals from the Doppler and/or inertial/GPS system. Radio altitude will be maintained using the basic altitude hold mode. Certain autopilots have an additional mode that programs the helicopter from higher altitudes, up to 2000 ft, and higher speeds to a flight condition from which the basic autotransition mode can be engaged.

Hover position hold. Hover position hold is achieved through the pitch and roll channels. The sensors used to generate the error signals will depend on the type of hover required. Plan position can be maintained accurately using a combination of signals from longitudinal and lateral accelerometers (translational acceleration), Doppler receivers (translational rate) and GPS (present position and translational rate). Hover position relative to a dipping sonar is often maintained using cable angle sensors. In some helicopters a rudimentary hover hold capability is provided using the basic attitude hold mode.

Translational rate command. Translational rate command (TRC), in the form of an auxiliary hover trim (AHT), has been available for a number of years on certain AFCS designed for SAR operations. Deflections of the inceptor, located at a crewman's station, commands a rate of horizontal translation. Centralizing the inceptor will cause the system to maintain the selected rate as sensed by the Doppler radio receiver. Full deflection usually commands a fairly modest rate, typically 10 kts, in keeping with the hover trimming function of this mode.

Automatic circuit. A relatively new autopilot feature, again provided for SAR operations, is the automatic circuit or Mark on Target (MOT) mode. The pilot overflies a desired point on the surface, usually the location of a survivor, and engages the mode. The existing flight path control functions of the AFCS then turn the helicopter downwind and initiate a deceleration and descent to the autotransition gate conditions. Some time later, using Doppler information, the helicopter turns into wind and at the appropriate point the autotransition mode is engaged and the aircraft descends and slows to a hover at an altitude pre-selected by the pilot. The navigation system will usually arrange for the hover to be established a short distance from the MOT point so that the AHT mode can be used for final plan positioning.

Dip-to-dip. Another relatively new AFCS mode is the so-called 'dip-to-dip' function. This uses a combination of hover-position hold, automatic transition and Doppler/IN/GPS based navigation to move the sonar equipped helicopter from one dipping location to another. The aim of this mode is to move the helicopter in the shortest possible time commensurate with the available performance and the ambient wind conditions.

Chapter 7
Systems Testing

7.1 METHODOLOGY

All helicopters, whatever their role, incorporate a number of systems to enable the aircraft to fly. Even the simplest of helicopters have flight controls and powerplant systems. With modern military rotorcraft it is not an exaggeration to say that the aircraft is merely a means of taking the various systems to the required location. The recent policy of awarding prime contractor status to system integrators rather than helicopter manufacturers illustrates this last point.

The systems onboard a rotorcraft can be divided into two major categories. The first category contains those systems that are required to allow the helicopter to fly but which do not require significant interaction with the crew other than monitoring. Examples of this category include engine lubrication, hydraulics, and electrical systems. Testing of these types of systems is beyond the scope of this book. The second category comprises systems that do require significant interaction with the crew to enable the aircraft to accomplish the mission. Examples of major types of system from this category include:

- flight controls,
- stability augmentation,
- cockpit displays and controls,
- air data,
- powerplant(s),
- navigation,
- weapons,
- sensors.

The first system on the list, flight controls will not be covered further in this chapter as the test techniques have already been described. The last three systems are examples of systems that are not required for flight but are needed if the mission is to be conducted safely and successfully. The principles involved in testing these *mission systems* are covered here but a detailed examination is again beyond the scope of this book. Although the function, design and operation of aircraft systems may vary widely it is possible to define some common principles to employ when evaluating any system. It is convenient to divide these into preparation activities and test activities Figure 7.1 shows the principles.

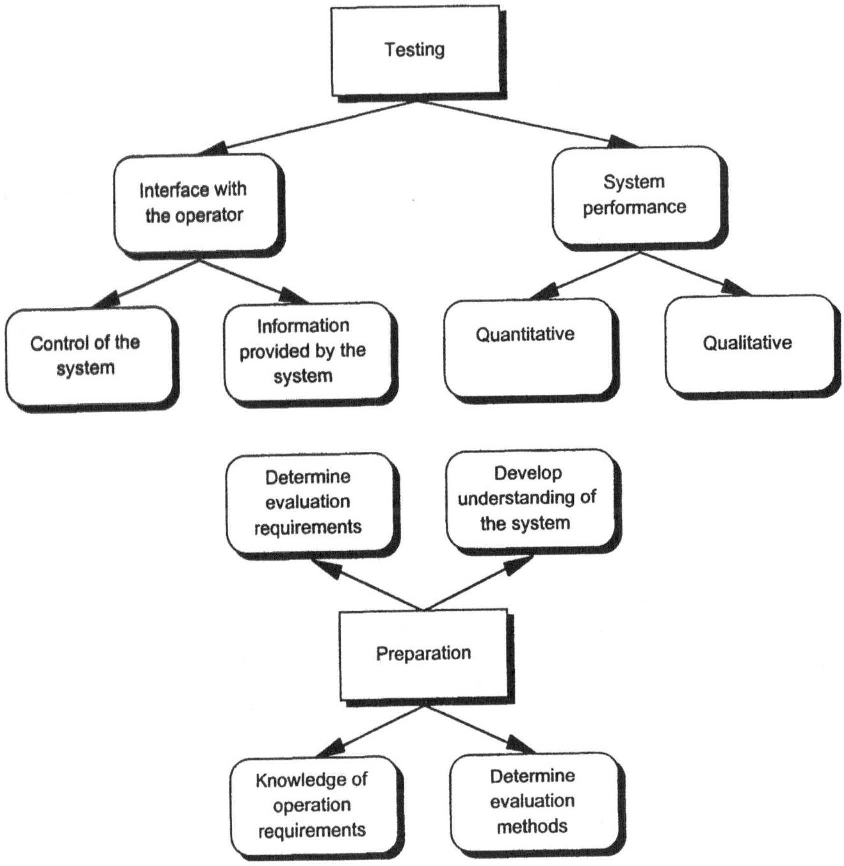

Fig. 7.1 Systems testing methodology.

7.1.1 Preparation for systems testing

The first, and most important, stage of systems testing is to determine the exact operational requirements. In other words what precisely is needed from the system to allow the pilot or operator to achieve the task or mission. The word *precisely* is important here because it is not just a question of defining the function that the system must meet but also defining the degree of accuracy required. Taking the example of a navigation system, the test team needs to understand what information the pilot will need (heading to steer, time to go, ETA, track error), at what point in the mission each item of information will be required, what the crew need to do with the information, and to what degree of accuracy the navigational information will be required. Staying with the last point, it is clear that if the navigation system feeds into an on-board weapon system a higher degree of accuracy will be required than if the information is only required for steering information by the crew. Similarly the accuracy needed in an air data system will also need to be higher if it has an input into the firing solution of a gun system.

Once the operational need is clearly understood a thorough understanding of the system under evaluation is needed. Part of the process is for the test operator to become proficient in its use. Becoming fully 'worked-up' can prove difficult with new systems as the training material and system documentation are sometimes not available. However, allowing sufficient time to achieve proficiency is included in the test planning. The effect of partial and total system failure is also addressed during the preparatory stage by conducting a failure modes, effects and criticality analysis (FMECA).

Armed with an understanding of the operational requirements and knowledge of the system itself the test planning takes place. As with all testing the aim is to evaluate the system as comprehensively as possible under operationally realistic conditions. The facilities required to conduct the programme will vary according to the system but this type of testing can often require complex instrumentation and significant external resources such as ranges, airspace and radar targets.

7.1.2 Test conduct

Systems testing can be divided conveniently into three main areas of interest: the interface with the operator, performance, and system integration. The interface with the operator breaks down in turn into the areas of information and control. Information covers the data input that the system needs from the operator, evaluating the ease with which it can be provided, and also covers the information provided by the system to the operator. In the later case it will be necessary to answer such questions as:

- Is the information presented in the required format?
- How easy is it to assimilate?
- What workload is required to monitor system performance?
- Are failures indicated adequately?

Assessing control aspects involves examining the range, degree and ease of control provided. In essence it answers the question 'Can the operator make the system perform the required function within its capabilities and can this be achieved easily?' It should be remembered that there may be limits to the amount of control that an operator should be given. For example it would not be desirable to allow a pilot the option of completely disabling visual or audio warnings. Another example would be providing individual lighting controls for each cockpit instrument, this would give the pilot complete control but it would be difficult and time consuming to exercise it.

Clearly evaluating the adequacy of control and the operator interface are largely a subjective process, while evaluating performance can often be an objective process. The system performance can be broken down into quantitative and qualitative performance. In the case of a navigation system the quantitative performance or accuracy is determined by comparing actual position measured from accurate maps or pre-surveyed points with displayed position. For a weapon system the accuracy of firing tests is measured against required accuracy. With other systems, such as displays and piloting vision aids, assessing performance relies to a large extent on the qualitative opinion of the pilot or operator. When dealing with qualitative aspects of systems testing it is important to describe fully the way that the performance impacts on the conduct of role tasks. As in all testing the fundamental question is 'Does the system contribute sufficiently to the efficient conduct of the mission?'

The last area to be discussed is systems integration. In the past most aircraft systems were stand-alone and the pilot was required to monitor the status and output of all the systems individually: the only integration which took place was in the pilot's head! This led to a high level of pilot workload. As more and more systems have been added to rotorcraft and operational tasks have become more demanding the need to off-load the crew has become increasingly important. The systems on board should share information so that the pilot or operator is not required to enter information more than once, pass information from one system to another or make unnecessary control selections. Examples of good integration might include automatic display of a cable hover screen when the sonar is armed or display of a pre-landing checklist as the final point on the navigation plan is approached. When assessing the adequacy of integration within an aircraft realistic simulated missions are conducted and the actions required of each crew member analyzed with regard to the systems. If any of the actions could be eliminated or made easier then the integration is deficient.

7.2 COCKPITS

It is hard to overestimate the importance that the design of the cockpit has on a pilot's perception of the qualities of a helicopter. It is not only the place where the pilot is accommodated but is also where he or she exercises control over the craft and all its systems. In addition all sources of information from both the aircraft itself and the external environment find their focus in the cockpit. Like early aeroplanes the first helicopters had very simple cockpits with little more than the flight controls, some basic instruments and simple engine controls. As the number of systems that the pilot had to interact with grew, helicopter cockpits became complex areas of controls and instruments. It is true to say that the ergonomic aspects of design have not always been addressed with the importance that they deserve and as a result rotary-wing pilots have had to cope with major deficiencies in the design of their cockpits.

7.2.1 Assessment methods

When assessing a cockpit as a whole or any individual part of it, it is vital to keep the role of the aircraft firmly in mind at all times. Clearly this requires a detailed knowledge of the role and precisely how operational crews will conduct each aspect of the mission. There are two distinct approaches to conducting the assessment, both of which are taken in a full evaluation.

Initially each item in the cockpit is evaluated individually. Gauges are checked for size, markings, location, etc., while each switch is tested for ease of operation, labelling, provision of guards against inadvertent operation, and so on.

The second and more important approach is to conduct the assessment in the context of a realistic mission. This will discover if the individual cockpit features, when combined, are ideally suited to the role. The various mission profiles are broken down into individual tasks and the actions that each crew member will need to make are determined together with the information they will need to receive. In other words the entire interaction of the crew with the aircraft and its systems is determined for each phase of the mission or missions. These factors are often considered before entering the cockpit so that the controls and displays *provided* do not influence the consideration of what is *required.*

A thorough evaluation of all aspects of the cockpit is made on the ground before the airborne assessment is conducted. This ground assessment should be conducted in a variety of lighting conditions. If a simulator is available this is also used to reduce the amount of flight time required. It must be stressed, however, that operating the aircraft in flight, under realistic environmental and operational conditions is essential if all cockpit deficiencies are to be identified correctly. All the equipment that may be worn or carried by operational crews is used during the assessment. For example, NVG, body armour and life preservers are worn and maps, respirator cases and personal weapons are carried and stowed. It is often the case that the designer has insufficient knowledge of what clothing and equipment operational crews will use.

7.2.2 Entry and exit

A logical place to start any cockpit assessment is with the entry and exit. The aim of this part of the evaluation is to determine the ease and safety with which the crew can enter the cockpit and also exit it under both normal and emergency conditions. As always a vital consideration is the way that the aircraft is likely to be operated. For example, if operations from field sites are likely then the effect of crews having wet and muddy footwear is considered. Similarly the effect of deck motion is taken into account when evaluating a naval rotorcraft. A slip by a crew member could result in injury or possible interference with the flight controls during rotors running entries and exits which could be disastrous. The security of the cockpit door when opened and the ease of controlling the rate of closure in high winds are also checked. When presenting the results of these tests the actions the pilot took when entering and exiting his or her station are described in some detail; photographs are often the best supporting data.

Exiting the cockpit in an emergency can literally be a matter of life and death to operational crews and therefore a thorough assessment is made in this area. From a normal seated position with the safety harness locked a simulated egress is made and timed. Any difficulty with operating the jettison control or any danger of becoming snagged on items in the cockpit is recorded. It should be remembered that following an accident the fuselage may come to rest on its side or roof and egress under these conditions is considered. For example, if a large, side-by-side seat aircraft comes to rest on its side, the crew member in the lower position may have difficulty reaching an available exit. In the case of ditching it is important that a handhold is provided at the exit to allow the pilot to remain orientated without visual cues. The force required to operate jettison controls is measured using spring balances; these forces should be light enough to allow easy operation under all conditions, even under water for example. It is essential that all controls can be operated with a single hand. Emergency ingress facilities should also be provided so that ground personnel are able to gain access to the cockpit in order to rescue an injured crew member. In a utility helicopter the safety of the passengers must not be forgotten and due consideration is given to entry and exit from the cabin.

7.2.3 Field of view

Documenting and assessing the field of view (FOV) available from the pilot's station is an integral part of a full cockpit assessment. The FOV is measured from the aircraft design eye position (DEP) which is the point in space where the manufacturer expected the pilot's eyes to be. The manufacturer will have designed the entire cockpit around this point so that the pilot should be able to see all the necessary items in the cockpit as well as having the best possible view of external references. Where a DEP is not available another point, known as the reference eye position (REP), is nominated from which all measurements are made. This will be the point where the assessing pilot's eyes are with the seat adjusted to his or her normal position for flight.

Measuring the FOV involves measuring the angular position of obstructions from the DEP or REP in both azimuth and elevation and then recording this information on a chart. As each measurement is added to the chart a picture of the obstructions to the pilot's FOV is built up. There are two main ways of presenting the FOV. The first type, shown in Fig. 7.2, employs an approximation to the Mollweide projection [7.1] and is the most commonly used presentation as it contains all the measurements made during the test and can show exactly which cockpit items are causing obstructions. It also has the advantage that the total area of obstruction can be seen at a glance. This type of presentation does have the disadvantage that it can be difficult to interpret, especially for people who do not have experience of using such charts. The second type of presentation uses a photograph using a fish-eye lens taken from the REP or DEP. The first stage when conducting the assessment is to determine the REP if a DEP is not available. For this the pilot sits in the seat which has been adjusted to his or her normal flight position. Then a marker is hung from the roof or canopy such that it is positioned between the pilot's eyes. A minimum of three, ideally othogonal, measurements are made from fixed parts of the cockpit structure to the REP to define the point for recording in a report. A mark is then made on the forward transparency, parallel to the fore and aft axis of the aircraft and in line with the REP. This mark is used as the zero degree of azimuth point. From the REP marker, the angles to obstructions in both azimuth and elevation are measured using an inclinometer and a protractor.

Measuring the FOV on the ground provides quantitative data to support the test pilot's qualitative opinion of the FOV during role manoeuvres: it is this latter part of the assessment process that is the more important of the two. The UK Defence Standard 00–970 [7.2] contains guidance on the minimum standards. It is extremely rare that a test programme includes dedicated flights purely to assess the FOV, therefore the test pilot has to evaluate this aspect during all test flights.

The aircraft designer faces a dilemma when planning the cockpit as he or she needs to accommodate all the controls, displays, sights, etc., but in addition must provide the pilot with the best possible view in each direction. It is worth remembering that unlike conventional aeroplanes, helicopters are not restricted to keeping the flight path close to the longitudinal axis of the aircraft and this increases the importance of having a good all-round FOV. A poor FOV can have a major influence on the operational pilot when conducting role tasks, as it will affect every manoeuvre that he or she makes. The FOV requirements of an aircraft will in turn be dependent on the role of the aircraft. For instance an attack helicopter that operates at high speed close to the ground will require a much better all-round FOV than an anti-submarine naval

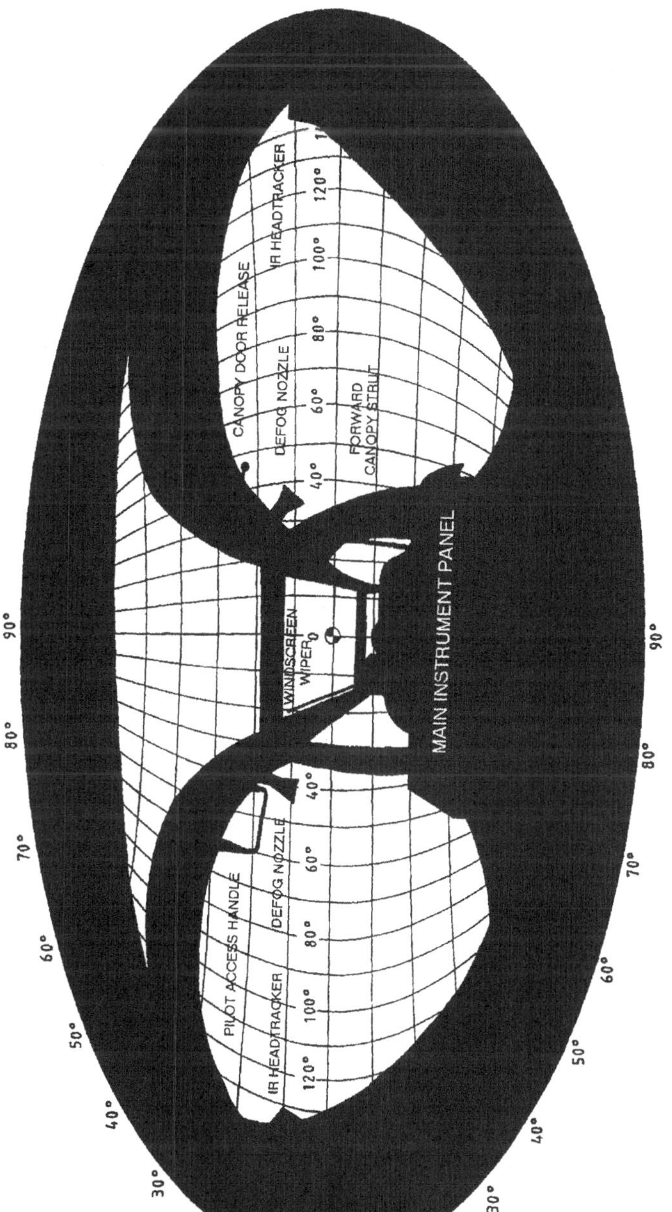

Fig. 7.2 Field of view diagram using an approximation to the Mollweide projection.

helicopter. However, the naval helicopter will still require a good FOV in certain arcs to allow the pilot to judge his position and relative motion to the deck when landing on a ship. Problems with the FOV often arise when a helicopter has not been designed for the specific role in which it is used. During many mission tasks the FOV may be the factor which limits the aggression and speed with which the task can be completed.

Once the FOV of an aircraft has been assessed on its initial entry to service there is a continuing requirement to conduct re-assessments whenever the cockpit is modified. Over the life of an aircraft these cockpit modifications can be very significant and invariably they result in a deterioration of the FOV rather than an improvement. Even if the FOV does not change it is often the case that the role of the aircraft or the way in which it is employed will change requiring a re-assessment to be conducted.

7.2.4 Anthropometrical considerations

When designing a cockpit the manufacturer needs to ensure that the pilot can be seated comfortably, can reach all the required controls and can fit through the normal and emergency entry and exit routes. This presents a considerable challenge to the designer because of the extreme variability of the human body. None of us are identical (some twins excepted!) and even when individuals are of the same height they may have very different measurements for body parts such as thigh length, functional reach, etc. A considerable amount of research has been conducted into the anthropometrical measurements of pilots. For example, surveys have been conducted of RAF aircrew [7.3 and 7.4] and these have been used as the basis for the design requirements of UK military aircraft [7.1 and 7.5]. It is worth noting that there is also considerable variation between different groups in a population, and between the populations of different countries so that cockpits designed to suit the mean of one ethnic group may not be ideal for another.

The designer could use the average measurements of the intended pilot population for each body part in his calculations. This would be better than designing for one or other extreme of size but would still mean that pilots larger or smaller than the average (in effect just about everyone!) would have difficulty in fitting into the cockpit or reaching the controls. To overcome this problem the designer usually provides a range of movement for the seat and yaw pedals. On some occasions an extreme percentile determines the design, for example the door opening must allow the largest pilots to fit through. The range of sizes that the designer must cater for has increased in recent years as more women, who are generally smaller, have entered the previously exclusive domain of male pilots.

7.2.5 Controls

The helicopter cockpit incorporates a plethora of switches and controls, each of which must be designed carefully to assist the pilot when operating the aircraft. Switches must be easy to locate, identify and operate under all conditions. There are a number of means at the disposal of the manufacturer to achieve these design aims. Switches can be grouped together by function, such as locating all electrical controls on a single panel. Alternatively they may be grouped according to task, such as having all the controls required for engine start located together and arranged in the order in which they will be used. The frequency with which a control is used is also a factor in deciding its location, so that frequently used items are located where they can be operated by the pilot most easily. Less commonly used items can then be located in the less convenient positions, however, some infrequently used controls are of great importance and must therefore be particularly easy to locate. Controls associated with emergency actions are the prime examples of this. For example, fuel cut-off levers and engine fire extinguisher buttons may be used only very rarely but must be as easy as possible to locate, identify and operate in the case of an emergency.

Identifying the function of a switch or control and identifying the purpose of its selection positions are important aspects of design and assessment. The pilot must be able to identify a control quickly and accurately even when he or she is tired, busy or frightened. Furthermore the effect of selecting each position on the switch or control must be immediately apparent. To achieve these aims, switches and controls have to have correct labelling which can be read under all circumstances. For example, if the aircraft is to be operated at night the labels must have adequate illumination. It is also often necessary to provide more feedback about the status of a selected system than merely the position of a switch. This can take the form of lights or indicators to show that the system has been selected and is operating. The marking of controls associated with emergency actions such as door jettison handles and an underslung load jettison switch is an area where clear marking is essential. For this reason all specification documents that cover cockpit layout lay down in some detail the size, colours and position of markings for these controls.

The designer must give the pilot control over all the aircraft systems which means that he or she can close down engines, discharge fire extinguishers, disengage flight computers and jettison parts of the aircraft or its cargo. Clearly a lot of design effort must be made to ensure that these events only happen intentionally. Aviation history is littered with examples of aircrews who have made inadvertent or incorrect switch operations, often with disastrous results. There are a number of steps that can be taken to minimize the chances of this happening. Where inadvertent operation is likely to have a serious consequence controls may be provided with gates or guards which require a conscious action to release or pass and operate the control. Electrical interlinks to disable a switch if the system configuration is not correct can also be employed, such as weight-on-wheels switches to disable undercarriage selectors when the aircraft is on the ground. Other methods of minimizing inadvertent operation include covers, catches, recessing, wire locking or any combination of these.

Designers can also ensure that it is easy to discriminate between switches by using different shapes, colours, sizes and even textures. A cockpit full of identical switches may look good in the manufacturer's brochure but it will spell disaster for the operational pilot. A good example of the use of shapes is the Westland Sea King autotransition panel, which employs a triangle, a cross and a bar to allow discrimination between three controls at night.

The distance between controls and their method of operation are also important factors that must be considered. For example, locating a heater control next to a fuel jettison lever would clearly not be sensible particularly if both controls operated in the same sense.

A well designed system will give immediate feedback of the system status independent of switch position. For example, if a fuel boost pump is selected on, an indicator could change from red to show a white line completing that part of the fuel system schematic. On a less sophisticated system the pilot has to use the selection position of the switch itself to gain information about the system status. This can add significantly to the pilot's workload and can also lead to errors, particularly for switches with multiple selections and those located further away from the pilot. Illumination of remote or integral lights and the use of indicators are the most common way of achieving status feedback.

7.2.6 Display of information

Having looked at the factors concerning the way the pilot is able to control the aircraft systems the next major area to consider is the way that information is displayed to the aircrew. Although the human being is extremely adept at processing large amounts of information quickly, pilots often reach saturation point in flight where the amount of information and the way it is presented make it difficult for him or her to analyze it. The type of information the pilot is required to receive and process is extremely varied. He or she must constantly monitor the aircraft's state in relation to the external environment encompassing such factors as velocities in all axes, altitude or height, separation from obstacles, heading and attitudes, etc. This may need to be achieved using external cues, instrument displays or most commonly a combination of both. In addition all the aircraft systems need to be monitored, some, such as the transmission torque and rotor speed, may require careful attention during manoeuvring. On top of all this there may be a requirement to take in and act upon the information presented on tactical and navigation displays.

There are a number of factors that affect the precision and speed with which the pilot is able to obtain and process the information presented on the aircraft displays. Instruments need to be positioned within the pilot's normal field of regard (the area of the cockpit that the pilot is normally viewing) so that they can be seen easily when operating the aircraft. In addition, the position of the instruments in relation to each other is an important factor. The pilot will often have to obtain information from several systems in a short space of time to perform a task such as starting an engine or performing an autorotation; a poor instrument layout will add significantly to the difficulty of the task. Associated with the position of instruments is the ease with which the correct instrument can be located and identified. If all instruments are of the same size and general appearance then it can be difficult to find the correct instrument quickly, particularly if there is no logical structure to the layout. Even worse, the pilot may misidentify the instrument during the stress of an emergency and shut down a serviceable system. Thus instruments must be easy to view and identify while conducting role tasks.

The instruments used for controlling the flight path of the aircraft are the most important instruments in the cockpit and therefore their layout has come in for particular study. Because of their importance they are always given the most prominent position on the instrument panel directly in front of the pilot. Following the example of their fixed wing predecessors the flight instruments of nearly all helicopters are arranged in the classic 'T' layout which puts the artificial horizon or attitude indicator (AI) in the centre with the directional indicator below. The airspeed indicator is located to the left of the AI and the altimeter to the right. This arrangement is usually laid down in specification documents, such as the Ministry of Defence Standard [7.2] and Joint Airworthiness Requirements [7.6]. Other instruments such as the vertical speed indicator and the radar altimeter are located below the 'arms' of the 'T'. This layout allows the pilot to build a viewing strategy known as a 'scan' that is centred on the AI; each instrument is scanned in turn with a scan of the AI in between. Thus when flying solely by reference to instruments the pilot is able to monitor the most important of all parameters, the aircraft attitude, frequently enough to prevent minor deviations from becoming larger errors.

The suitability of the layout of the flight instruments will naturally be affected by the requirements of the role. For example, an aircraft that is required to operate at night, low level over the sea will need radar height displayed more prominently than barometric height. For this reason some naval helicopters, such as the Westland Sea King, have the radar altimeter positioned in the place normally reserved for the barometric altimeter.

The position and distinctiveness of instruments is only part of the story. The pilot must also be able to interpret the information displayed easily and for this the size, markings and scales used must all be correct. The ease with which the pilot is able to interpret the information displayed by an instrument is often referred to as 'readability'. The effect that size has on readability needs little amplification but when assessing the size of an instrument consideration must be given to the distance it is from the pilot's eye, the frequency with which the pilot will need to interrogate it, and the precision with which information will need to be gathered. During an instrument approach, for example, the pilot will need to monitor the direction indicator frequently and to an accuracy of one or two degrees. This is in contrast to an oil temperature gauge that will need only infrequent scanning with far less precision.

The markings and scales used on instruments can be a more complex aspect of assessment as they are dependent on the way in which the pilot is required to use the information presented. For example, when considering a gauge displaying rotor speed (N_R) it is clear that the pilot will be interested primarily in that portion of the scale which relates to rotor speeds that will be seen in flight. If the full range of rotor speed is shown on the gauge from zero to the maximum permitted, the size of the portion corresponding to flight values is likely to be very small. This problem is sometimes tackled by using scales with expanded portions as in the case of radar altimeters where the scale is larger at low altitudes where greater precision is required. The test pilot must make the decision as to whether or not the choice of scale provided is the optimum for all mission tasks. Care must also be taken with non-uniform scales to ensure that rate of change information on the parameter displayed does not lead to confusion as a constant rate of change will not lead to a uniform rate of pointer movement.

7.2.7 Electronic cockpit displays

As in all new aircraft, helicopter cockpits are increasingly incorporating electronic displays in place of mechanical, analogue dials. These can range from large multi-function displays (MFD) used to present flight information to small light emitting diode (LED) displays for an individual parameter. The electronic display offers a number of advantages over the mechanical dial, such as reduced maintenance costs, flexibility, clarity, and significantly smaller space requirements. The considerations for the display of information already stated in this section apply equally to electronic displays, however, there are a number of specific considerations that must be included in any assessment of these more modern instruments.

As the designer can place information on an MFD very easily there are sometimes problems with size and clutter. Each individual number, letter and symbol must be large enough to be distinguished easily and quickly. In addition the pilot must be able to locate and identify the parameter required with rapidity and with precision. Some MFDs provide a great deal of information in a very small space but do not reduce the pilot's workload. When assessing these types of displays an important element is determining if the size and number of display elements helps or hinders the presentation of essential information for each phase of the mission.

There are a number of steps that the designer can take to improve this situation. Firstly the crew can be given some control over the displays to select the items of information they would like presented. Clearly this brings its own problems and will require the test pilot to check that essential information is never missing. Another approach commonly used in modern aircraft is the 'black cockpit'. In this the pilot is only given information on items such as system status when he or she selects it or if the system detects an unusual change in a parameter. For example, engine oil pressure would not be displayed if the value lay within a pre-determined band and the rate of change was below a certain value. Thus by reducing the total amount of information presented and automating the systems monitoring task the designer can reduce the crew's workload. Many pilots have resisted this approach, however, preferring instead to decide for themselves what to monitor and believing that they are better at detecting subtle changes in system status than an automated system. Another solution to a surfeit of information is to automatically de-clutter a display in certain circumstances. For instance heading and navigational information can be removed from a primary flight display if the helicopter attitude exceeds a certain value. This allows the pilot to concentrate on attitude information to effect a recovery to a normal flight condition.

The presentation format also requires careful assessment. Unlike conventional instruments the electronic display gives the manufacturer an almost infinite number of ways to present information. It is important that the manufacturer chooses the optimum format and it is often in this area that the test pilot with an extensive knowledge of the role requirements can make a significant contribution. The display format should not merely reproduce conventional instruments in electronic form but should also employ the greater flexibility offered by this technology. For example, airliners such as the A320 provide trend arrows showing the predicted airspeed in 10 seconds in addition to airspeed limit markings that change with aircraft configuration. The provision of flight path vector information is another common example. Many electronic displays use digital formats to present numerical information. This type of presentation has the advantage of providing information to a high degree of accuracy and does not require the pilot to interpret a pointer position. However, these displays are very poor at presenting trends as the pilot has to interpret rapidly changing numbers. Data sampling rates can also pose problems. If this is set too high the presentation will be changing constantly, if it is set too low then it will introduce significant lag. Sometimes cockpit designers will combine digital displays with other displays to attempt to get the advantages of greater accuracy but without the disadvantages of poor trend information. One example of this is the VIDS display in the Sikorsky Hawk (UH/SH-60) family of helicopters which uses strips of coloured LEDs together with digital displays at the top of each strip. Unfortunately strip instruments themselves are poor at displaying rate of change information. When assessing any display format, but especially digital or strip formats, conducting tasks that result in rapidly changing parameters is always important.

Any evaluation of electronic displays also includes an assessment of the appropriateness of the information displayed. The greater flexibility offered by these displays should be utilized to reduce the pilot's workload by displaying information that is appropriate to the flight phase or mission task that the aircraft is in. Perhaps the best example of this approach again comes from the civil airliner world in the form of the Airbus aircraft. The A320 for instance automatically displays different systems pages depending on cabin door position, engine condition, airspeed, etc. System malfunctions result in the display of the appropriate systems page and automated checklists. A rotary example comes from the Merlin helicopter where raising the sonar body changes the cable angle display to a Doppler hover presentation. In all cases the test pilot uses his or her experience of the role to decide if the designer has produced the most appropriate display. Where displayed limitations change with aircraft state it is important that the change happens quickly. On some helicopters it has been found that the more restrictive Power-On rotor limits do not change quickly enough to the Power-Off limits when entering FIG leading to spurious N_R overspeed warnings.

Arguably the most important consideration when dealing with electronic displays is the ease with which the pilot can see the information presented. This will be affected by a number of factors such as viewing angle, colours used, contrast ratio, refresh rate, and display brightness. With the earlier type of LCD the viewing angle could present problems for centrally mounted displays in aircraft with side-by-side seating. This has become less of a problem with active matrix LCDs. Colour is used extensively in cockpit displays both to present information and to warn or advise of changes in state such as entering a 5-minute rating band or arming of a missile. During an evaluation the appropriateness of the colours used is determined. Using colours to alert the pilot to approaching limits may not always be enough. On the EHI Merlin, for example, it was found that despite changing the torque display to red at high torque values a master caution caption had to be added to provide greater attention-getting qualities. All civil airworthiness authorities dictate which colours are permitted on flight displays and define the appropriate purpose for each colour. This ensures commonality on civil aircraft. Night operations can be a critical area for colour displays as, with the brightness dimmed down, a display that is satisfactory during daylight may be unsatisfactory at night. This is due to the colours becoming indistinct at low intensity and therefore any night evaluation concentrates on areas where colour is used as a primary source of information. The brightness of a display is assessed in the full range of lighting conditions including direct sunlight. In addition the control of brightness is checked. Associated with the overall display brightness is the contrast ratio, this is the ratio between the luminance of any display element and the background. The contrast ratio should be such that throughout the range of available screen brightness it is possible to identify the individual elements quickly and easily.

7.2.8 Night lighting assessment

Very few helicopters are restricted to operations purely during daylight hours therefore assessing the efficiency of the internal lighting arrangement forms an important part of any cockpit evaluation. Illumination is provided in the cockpit to assist the pilot to locate and identify switches and controls in addition to allowing information to be read from cockpit instruments and displays. However it must be remembered that in most tasks the pilot's primary visual task will involve looking outside the cockpit for external references and the internal lighting should not make this more difficult. The assessment method for night lighting follows the normal cockpit assessment process of conducting a ground evaluation first and then moving on to airborne evaluations. The ground assessment can be conducted using covers to create a fully dark cockpit but further flight assessments under a range of ambient light conditions are essential.

The first task conducted on a night lighting assessment is an evaluation of the illumination of all controls and instruments. Floodlighting often leads to parts of the cockpit being completely in shadow or at best poorly illuminated. Even a small shadow across an instrument can cause problems in extracting information from it. The difficulty of providing balanced and well-controlled illumination can even result in some instruments and controls not being illuminated at all.

The next considerations are control and balance of lighting. With control the designer has to match two opposing requirements. Ideally the pilot would like to be able to control the intensity of illumination for every switch, control and instrument individually to ensure an even balance. On the other hand this would make it difficult and time consuming to switch the lighting on and off and to set the required intensity. The universal solution to this problem is to group lights onto lighting circuits each controlled by a single switch. These groups can be arranged either by cockpit location, such as the complete overhead console, or by function, such as the engine controls. If the balance of lights across a group is not correct it can be problematic; either the brightest instrument can be at the right intensity and the dimmest unreadable or the dimmest readable and the brightest distracting. The balance of lighting is checked to ensure that it is even across individual instruments. The precision with which lights can be controlled is also important as the aircraft will have to be operated in the whole spectrum of environmental light conditions from brightly lit dispersals to near complete darkness. For lights that perform a warning or advisory function it will normally not be advisable to allow the pilot full control over the intensity as it could lead to the light being dimmed to the extent that it cannot be seen.

A major problem with cockpit illumination is unwanted reflections from the variety of internal lighting sources. The position of these reflections is determined and if they cannot be eliminated completely then a judgement is made on the effect they will have during the operational role. External lights such as landing lamps can also create reflection problems. Although not part of a cockpit assessment the effectiveness and control of external lights is usually evaluated during the night lighting assessment.

The provision of lighting during emergencies and following system malfunctions is also assessed. Items such as door jettison handles require luminous markers to be placed on them to provide illumination that is totally independent of the aircraft electrical system. The lighting for critical instruments and displays should not be dependent on the electrical generation system but should be on a busbar supplied directly from an aircraft battery.

The first stage in the testing process is to conduct a failure modes, effects and criticality analysis to determine where problems may lie. The next stage, which is conducted on the ground, is to physically fail or deselect the electrical supply. This is done even if a failure modes analysis indicates that there will be no problem. Following ground tests a carefully planned airborne test is conducted in an incremental manner. The test pilot ensures that the items illuminated and the intensity of illumination are satisfactory for all possible conditions and tasks. During the lighting assessment consideration is also given to engine starting where the power drain on the battery may lead to loss of effective lighting for the engine instruments.

The provision of utility lighting for illuminating piloting documentation such as maps, kneeboards and checklists is also considered. This illumination is usually achieved through 'wander lamps' or 'stalk lights'. Assessment criteria here are controllability, effectiveness and rigidity once adjusted to a position. Checks are made that stalk lights cannot be placed in a position where they could interfere with the flight controls.

7.2.9 Warning and alerting systems

The most important aspect of evaluating warning and alerting systems is to determine the effectiveness of the system in capturing the attention of the crew in a timely and appropriate manner. An effective warning will posses adequate attention-getting qualities so that it can gain the crew's attention even during high workload situations where there may be little surplus mental capacity to register new information. In rotorcraft the main means of providing warnings is by lights usually on a centralized warning panel (CWP). The effectiveness of these illuminated warnings depends on factors such as the intensity of illumination, the position of the CWP in relation to the pilot's normal field of regard, and the size of the light. Often separate lights known as 'attention getters' are used to increase the chances of capturing the crew's attention and causing them to look at the CWP. Warning lights may be placed on the instrument of interest as is the case of the Aerospatiale Gazelle overtorque warning. To test the effectiveness of warning lights the aircraft is operated under a range of lighting conditions and, in particular, with bright sunlight on the instrument panel. The conversion of some aircraft to NVG compatibility has caused problems with filters being placed over CWPs resulting in poor daylight readability. Testing the suitability of the position of warning lights ideally involves unannounced illuminations during high workload tasks.

All warnings should be appropriate to the emergency or malfunction that they indicate. Major warnings requiring immediate intervention by the pilot are normally coloured red and are often supplemented with audio tones while lesser warnings are indicated by an amber colour. Advisory lights are usually coloured blue or green. Where warnings are used to indicate the approach of a limit it should activate sufficiently early to aid the pilot in respecting the limit. Audio warnings are checked to ensure that they can be heard even when radio calls are being received and crew members are talking over the intercom. If multiple audio warnings are fitted then the system of allocating priority is checked.

7.3 PRESSURE ERROR MEASUREMENT

7.3.1 Sources of pressure error

The altimeter, airspeed indicator and vertical speed indicator are designed on the assumption that they are fed with pressures from the undisturbed freestream. However, as a helicopter flies it disturbs the air mass and in so doing generates a pressure field around the vehicle. In addition to true airspeed this pressure field is affected by factors such as: the downwash from the main rotor, the carriage of external stores, and the configuration of hatches, doors and movable weapons. Since it is common practice to use either a fixed pitot-static probe or a fixed pitot probe with a fuselage mounted static source then errors in measured air data will also arise as a consequence of sustained or transient changes in the angles of attack and sideslip of the fuselage. These changes can be caused by:

- low-speed out-of-wind manoeuvres, transitions, turns, climbs or descents;
- variations in all-up-mass (AUM) and centre of gravity (CG);
- carriage of underslung loads;
- operation of movable aerodynamic surfaces.

Errors associated with feeding pressures, other than the freestream values, to the pitot-static system are called *pressure errors.* The action of the main rotor and the large fuselage angles of attack and sideslip that are commonplace in helicopter operation can cause errors in both total and static pressures. Whereas a static pressure error (ΔP_S) is applicable to both the ASI and the altimeter, a pitot (or total) pressure error (ΔP_P) is applicable to the ASI only.

Although there are several different methods for determining pressure errors they fall into either of two distinct groups. Those that compare the aircraft instrument readings with air data obtained from the actual freestream conditions in the vicinity of the aircraft, and those that make a comparison between aircraft instrument readings and data from independent external sources. Pressure errors are measured for a number of reasons and at various times during the life of an aircraft. During development, tests are made to determine the best position for the pitot head(s) and static vent(s) to minimize the errors over the full flight envelope of the helicopter. Due to the complexity of the flow field around the fuselage of a typical rotorcraft this is usually accomplished by fitting several probes and vents and by comparison selecting the location(s) that generate the smallest pressure errors. Prior to release to service formal trials are conducted to obtain data for publication in operating manuals and aircrew manuals so that the rotorcraft can be operated safely with changes in CG, AUM, rotor speed, power and external configuration. Knowledge of the pressure errors will be required to convert any true airspeed or pressure altitude restrictions into limits based on cockpit indications. Likewise, special operating techniques may be necessary to compensate for excessive pressure errors. Detailed knowledge of the pressure errors is also a pre-requisite for other flight tests such as performance evaluations and stability testing. Later in the service life of the rotorcraft, if external store configurations other than those originally fielded are considered, it may be necessary to re-evaluate the pressure errors.

7.3.2 Tower fly-by or aneroid method

The tower fly-by method is used to determine the altimeter pressure errors by comparing the cockpit indication with an external measurement of the aircraft's true pressure altitude. This altitude is found by the summation of the pressure altitude of a fixed datum (base or top of the tower) with the tapeline height of the aircraft above that datum suitably converted to a pressure height. Aircraft height may be measured using a calibrated radar altimeter, a space-positioning system (such as kinetheodolites) or photographically. The method can be improved by connecting a sensitive aneroid to the static system of the aircraft thus identifying errors associated with the cockpit instrument. Care needs to be exercised when handling these aneroids as they are often capable of only tolerating a small altitude range above sea-level and are susceptible to aircraft vibration. Using a radar altimeter is a simple technique that can be used when pre-surveyed facilities are not available. Care must be taken to eliminate the effects of lag and the helicopter should be flown over a smooth level surface to avoid transient errors.

The actual tower fly-by method is normally only used for level flight and hover testing. The aircraft is flown a pre-determined distance from the tower, perhaps down a runway centreline, and is photographed just as it passes the tower. A grid superimposed on the photograph allows determination of the aircraft height. It is common practice for aircraft conducting fly-by testing to have the vertical position of the static ports and any associated transducers or instruments clearly marked as an aid to calculating the true height of the aircraft. Space-positioning equipment may be employed during steady climbs and descents provided a means of synchronizing the data taken in the air and on the ground is available.

Fly-bys are conducted throughout the speed range of the helicopter. It is important that the flight condition is stabilized before data is gathered. Thus an adequate run-in distance must be allowed so that the test height and airspeed can be well stabilized before the instant of measurement. If photography is being used it is important that the aircraft is flown at the correct horizontal distance from the tower otherwise trigonometric errors will be introduced during interpretation of the photographs. As the helicopter approaches a position opposite the reference point the observer in the aircraft records the altimeter reading and radar altimeter indication if appropriate. At the same time a ground observer, at the reference point, notes the static pressure reading from a sensitive aneroid, and photographs the helicopter. A radio link is therefore highly desirable. Both observers keep a record of the runs made so that the test results can be matched at the end of the flight. The aircraft OAT and fuel state should be recorded at the beginning and end of each series of runs. At low IAS singled-engined helicopters may well be operating well inside the avoid curve so time spent in such a condition should be minimized. Tower fly-bys frequently involve use of the primary runway at test establishments and therefore close co-ordination with other activities is required to minimize the risk. As stable weather conditions (low wind and absence of thermal activity or turblence) are a must for accurate testing it is quite often necessary to perform tower fly-bys at dawn or dusk thus easing the problem of deconfliction with other air traffic.

Post-flight data reduction usually uses the convention of relating all the data to the same reference height. Thus the altimeter pressure error is obtained by subtracting the pressure altitude observed in the aircraft, corrected to the reference height, from the pressure altitude of the reference point itself:

Δh_P = reference pressure altitude − corrected aircraft pressure altitude

Typically the aneroid placed in the aircraft is compared with a *similar* device placed at the reference height. Obviously each individual aneroid will have a calibration curve that relates the gauge indications to corrected pressure altitudes and these are applied before the pressure error is determined. Errors due to mechanical differences between the aneroids are removed by taking readings from each aneroid when at the same height before and after the test. This so-called ground correction is added to the pressure altitudes recorded in the aircraft:

$$\begin{aligned}\text{aircraft pressure altitude} &= \text{corrected aneroid reading} + \text{ground correction}\\ &= h_{P_a} + (h_{P_r} - h_{P_a})_G\end{aligned}$$

Although the pilot will endeavour to fly past the tower at the correct height there will often be a small tapeline error, Z_c. This error is quantified photographically and if the aircraft is above the camera datum the error will be added. It is important that the tapeline error, or the error *plus* the height of the tower if the ground aneroid was placed at its base, is converted to a pressure height before the altimeter PE is determined. The correction is achieved by noting the temperature of the air at the reference height and ratioing the tapeline height using the sea-level temperature on the day of the test, assuming an ISA lapse rate, and the ISA sea-level temperature of 288.15 K. Thus:

$$\begin{aligned}\Delta h_P = h_{P_r} - (&\text{aircraft pressure altitude}\\ &+ \text{pressure height of aircraft above reference point})\\ = h_{P_r} - [&h_{P_a} + (h_{P_r} - h_{P_a})_G + h_{P_c}]\end{aligned}$$

$$\Delta h_P = h_{P_r} - \left[h_{P_a} + (h_{P_r} - h_{P_a})_G + Z_c \frac{288.15}{T_c + \delta_c}\right]$$

where T_c is the temperature recorded at the reference point on the day of the test and δ_c is the temperature difference between the reference point and MSL, assuming a standard lapse rate.

7.3.3 Speed course method

The level flight ASI pressure error can be determined by timing the aircraft over a measured course. The course length is chosen to give a data run lasting between 30 and 60 seconds. EAS is calculated from the measured ground speed, after allowing for wind, pressure and temperature, and this is compared with the aircraft IAS. This method may be used at low altitude using visual timing or at high altitude using radar tracking or Doppler to measure the ground speed. Although the low level visual method is used at most test establishments it is reserved for level runs and requires a surveyed and marked ground course, and a stopwatch. It is highly desirable to locate an anemometer in the centre or, preferably, at each end of the course so that local winds can be noted. If timing back-up is to be provided by ground personnel then a radio link is usually employed. Alternatively radar tracking, Doppler or electronic timing may be employed. Where possible a trace recording of the airspeed flown is used to determine the accuracy of the trim state of the subject helicopter.

Obtaining quality data from the speed course method requires that the helicopter be accurately timed over a course of known length whilst being flown accurately (± 1 kts and ± 20 ft). If the course is flown in both directions a correction can be made for the wind (assuming it was of consistently low strength and constant direction) by averaging the ground speed for each run. This speed is then assumed to equal TAS. Errors are minimized by selecting a course that is oriented into the local wind. If a steady crosswind is unavoidable, the aircraft heading is maintained along the course track and the aircraft allowed to drift. If the heading were continually altered to ensure that the ground track matched the speed course then during the test the helicopter would be subjected to undefined amounts of sideslip which would affect the accuracy of the PE measurement. Sometimes it is necessary to displace the aircraft upwind of the start of the course to ensure that the lateral displacement from the end of the course is not excessive and accurate timing can still be accomplished.

Accurate results rely on both ground and aircrew observers avoiding timing errors due to parallax. In the aircraft this is typically achieved by sighting the ground markers across the same two fixed points on the airframe (the bottom of the instrument panel and a window frame for example). Longer course lengths should produce greater accuracy but, in practice, the optimum length for a test will depend upon the ease with which a constant IAS can be maintained. In calm conditions, this will depend upon the overall flying qualities of the helicopter. For the stopwatch and ground markers technique, a course length which gives a run time of approximately 1 minute is employed, providing the test IAS can be held to within ± 1 kts (brief excursions to ± 2 kts are usually acceptable). Pilot workload and fatigue may be reduced by accepting shorter run times, down to 30 seconds, if difficulty is experienced in sustaining a stable condition. Obviously, calm conditions are desirable but acceptable results have been obtained in steady wind speeds up to 10 kts. Turbulence must be avoided and therefore, depending on the ground clutter around the speed course, flight with a crosswind component may be precluded. The ground course can be overflown at any height but between 50 ft and 200 ft is usually recommended. At lower heights timing errors may be introduced due to the short time that the marker is within the timer's field of view and there may also be additional pressure errors due to ground effect. Above 200 ft parallax errors often become too significant. Despite operating at a height within the range of typical radio altimeters it is usual for the pilot to maintain height with reference to the barometric altimeter. This is because the pitot-static system is the system under test and use of a radio altimeter may introduce additional errors as the pilot attempts follow undulating ground.

Post-flight data reduction is relatively straightforward. First the airspeed indicator reading (ASIR) is corrected for instrument error to yield the IAS:

$$V_i = \text{ASIR} + \text{instrument error}$$

Then the timing for the two runs (upwind and downwind) are converted to ground speed, which is assumed to equal the true airspeed:

$$V = \frac{1}{2}\left(\frac{D}{T_{up}} + \frac{D}{T_{dn}}\right)$$

Using knowledge of the air temperature and static pressure at the altitude of the test aircraft, the equivalent airspeed is determined. This is assumed to equal the calibrated airspeed since the scale altitude correction is usually zero for rotorcraft operating at low level:

$$V_c = V_e = V\sqrt{\sigma}$$

Finally the ASI pressure error is found by subtracting V_i from V_c.

7.3.4 Formation method

In the formation method, the test aircraft is flown in concert with another (pace) aircraft whose pressure errors are known. This method has several advantages over others:

- Both level flight altimeter and ASI pressure errors can be determined simultaneously.
- The method can be used for climbs and descents.
- Since the PE calibration is made by comparing the altimeter and ASI readings of the two aircraft in the same air mass there is no need to calculate TAS.

The pace aircraft is normally equipped with calibrated pitot and static systems fed from boom mounted sensors to eliminate the effect of main rotor downwash. The pitot head is usually swivel-mounted to reduce the effects of α and β.

The distance between the test aircraft and the pace aircraft is carefully chosen so that it is sufficiently close to aid holding a steady station yet far enough away to ensure no mutual interference. A spacing equal to two rotor diameters of the larger helicopter is often used. Once again the required test conditions are maintained for about one minute while the observers in both aircraft record IAS and height every 15 seconds. Clearly IAS comparisons can only be made if the pace aircraft can cover the speed range of the test helicopter. However, this is not necessary for static errors as the test helicopter can be flown past the pace aircraft (or vice-versa) and altimeter readings compared. As with any formation flying, a comprehensive briefing, which deals with all aspects of normal and emergency operations, communications and safety procedures, is essential. There is often a temptation to fly too close together, which must be resisted if good data is to be gathered with minimum risk to either aircraft. Since the pilot will spend the majority of the time 'eyes out', the co-pilot or FTE usually assumes the responsibility for monitoring aircraft systems as well as advising the pilot of slip-ball position. It is important that all changes in flight path are made gently with due consideration to any performance differences between test and pace aircraft otherwise time will be wasted unnecessarily in re-establishing a tight formation.

Reduction of data gathered using the formation method is perhaps the simplest of any method currently used to determine pressure errors. After adjustment of all readings from both the pace and the test aircraft, to take account of instrument errors, the pressure errors are found simply by subtraction. If the airspeed and altitude from the pace aircraft come from test gauges fitted to a boom mounted system operating in the freestream there is usually no need to be concerned with the effect of the pressure field surrounding the aircraft. So:

$$\Delta V_i = (\text{ASIR} + \text{instrument error})_{\text{pace}} - (\text{ASIR} + \text{instrument error})_{\text{test}}$$

$$\Delta h_p = (\text{AltR} + \text{instrument error})_{\text{pace}} - (\text{AltR} + \text{instrument error})_{\text{test}}$$

7.3.5 Trailing pitot-static method

Freestream total and static pressures can be measured by suspending pitot and static sources on a cable attached to a suitable strong point fitted with an emergency jettison facility. Pressures are transmitted through two tubes to the helicopter where they are converted to accurate pressure altitudes and equivalent airspeeds by sensitive calibrated instruments. Since the pressures are transmitted to the helicopter prior to conversion to airspeeds and altitudes, no systematic error is introduced by trailing the sources below the helicopter. The system is usually calibrated in a wind tunnel prior to installation and the cable is around two rotor diameters in length. The trailing pitot-static (TPS) typically comprises a streamlined body equipped with a drag cone or tail fins to align it with the relative airflow. Pitot and static ports are placed at suitable locations on the body. A suspension point over the CG incorporates a swivel so that the cable cannot exert a moment on the body. Test instruments in the aircraft cabin display, once their own instrument errors have been taken into account, calibrated airspeed and true pressure altitude. Tubes connect the trailing sources to the test instruments via breakaway links and the support cable is attached to the underslung load hook or, if none is available, to a special-to-type release bracket. Experience has shown that it is possible to deploy a TPS from the aircraft cabin in low-speed flight if OGE hover performance is not available at take-off.

As before, level flight runs are maintained for approximately one minute with the observer recording both TPS and aircraft IAS and pressure altitude every 15 seconds. For climbs and descents measurements are made over a height band 500 ft either side of the nominal test altitude. The same parameters as for level flight are measured on entering the test band, at the test altitude and when leaving the test band. For accurate data it is important that the run is started at sufficient height above or below the test band to ensure that the helicopter is fully stabilized (speed, power, N_R and rate of climb/descent) when the test band is entered.

7.3.6 GPS methods

Since the advent of GPS, several different flight test techniques have been developed to make use of this system in determining pressure errors. There are good reasons why GPS is attractive for such tests:

- No special equipment is needed, such as trailing cones, kinetheodolites, cameras.
- There is no need to keep and maintain a calibrated pacer aircraft.
- A surveyed ground course is not required.
- Flying close to the ground is not required.
- There is no requirement for radar tracking.

The largest drawback to these methods is that results typically show high data scatter. The paragraphs below outline two multi-track GPS techniques that have been developed.

7.3.6.1 NTPS method

In this method, three orthogonal headings are flown at the same IAS and altitude (see Fig. 7.3). The ground speed is read about 15 to 20 seconds after the IAS is stable,

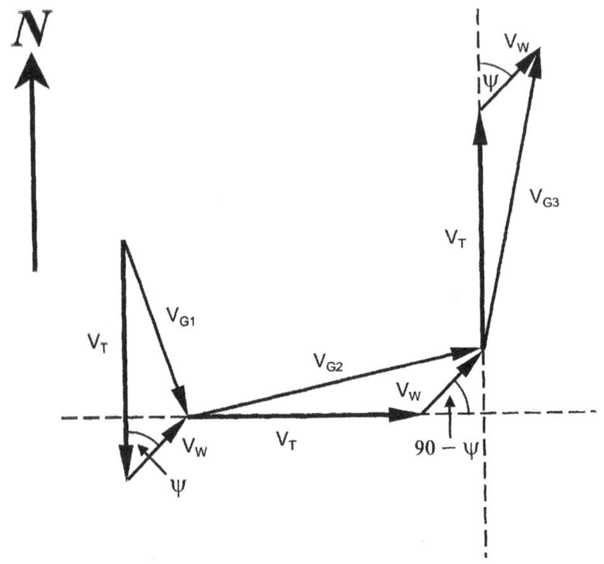

Fig. 7.3 Flight path for PEC measurement using GPS (assuming constant wind strength and direction).

allowing time for the GPS data to stabilize. From the three groundspeeds, the wind velocity, wind direction and true airspeed can be determined using basic trigonometry [7.7]. The resulting angle for the wind direction is the angle clockwise from north, assuming that the original heading was south. For other initial headings, the wind angle must be adjusted. Knowledge of wind direction and velocity is not strictly necessary to determine the PEC as they can be eliminated from the equations. However, a review of the wind data might prove useful in determining if the test data is reasonable.

7.3.6.2 AFTTC method

This method was developed in the early seventies at Edwards AFB, and involved the aircraft performing consecutive turns through a common airmass. It was dubbed the 'cloverleaf' method due to the pattern the aircraft described in the sky. The aircraft flew three passes at 120° apart from each other at the same indicated airspeed. Although pre-dating the first method by some margin it fell out of favour in its original form as it required radar tracking which resulted in it being expensive and difficult to co-ordinate from the ground. However, with the advent of GPS, this method has now become much more economical. Recently Olson [7.8] re-introduced the method and presented a non-linear mathematical solution that removes the need to fly orthogonal headings.

7.3.7 Data presentation

Test results are reduced and presented in the form of IAS against altitude or airspeed *pressure error correction* (PEC) and it is this correction that must be *added* to the

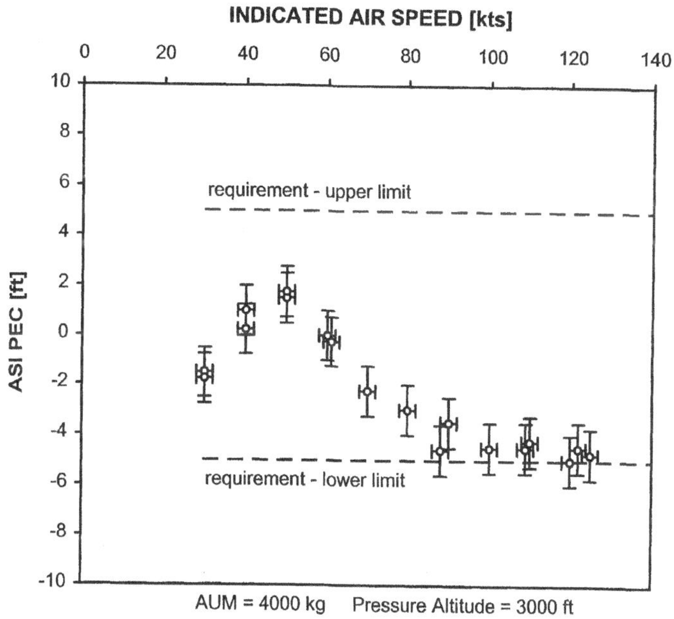

Fig. 7.4 Altimeter PEC data (obtained using the trailing pitot-static method).

indication to yield the calibrated value. Level, climbing and autorotative flight PECs are normally presented on one graph for one aircraft configuration. Airspeed indicator PECs with sideslip are normally plotted in the form of PEC against sideslip angle for specific flight conditions (airspeed and power) for one aircraft configuration. Figures 7.4 and 7.5 show test data obtained for a medium sized twin engine helicopter and compares it with the specification requirements.

7.3.8 Managing pressure errors

The Royal Navy developed one method of compensating for the deficiencies of airspeed indicators at low airspeeds. When Westland Lynx helicopters were taking off from ships' decks it was observed that the pilot did not have reliable airspeed indications until the aircraft had achieved in excess of 40 KIAS. This was due to a combination of the low dynamic pressure itself and the downwash of the main rotor at high power settings. Consequently the pilot had no reliable means of judging when the helicopter had reached an airspeed from which it would be able to perform a flyaway in the event of an engine failure since the minimum speed for the flyaway was typically below 40 kts. The solution to this problem was to determine the relationship between airspeed and time when flying a standard take-off technique. This then allowed pilots to use time rather than airspeed to determine at which point a flyaway should be made rather than ditching conducted. The pilot was provided with charts that accounted for aircraft weight, OAT, height above water, and windspeed thus producing a critical time, known as T_{crit}. Provided that the aircraft had passed T_{crit} and the total

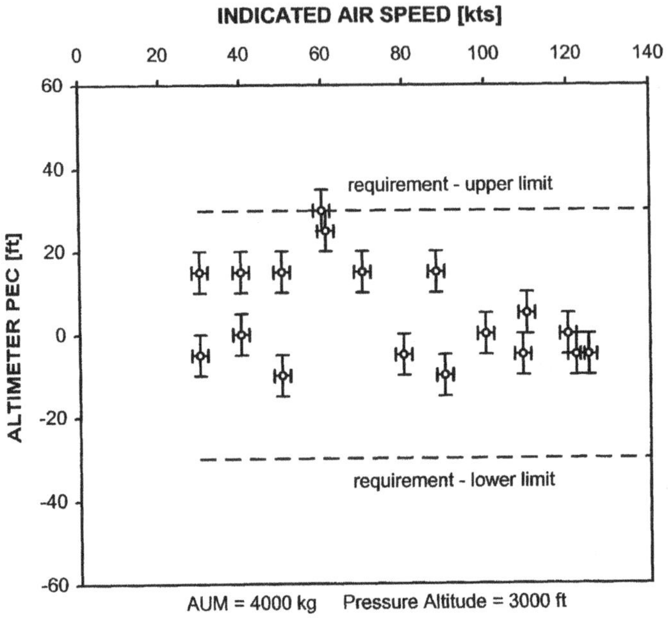

Fig. 7.5 Airspeed indicator PEC data (obtained using the trailing pitot-static method).

intervention time did not exceed two seconds, then pilots were assured of being able to fly away with a minimum separation of ten feet from the surface.

Another area of the flight envelope where PEs can have a major influence is during instrument approaches. When involved in a precision approach, such as a GCA, the pilot is told the decision height (DH) for the runway he intends to approach. To this figure he may need to add the helicopter type allowance (see below) and a further increment depending on his instrument rating. In the UK it is usual to reduce the DH reported by the air-traffic controller by 50 ft. This takes account of the fact that the helicopter can transition to a lower speed, ultimately the hover, rather than having to perform an 'S' manoeuvre to align with the runway centreline before touching down. In the UK the minimum is 150 ft for a CAT 1 approach (a FW minimum of 200 ft reduced by 50 ft). This minimum may also be considered as being made up of a height to avoid obstacles, the *dominant obstacle allowance*, and a factor to account for both errors in the pitot-static system and the anticipated height loss involved in a missed-approach procedure, the *aircraft allowance*. Work done some 25 years ago [7.9] identified that for medium to large twin-engined helicopters using a 3° glide slope the aircraft allowance (AA) was typically 100 ft. This allowance was the summation of a number of factors and error sources associated with the static pressure system that feeds the altimeter. One source was the maximum permissible variation in the PEs across a fleet of similarly equipped aircraft.

The actual PEs associated with a particular model of rotorcraft in descending flight are dealt with separately in the helicopter type allowance (HTA). Although the runway DH plus the instrument rating factor is usually reported back to the GCA controller the effect of the HTA remains within the cockpit. Thus the HTA represents the increment in pressure height required to ensure that the true pressure height at the decision point does not fall below the minimum allowed. The HTA is determined by measuring the altimeter PEs whilst performing a GCA profile, typically 80 to 100 kts with a 500 ft/min RoD. This basic PE is increased by a factor to account for the stiction and lag in the instrument associated with responding to a steadily increasing static pressure, provided these errors have not already been included in the AA.

7.4 ENGINE AND ROTOR GOVERNING

The engine and rotor governing system of a helicopter is designed to regulate the flow of fuel to the engine(s) to provide the pilot with the correct amount of power at a suitable rotor speed for the demands being made with the flight controls. In addition the system is required to provide control over the starting and stability of the combustion process. The aim of the test programme therefore, is to assess the operation of the governing system over as wide a range as possible of operationally relevant conditions. Since the main focus of this book is flight testing, the qualification testing of gas turbine powerplants [7.10] will not be covered in any detail. The results from flight tests are used to make recommendations for improvements to the system as well as defining system limitations. These limitations include restrictions on the rate of conducting power changes, residual engine temperature limits for starting, and altitude and airspeed limits for engine re-lights whilst airborne.

7.4.1 Trials planning

To conduct a full engine and governing system assessment a comprehensive instrumentation package is fitted to the aircraft. The parameters recorded typically include:

- the state of the aircraft (airspeed, altitude, OAT, RoC/RoD, sideslip);
- the position of relevant aircraft controls (collective lever, SSL or 'beep' control, throttle);
- the state of the engine(s) (compressor speed, power turbine speed, engine temperatures, engine pressures, rotor speed, positions of any anti-surge devices).

For trials involving digital engine control systems the inputs to the digital engine control unit (DECU) and the outputs from it are recorded. For more limited trials where it is not practical to fit a full instrumentation suite, it is possible to obtain useful data by using a video camera trained on the instrument panel. As part of the trials planning process the environmental conditions required and how they will be achieved need to be decided upon. For example, it is necessary to conduct engine starting tests with the aircraft both hot- and cold-soaked. This can either be achieved by placing the aircraft in an environmental chamber or by conducting the tests during cold weather and/or hot and high trials in suitable locations.

Before the trial begins the aircraft is checked to make sure that the engine(s) and the rotor are set-up in accordance with the manufacturer's instructions. This may include such tests as determining the engine power output and FIG or autorotative descents to establish that the rotor has been correctly rigged. As the trial progresses these settings are checked periodically to make sure that engine and rotor have not shifted excessively from these baseline settings. Usually the engine(s) are bench tested and calibrated both prior to the trial and post-trial.

7.4.2 Cockpit assessment

The controls and indicators associated with the engine(s) and rotor governing system are assessed in the same way as any cockpit assessment. Particular emphasis is placed on the sensible grouping of items so that tasks such as engine and rotor starting or stopping are easy to accomplish. For example, it is common to find that the start switch is remote from the igniter indication which is also remote from the engine temperature gauge; this can make it difficult for the pilot to monitor the start adequately and may lead to late intervention during a hot start. Emergencies that require the use of the engine(s) and rotor controls are also studied to determine if all the controls can be reached and operated easily. For these evaluations the normal and emergency drills contained in the aircraft checklists are commonly used.

7.4.3 Engine(s) and rotor starting

Tests are made to assess the procedure for starting the engine(s) and engaging the rotor; this will include determining the starting envelope. Eventually, successful engine starting will have to be demonstrated throughout the anticipated operational range of environmental temperatures. Tests include internal battery starts of cold-soaked engines and restarts of engines with high residual temperatures. The airborne re-light envelope and recommended airspeed are also established for each engine in a multi-engine aircraft. This re-light envelope is often smaller than the aircraft operational flight envelope as at high altitude there may insufficient air density to windmill the engine properly or insufficient oxygen to ensure adequate ignition. The drain on the electrical system caused by starting an engine can sometimes result in other aircraft systems being affected, for example, the AFCS can drop out, lighting circuits can be switched off and navigation systems can close down. For normal ground starts none of these effects may matter as the engines are usually started before these systems are engaged, however, for airborne starts the implications may be serious and must be investigated.

The starting sequence is assessed to determine if the actions required of the pilot are easy to perform and the order is logical. Many engine starting sequences are extremely complex and require a high degree of concentration, although with the advent of FADEC systems this problem has become less common. In addition to a subjective opinion on the starting process, quantitative data is collected on time to light-up, rate of oil pressure rise, and the temperature rise profile. This data is normally presented in reports in the form of time histories and an example is shown in Fig. 7.6. The maximum rate at which the rotor can be accelerated to the normal operating RPM is found by incrementally increasing the rate at which the throttle is advanced until a limiting condition is reached. The total time required to complete a start from commencing the pre-start checks to being ready for take-off is also established; for certain roles, such as search and rescue, air ambulance and special forces operations, this time may be a major factor in determining the acceptability of the aircraft.

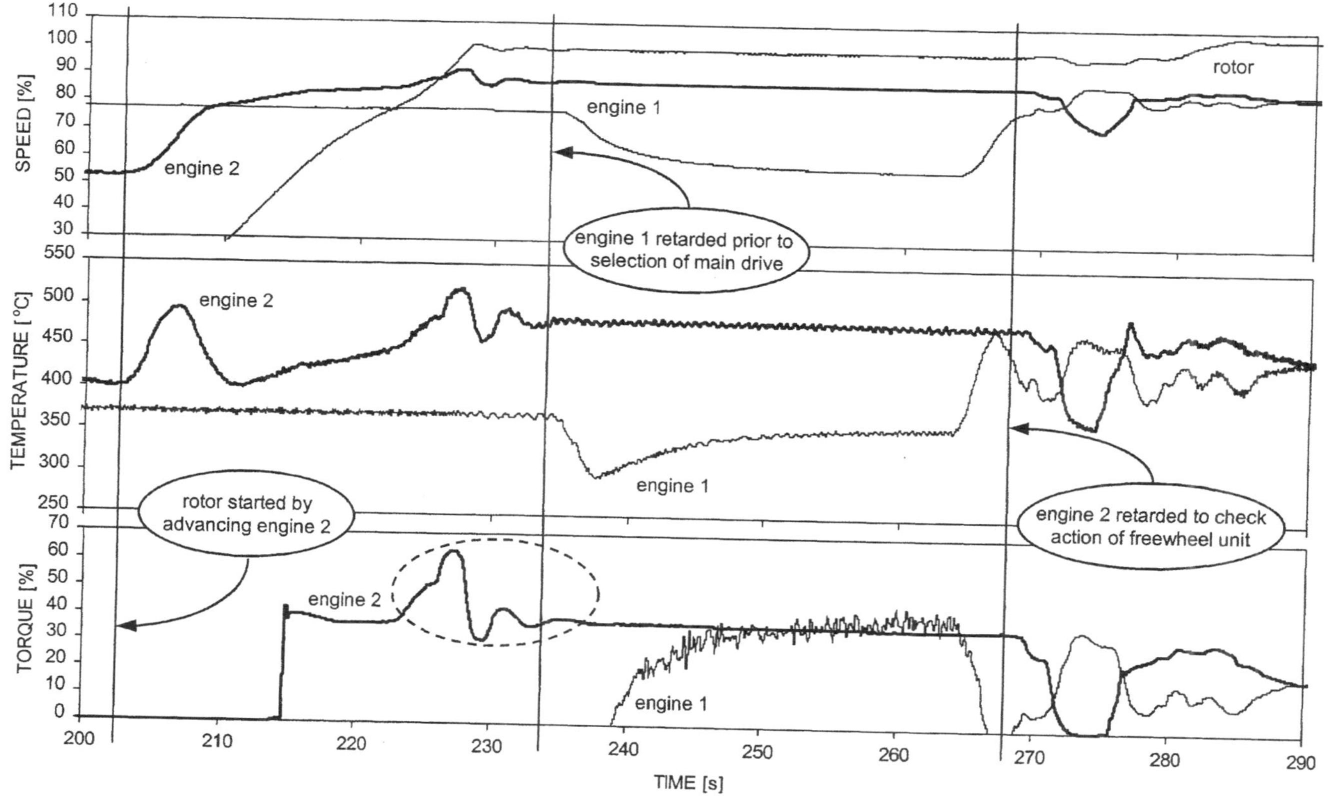

Fig. 7.6 Engine data – time history of rotor start (note the torque oscillation, caused by FPT governor, as rotor first reaches normal operating speed).

The engine and rotor shutdown procedure is also assessed to determine if it is easy to perform. The lack of a rotor brake or an inefficient one may constitute a deficiency in some roles where having the rotor turning for prolonged periods at low RPM may create a hazard to the aircraft or personnel. In addition, any requirement for long engine cooling periods at ground idle may be considered a deficiency.

7.4.4 Static droop measurement

Measurement of static droop is achieved by setting the nominal rotor speed on the ground or in flight and noting the variation of N_r with power setting at a variety of airspeeds. For each test point the aircraft is flown accurately and the condition allowed to stabilize before data is recorded. For the zero airspeed point the N_R value for a series of power settings is recorded as the aircraft is raised into an OGE hover and then higher powers are achieved during vertical climbs. If the aircraft has an avoid area then the OGE hover and vertical climbs are performed above the danger zone. An alternative to the zero airspeed point is to use the tethered hovering technique (Section 3.5.6.3) which will permit the full range of power settings to be tested without climbing. Forward flight tests are flown using a saw-tooth profile around the datum altitude starting at the power for level flight for the chosen airspeed. Static droop testing is often combined with aircraft performance and static stability testing. Test results are usually presented in the form of a plot of power or torque against N_R for each speed condition, as shown in Fig. 7.7.

For aircraft without droop cancellation the rotor speed will depend only upon the power setting. The amount of droop will not be affected by the collective lever position and therefore a single test airspeed will be all that is required. For systems that

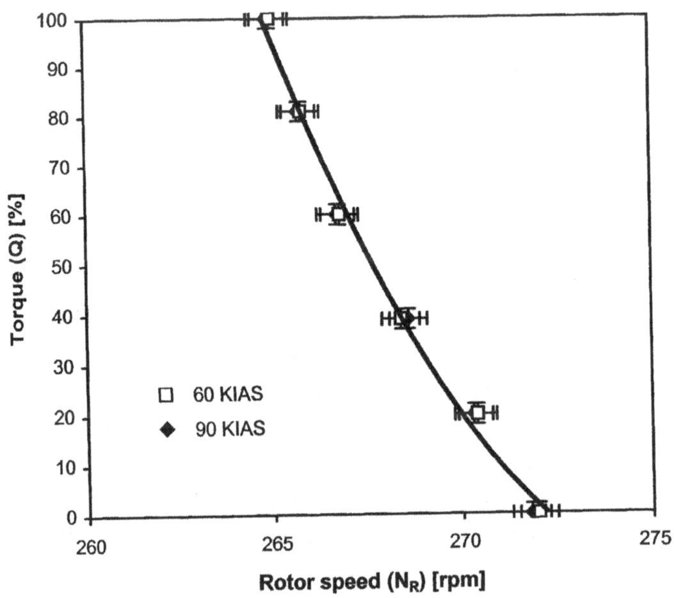

Fig. 7.7 Static droop test data.

incorporate droop cancelling, airspeed becomes a factor as it will affect the collective position for any given power setting due to the different inflow through the rotor. In the same way the collective position will also be affected by the rates of climb and descent achieved for any given power setting which means that the aircraft weight will affect the results. To fully document the static droop of such a system requires, in theory, a large matrix of test points at various density altitudes, weights and airspeeds; in practice however, a zero airspeed, minimum power, and V_H point at low and high density altitudes normally suffices.

To the operational pilot static droop is never a desirable system characteristic. A large amount of static droop at high power settings will cause the tail rotor to operate at a reduced RPM and may lead to a loss of tail rotor effectiveness. Some systems require the pilot to compensate for droop by providing a manual 'beep' control. However, this has the serious disadvantage of requiring the pilot or co-pilot to direct his or her attention inside the cockpit for relatively long periods often at critical moments such as when transitioning to the hover. Even if the pilot is not required to compensate and the aircraft handling is not affected, large variations in N_R with power demand are a distraction for the pilot and should be considered to be a deficiency. A series of role manoeuvres that requires large changes in power setting such as lifting external loads or rapid transitions to the hover are flown to determine the effect of any droop.

7.4.5 Engine matching

In multi-engined installations each engine will typically operate to an individual droop law due to different mechanical tolerances within the engines and their control systems. To ensure that the maximum power is available from all the engines a means of power matching is usually provided. The pilot will normally match the engine torques but engine temperatures may also be used if they become the limiting variable. Pilots are sensitive to torque mismatches at all power settings as this can be the initial indication of an engine failure or governor malfunction. Consequently some automatic alerting systems use torque mismatches, above a certain threshold, as the trigger for activation of an engine failure warning. These have not proved popular due to the difficulty of eliminating false warnings.

Testing of static power matching can be combined with static droop testing by matching engine torques at a datum power setting, normally minimum power for rotors at pre-take off N_R on the ground or maximum continuous torque in flight, and then noting the mismatch throughout the power range. Mismatches during rapid power changes are also documented during transient overswing and engine acceleration tests. The additional workload placed on the crew in dealing with power matching will determine if a deficiency exists. This will depend not only on the size of the mismatches but also on the characteristics of the manual power matching system. For example, an aircraft which suffers from a large power mismatch on take-off and has the manual matching control mounted off the flight controls may significantly increase the crew workload. For obvious reasons the engines must be correctly adjusted before these tests.

7.4.6 Transient droop, transient torque response and overswing

Once the static variation of N_R with power setting has been established tests are made to determine the dynamic characteristics of the engine(s) and governing systems. Transient droop, transient torque response and overswing testing involves assessing how the power supply system reacts to sudden changes in power required.

7.4.6.1 Transient droop

The amount by which the N_R droops following a rapid collective pitch increase depends on the rate of collective increase, the rotor inertia, the acceleration capabilities of the engine(s) and the speed of reaction of the engine(s) and rotor governing system. For the operational pilot a large amount of rotor droop following a collective lever pull could have serious consequences; it might not be possible to arrest a rate of descent as quickly as required and may also lead to problems with control response at low rotor speeds.

Testing transient droop involves collective lever pulls from a low power position to a high power position at incrementally increasing rates until a limiting factor is reached or a satisfactory result consistent with the aircraft role is obtained. The low power position should ideally equate to zero torque but depending on the design of the gearbox it may be necessary to 'join the needles' by matching the speeds of the power turbine(s) and the rotor to avoid damage to the transmission system. The higher collective position is chosen such that any torque spikes remain within the transient limits; typical values are collective positions that equate to 90% or 95% of maximum continuous torque. Some means is used to block the co-pilot's collective to prevent the maximum test value being exceeded; this can either be the flight test engineer's hand or a fixture held by him or her.

The test is conducted by setting the maximum collective position and establishing the block. The collective is then lowered to the low power position previously established. A slow collective ramp input is made over a period of at least 5 seconds until the collective contacts the block. The timing is achieved by one of the crew members conducting a cadence count. It is important for the control to be moved at a constant rate throughout the pull and no attempt should be made to vary the rate of application simply to meet the target time. In addition to noting the N_r droop, the maximum torque value and the maximum engine temperatures are recorded. The crew must also be alert for any sign of instability within the power system and for other effects such as large yaw rates occurring. If the test team conclude that it is safe to increase the rate of collective application then the test is repeated using a count reduced by one second. Although this incremental approach is the safest way of approaching rapid lever rates experience has shown that large variations in peak torque may occur with even small increases in collective rate.

Once the 'academic' tests have been completed a series of mission tasks should be flown which are designed to identify any problems with transient droop. These may include jump take-offs, baulked landings, recovery from flight idle glide and the final stages of a quick stop. For naval aircraft, landings are made onto a heaving deck or this task is simulated. Aircraft with collective anticipators fitted may have no droop problems with collective inputs, however, droop may occur with inputs made with the cyclic or yaw pedals. In these cases tests are also made of rapid rolling in forward flight and of rapid yaw inputs in the hover.

7.4.6.2 *Transient torque response*

Allied to the transient droop characteristics of a rotor system is the transient torque response to power demands made by the rotor system. A rapidly reacting governing system may effectively prevent transient droop but may create a problem by producing a large torque spike. An excessive torque response may restrict the agility of the rotorcraft as fully as excessive transient droop would. Part of the assessment of the torque response is to assess any lag in the torque indication system. If the indication lags the actual value significantly then this can lead to the transmission limit being exceeded when the pilot selects a high power setting. This problem can often lead to operational pilots being unwilling to manoeuvre the aircraft with any more than moderate aggression which may have a serious effect on the conduct of the role. The test techniques and mission tasks used to test the transient torque response are identical to those used for transient droop testing.

7.4.6.3 *Transient overswing*

Transient overswing testing is conducted to examine the rate at which the governing system is able to reduce the fuel flow to the engine(s) with rapid decreases in collective pitch. If the system is not able to react quickly enough the power output from the engines will momentarily exceed the power required by the rotor causing the rotor speed to increase.

Transient overspeed testing is conducted in a similar way to transient droop testing. Rapid collective lever lowering is made from the collective position which gives 95% of maximum continuous torque to the lever fully down position. As in transient droop testing the rate of lever movement is increased incrementally starting with a minimum of a 5-second lowering and then reducing by one second for each subsequent test point. The maximum transient rotor speed is recorded together with the minimum normal acceleration achieved. The decision on whether to proceed to the next test point depends on the proximity to the rotor speed and '*g*' limits. Once the transient rise in rotor speed has been observed the pilot may have to raise the collective lever to control any subsequent rise in N_R caused by entering flight idle glide.

The operational effect of transient overswing is assessed by conducting mission tasks which require rapid lever lowering such as practise forced landings, entry to quickstops, and re-masking manoeuvres. Excessive overswing will lead to excursions above the rotor speed limit or may lead to restrictions on the rate of lever lowering that the pilot is able to employ. Of course there may be other reasons for imposing limitations on the rate of lowering the collective lever such as excessive coupling in the pitch or yaw axes. In addition, the need to respect the minimum normal acceleration limits may prevent the pilot from lowering the lever rapidly. Figure 7.8 shows a combined plot illustrating the effect of lever rate of transient droop and overswing.

7.4.6.4 *Governor stability*

While transient droop and overswing testing is an examination of the governor system characteristics during rapid changes in power demand, governor stability testing is concerned with the reaction of the system to high frequency, small amplitude collective lever inputs. If the system has low bandwidth then the engines will not be able to respond to higher frequency collective inputs leading to rotor speed variations occurring out of phase with lever movements. On the other hand if the gain of the droop law is too great it is possible to drive the system unstable.

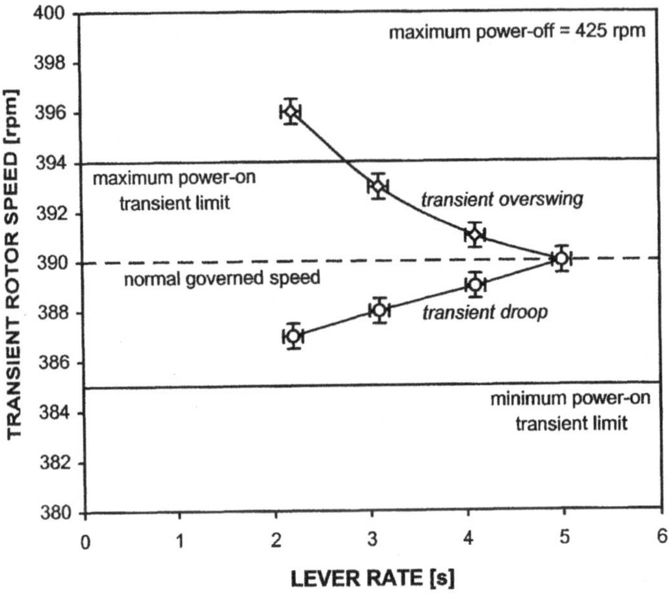

Fig. 7.8 Transient droop/overswing test data.

The test method is to conduct a collective frequency sweep using small amplitude ($\pm 10\%$ to $\pm 20\%$) collective inputs starting at a low frequency. The frequency is gradually increased to a maximum frequency of approximately 2 Hz: higher frequencies are unlikely to be generated during operational tasks. Any tendency for the response of the engine(s) and rotor speed to become out of phase with the collective input is noted. The reaction of the governing system once the inputs are stopped is also observed to ensure that oscillations of the power system do not continue. Like all testing which involves high frequency control inputs considerable care is required to prevent fatigue damage to the aircraft. Significant lags in the torque indication system will also be highlighted in this testing. Following the frequency sweep tests, mission tasks that involve high frequency collective inputs are conducted. These may include positioning for a deck landing, precision landings, and setting down underslung loads.

7.4.7 Flight idle power contribution

When helicopter pilots complain about the powerplants and governing system of their aircraft it is usually because power is not supplied quickly enough to match demand. There are occasions, however, when more power is supplied than the pilot wishes. This may happen when the pilot has lowered the collective lever fully or to a low position to descend or to reduce speed rapidly. Any power supplied by the engine(s) to the rotor when the collective lever has been lowered fully or to the collective position used in power-off autorotation is known as the *flight idle glide power contribution*. This situation occurs because the minimum fuel flow of the engine(s) at flight idle is set too high to allow the engine(s) to back off fully. This high minimum fuel flow may be set by the manufacturer deliberately either to prevent the poor acceleration characteristics that would occur at lower engine speeds or to avoid problems with governor stability once N_R is higher than N_F. It is important to realize that once a needle split (N_R indication higher than N_F indication) has taken place then the rotor is turning faster than the engine(s) power turbine (in percentage terms) and there can be no power contribution.

A FIG power contribution can present some serious problems. For example, if the pilot wishes to reduce speed rapidly and lowers the collective fully the undemanded power from the engine(s) will combine with the autorotational force to cause N_R to rise. The pilot will be forced to raise the collective to prevent a rotor overspeed and this will significantly reduce the rate at which airspeed can be bled off. During a quickstop manoeuvre this will result in a much increased stopping distance. A FIG power contribution will also have operational implications for rapid descents and for forced landing practice, as it will reduce the rate of descent compared to the real power-off case.

Documenting a FIG power contribution is achieved by comparing the rates of descent for a variety of collective lever positions with the engine(s) at flight and ground idle. The test is usually performed at the speed for minimum rate of descent and is conducted by timing the descent at each collective lever position through a band centred on a datum altitude. The results can be presented in the form of a plot of torque, ROD and N_R against collective lever position, as shown in Fig. 7.9. Of course the power-on line will be the static droop line. The important point on the plot is the point at which the collective position equates to the nominal N_R for a true autorotation. The manufacturer normally only gives a range of permitted N_R, therefore the test team has to decide on the nominal rotor speed to use in autorotation. This will usually be a compromise between a speed which provides sufficient rotor kinetic energy for an engine(s)-off landing and that which maintains a margin below the maximum N_R limit to account for rises in rotor speed during manoeuvring. Comparing the ROD at the collective lever position that gives the nominal N_R in autorotation with the ROD at the same lever position in a FIG allows the power contribution to be quantified both in terms of torque and decrease in ROD.

7.4.8 Specific testing of FADEC systems

FADEC systems are becoming increasingly prevalent in new helicopters and understanding how to test them is an increasingly important skill for the rotary wing test pilot and flight test engineer. Essentially the testing of these systems involves the same techniques that have already been described, however, there are some areas where extra attention is required.

A key area of renewed interest is the instrumentation requirements. As FADEC systems are inherently of high bandwidth it is necessary to have a high sampling rate (>100 Hz) for data acquisition. In some systems this high bandwidth has caused the FADEC to generate damaging resonances within the drivetrain. Therefore modern instrumentation systems are designed with this possibility in mind. Extensive sets of transducers are fitted at key areas within the transmission system and there is often a capability to present the data in real time through telemetry. As the airborne environment is different from the bench, FADEC performance is checked against the results of bench tests at an early stage. To test failure modes of both hardware and

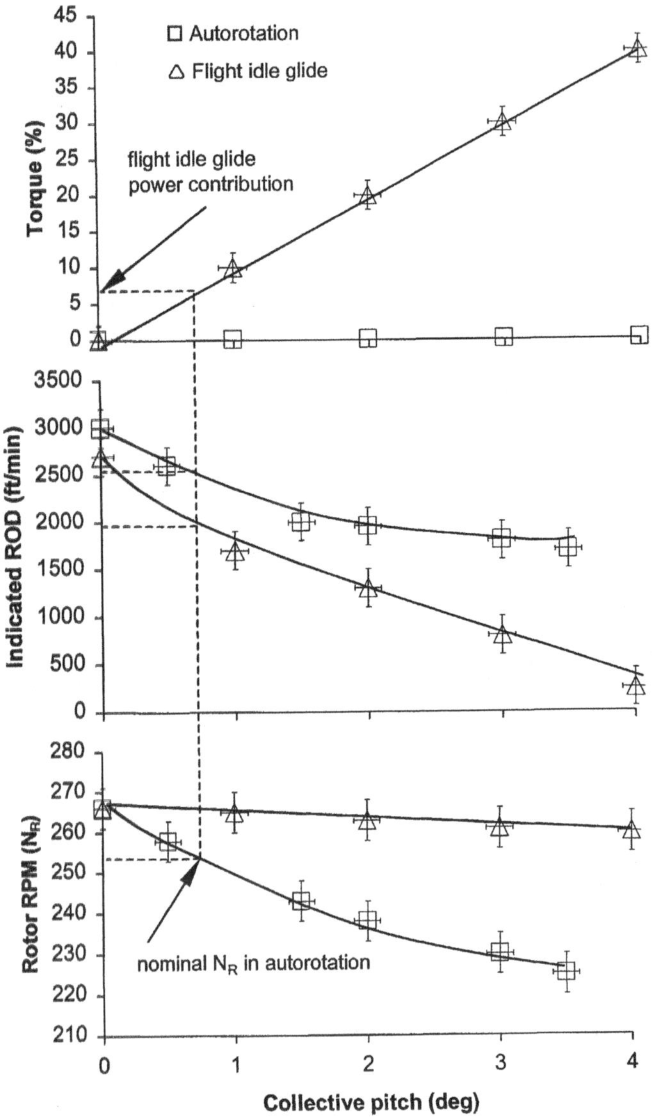

Fig. 7.9 FIG and autorotative test data.

software it is often necessary to build and test a fault injection system to simulate a range of failure modes.

Areas of specific concern for FADEC testing include:

- *Rotor acceleration.* The appropriateness of the acceleration schedule is checked to ensure that it does not lead to the aircraft yawing on icy or wet dispersals.
- *System stability.* Due to the high system gain a danger of instability exists in addition to the danger of drivetrain resonance already discussed. Collective doublets and frequency sweeps are made to assess these areas.

- *Torque spikes.* The rapid response characteristics of FADEC systems can eliminate problems with transient droop but may introduce problems with torque spikes if the gain is set too high.
- *Training mode.* Most FADEC systems incorporate a training mode to provide safe and realistic engine failure training. This mode is evaluated to ensure it meets its design aims. Many of these systems will restore the 'failed' engine to full power if N_R drops below a threshold value. The operation of this part of the system is checked with care to ensure that rapid acceleration of the engine does not lead to an overtorque.
- *Failure testing.* For correct operation a digital engine control system relies on software, processors and information supplied by a large number of sensors. Very extensive testing is usually required to determine the effect on the system of a failure or malfunction of any of these.

A final point to stress about FADEC testing is the need to have good configuration control over the software. During the development and testing phase a number of different standards of software may be in existence at different stages of testing. It is imperative that adequate control methods are in place so that the software version installed in the test aircraft can be tracked.

7.5 EFFECT OF AFCS ON HANDLING QUALITIES TESTING

In Chapter 6 the hardware associated with typical automatic flight control systems and the functioning of generic systems was described briefly. Although it is always important to distinguish between the functionality of stability augmentation and outer-loop or autopilot modes, a possible blurring of the definitions was pointed to when the hardware implementations of typical heading and height holds were considered. In terms of system architecture these channels tend to mimic basic pitch/roll attitude stabilization. In each case a primary sensor (compass or altimeter) is used to generate an error signal with the tightness of the hold being enhanced by rate feedback obtained from a gyro (for yaw) or integrated acceleration (for height). To provide a useful shorthand for the following discussion system based definitions of two generic forms of automatic flight control systems will be made. Thus a SAS is defined as an AFCS that provides pitch, roll and yaw *rate* stabilization (enhanced rate damping) and *short*-term attitude hold. An ASE or ATT mode AFCS is a system that provides pitch/roll *attitude* stabilization for enhanced stability augmentation and gust rejection via *long*-term attitude hold with pitch, roll and yaw rate feedback active at all times (unless disabled by the action of a trim release switch). The ASE will also provide heading hold at all speeds and possibly turn co-ordination in forward flight. It is interesting to note that these two definitions are sometimes used to describe modes provided by a dual functioning AFCS that overcomes the reduced agility commonly associated with ACAH by providing a SAS/ASE or SAS/ATT switch to restore rate command.

7.5.1 Stability augmentation systems (SAS)

7.5.1.1 Longitudinal long-term, LDO and falling leaf modes

The additional rate damping associated with a SAS will help to suppress the nuisance modes. The action of the SAS in applying a control input in opposition to any measured rate of change of attitude will reduce the ease with which any long-term mode is excited. In addition the pseudo-attitude hold available by integrating the rate signal will provide a short-term datum to which the aircraft can be returned following minor excursions arising from atmospheric disturbances. Due to the limited authority of the series actuators employed and the low quality attitude datum, it is unlikely that these long-term modes will be completely suppressed in all levels of turbulence although an acceptable reduction in pilot workload is usually provided.

7.5.1.2 ***Control response***

In addition to providing a reduction in pilot workload, during straight and level flight and hovering, a SAS can be employed to improve the control response. As mentioned earlier the increase in rate damping resulting from rate feedback will tend to increase control predictability by reducing the time constant (at the expense of steady state rate). Thus for any given discrete attitude change although the control will need to be displaced further the time taken for the rate of change of attitude to achieve a steady value will be reduced.

7.5.1.3 ***Longitudinal static stability***

Since a SAS only responds to rates of change of attitude it will have no effect on static stability data (apparent or collective-fixed). As there will be no pitch rate present when the test data is gathered the stick position will be a true reflection of the longitudinal cyclic pitch demand. A SAS will not, therefore, affect the ease with which a pilot can *select* airspeed as indicated by the trend from apparent static stability tests. The presence of a SAS will, however, affect the *interpretation* of collective fixed static stability (CFSS) results. Since in this channel the basic function of the SAS is to oppose pitch rates and hold a given pitch attitude (for a short period) it may also help to *maintain* airspeed. Thus any assessment of the helicopter's ability to hold airspeed, as gauged from CFSS, must be modified by an appreciation of the effect of the SAS on pitch attitude maintenance and any subsequent airspeed keeping. If, for example, there is a strong relationship between airspeed and pitch attitude, as measured during the apparent static stability tests, then the SAS may be quite successful in maintaining an airspeed for a short period of time.

7.5.1.4 ***Collective-to-pitch coupling***

Open loop trim changes with power (or collective-to-pitch coupling) are likely to be less 'vigorous' with the addition of a SAS. As the collective is raised and a nose-up pitching moment develops the resulting pitch rate will be sensed and opposed by longitudinal cyclic pitch at the rotor head. Since a pitch rate must develop before the SAS can act and it will quickly lose the original attitude datum it is likely that following a power change a residual pitch rate will exist. The SAS will not be able to completely eliminate the coupling since the nose-up pitching moment from collective will remain as long as it is above PFLF and the opposing control input from the series actuator will be directly proportional to pitch rate.

7.5.1.5 ***Manoeuvre stability***

Although maintaining a steady turn is a quasi-static situation (a point of dynamic equilibrium) a SAS can have a significant effect on the pilot's perception of manoeuvre

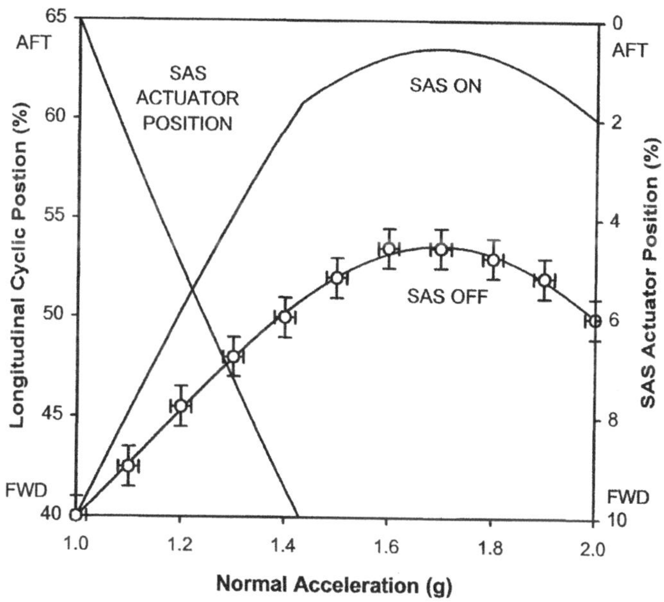

Fig. 7.10 Effect of SAS actuator on manoeuvre stability data.

stability since any flight at reduced or elevated load factor requires the development of a pitch rate. During a pull-up manoeuvre or steady turn, the resulting pitch rate will be sensed and opposed by the SAS and in order to maintain the desired load factor additional aft cyclic will, therefore, be required. Since additional aft cyclic will be required for any load factor above unity (and vice-versa) the SAS has the effect of increasing the manoeuvre stability as seen in the cockpit, although the amount of longitudinal cyclic pitch required at the rotor head, for a given value of '*g*', will remain unchanged. If the SAS has sufficient authority it can completely mask any tendency towards manoeuvre instability and can also prevent excessive control activity in regions when the helicopter is manoeuvre neutral/unstable by opposing the tendency to 'dig-in'. It is usual however for series actuators to saturate at quite modest load factors, typically around 1.5*g*, due to the low authority inherent in that form of actuation. Consequently although with the SAS engaged there will be a greater stick migration with increasing load factor than with SAS off, the underlying manoeuvre instability of the baseline aircraft is still evident, as shown in Figs 7.10 and 7.11.

7.5.1.6 Lateral and directional static stability

As with longitudinal static stability since steady heading sideslips (SHSS) are static tests a SAS will not affect the lateral cyclic and pedal deflection required for a given lateral velocity. The presence of a SAS will, however, affect the results of any turns on one control (TO1C) since roll and yaw rates are generated during these tests. Also any decision on the acceptability of the strengths of L_V and N_V cannot now be based solely on the results of SHSS tests since the LDO and spiral modes will be affected by the additional L_p and N_r afforded by the SAS. A SAS also prevents the initial phase of TO1C tests from being used to evaluate the strength of L_V and N_V, although TO1C-C may still be used to assess turn co-ordination. During TO1C-P the natural

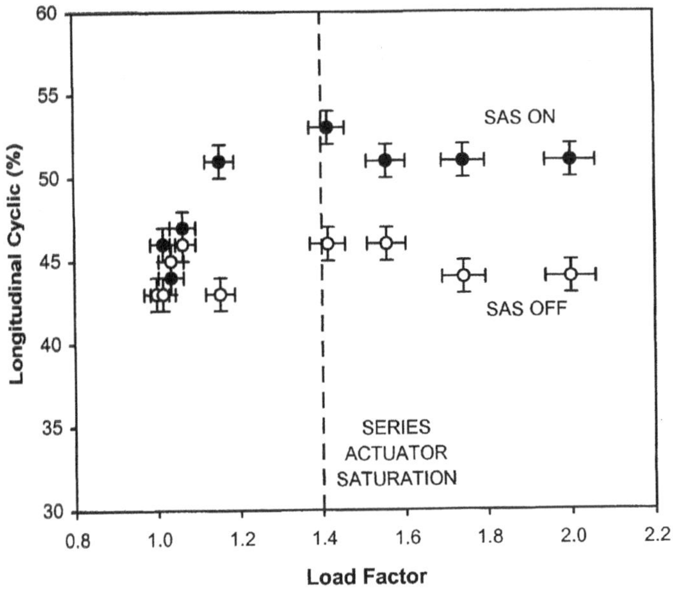

Fig. 7.11 Manoeuvre stability test data.

tendency of the main rotor to flap away from the sideslip will be opposed by the SAS as it applies lateral cyclic in the direction of the sideslip in proportion to the roll rate sensed. Also during TO1C-C tests the action of sideslip on the fin and tail rotor, giving rise to 'weather-cock stability', will be opposed by the SAS as it feeds in opposing tail rotor pitch control in proportion to the yaw rate sensed.

7.5.1.7 Spiral stability

Although the additional rate damping, and pseudo-attitude hold, associated with a SAS will tend to reduce the tendency of a helicopter to enter an uncommanded descending turn or spiral dive, it will affect the results of academic tests used to assess spiral stability. The position of the cyclic during a TO1C-C can still be used, however, as an indication of the sign of the spiral stability since in opposing the yaw rate the SAS should not apply lateral cyclic. The release to trim test will be affected since in rolling wings-level, or departing in roll, a roll rate is generated which the SAS will tend to oppose. The times to half or double bank angle will, therefore, be increased by the action of a SAS.

7.5.2 Automatic stabilization equipment (ASE)

7.5.2.1 Longitudinal long-term, LDO and falling leaf modes

The attitude stabilization provided by an ASE should completely suppress these nuisance modes. The attitude holds should ensure long-term datum holding with the tightness of this hold improved by pseudo-rate feedback obtained from differentiating the output from the vertical gyros. The magnitude of any underlying dynamic instability can be gauged by observing series actuator activity. What constitutes appropriate testing must be carefully considered since movement of the cyclic stick by the pilot in order to generate a disturbance may alter the attitude datum being used by the ASE. The attitude hold performance during protracted flight in role representative turbulence is by far the most appropriate test. Additional tests to document heading hold performance, and altitude hold performance (if operative), are also to be conducted.

It should be remembered that tight control of pitch and roll attitude may not give satisfactory handling qualities whilst hovering in turbulent conditions. If the helicopter is hovering in a wind of varying strength then even though the aircraft attitude may be being maintained it will drift relative to the ground and so plan-position keeping may be poor. To overcome this deficiency and alleviate the pilot from the workload associated with continual re-positioning of the rotorcraft during extended hovering it is common for AFCS system designers to substitute a Doppler-based hover hold for the attitude hold used in forward flight. Increasingly these auto-hover functions make use of a mix of Doppler, inertial and satellite signals. Assessment of such a hold would normally be part of an autopilot assessment and therefore the assessment of the attitude hold whilst in the low-speed regime needs to be approached with these additional tests in mind.

7.5.2.2 Control response

The rate feedback included in most ASE systems can be used to improve the control response characteristics of the helicopter. A step cyclic input in the cockpit will, however, cause a discrete attitude change rather than generate a rate response. Since most ASE systems use foot pressure within the unlock logic for the heading hold it is likely that in the low-speed regime a step yaw pedal input will still generate a yaw rate. The method used to unlock the attitude hold during pilot inputs may cause anomalies in the control response or lead to the adoption of non-standard piloting techniques. It is worth noting that the ACAH nature of the typical ASE may result in excessive control activity during aggressive tactical flight as the pilot exercises the controls in an attempt to generate sustained angular rates. In such situations a switch allowing a change to a SAS mode providing RCAH may be warranted.

7.5.2.3 Longitudinal static stability

Since the purpose of the pitch channel of an ASE is to maintain pitch attitude for long periods it should in theory have no effect on longitudinal static stability (LSS) data. If, however, the ASE features trim follow-up and the aircraft is significantly unstable, requiring continuous trim movement to maintain series actuator authority, it may be difficult to quantify CFSS using the conventional test. The validity of CFSS testing must therefore be considered. In normal operations the pilot will wish to select a given airspeed and maintain it with minimum workload. The quality of airspeed maintenance afforded by a typical ASE will depend on the tightness of its attitude hold (this can be assessed during 'long-term testing') and the variation of trimmed attitude with airspeed, which can be documented during a conventional trimmed flight control position (TFCP) test. Handling problems may arise during *airspeed selection* if, for example, movement of the stick to generate a new attitude causes the trim follow-up to activate, which in turn leads to oscillations in airspeed. Thus, precise selection of a given airspeed may be very difficult requiring inordinate pilot workload to match stick position and aircraft attitude with the desired ASI reading. A valid test of airspeed maintenance, provided the ASE does not feature trim follow-up and the cyclic has absolute centring, is the conventional release to trim test. It should be remembered that during this test the pilot is generating a pulse disturbance in *attitude* rather than airspeed. The ASE should quickly re-acquire the datum attitude, however, the helicopter may regain the original airspeed only very slowly, with many overshoots, or not at all.

7.5.2.4 *Cross-coupling*

Open loop changes in power should not cause any pitch/roll/yaw coupling if the attitude hold is sufficiently tight. Indeed this test can be used to assess attitude hold performance. The magnitude of the underlying cross-coupling can be gauged by observing series actuator activity. Large changes in power may cause the cyclic to move longitudinally through the action of the trim follow-up system.

7.5.2.5 *Manoeuvre stability*

Assessing the manoeuvre stability characteristics of an ASE equipped aircraft is difficult and less important than discerning the ease with which the pilot can turn the aircraft. If the role requires aggressive altitude changes (NOE flight) a qualitative assessment of the handling qualities during pull-up and pushover role manoeuvres, such as wire avoidance, should be considered. Formal academic PUPOs cannot be conducted, however, due to the suppression of pitch rates by the ASE. Thus, although level turns and PUPOs may be assessed qualitatively with ASE engaged the resulting cockpit control positions and activity should not be taken as indicative of the underlying characteristics of the helicopter.

Consider performing a level turn with altitude hold engaged and assume that turn co-ordination is active. The pilot will initiate the turn with lateral cyclic by selecting an appropriate bank angle. Since the thrust vector is no longer vertical more collective pitch will be required to maintain level flight and the lever will automatically trim upwards possibly generating a pitching moment. Any tendency to slip or skid in the turn will be eliminated by the yaw channel of the ASE. Movement of the cyclic stick in the longitudinal sense during the turn is dependent on many factors not least of which is the AFCS designer's perception of the most suitable stick cues for this phase of flight. The tendency of the helicopter to 'dig-in' in steep turns should be completely suppressed by the fast-acting series actuators so that no residual movement is seen at the stick, although excessive instability could cause a trim follow-up system to activate.

In balancing the various forces and moments associated with turning flight it is quite possible, however, that the pitch attitude required for a given airspeed in level flight is different to that necessary to maintain the same speed during a turn. The pilot may therefore have to move the cyclic fore or aft to re-acquire the airspeed. The requirement to trim forward in this situation is *not* necessarily indicative of manoeuvre instability but simply means that a slightly lower pitch attitude is needed to hold speed. In order to provide the pilot with good cues in turning flight some AFCS have been deliberately engineered so that an aft cyclic stick deflection or force is required to increase and maintain the load factor regardless of the actual longitudinal cyclic pitch angle required at the rotor head.

7.5.2.6 *Lateral and directional static stability*

The complex inter-relationship between the roll and yaw channels of a typical ASE make conventional lateral and directional static stability testing both difficult and unwarranted. It might actually be impossible to perform a steady-heading sideslip without over-riding part of the ASE system. For example, if lateral cyclic immediately initiates a turn the pilot will have to apply yaw pedal to hold a given sideslip angle thereby effectively disabling the yaw channel of the ASE and negating the test. For unaugmented helicopters measurement of L_V and N_V help to quantify the aircraft's tendency towards spiral instability, its natural turn co-ordination characteristics and its propensity to oscillate laterally and directionally. All these qualities are heavily affected by the ASE further negating any assessment of static stability. For helicopters without bank angle initiation of turn co-ordination, turn-on-one-control tests may still be appropriate as methods of assessing the performance of the heading hold and roll attitude hold.

7.5.2.7 Spiral stability

As with manoeuvre stability the actual spiral stability characteristics of a helicopter may be completely suppressed when the ASE is engaged. The amount of in-turn stick required during a level turn can no longer be used as an indication of the helicopter's natural spiral stability since the lateral cyclic deflection from trim may be derived purely from the requirements of the ASE system. Some systems are engineered to display *neutral* stability with the stick returning to the centre once a turn has been initiated. Alternatively other systems show the characteristics of *strong* stability, with the angle of bank and rate of turn being directly proportional to the amount of in-turn stick.

7.5.3 Autopilot testing

There are three main aspects to testing autopilots: the interface with the pilot, the performance of the system, and the safety of operation.

7.5.3.1 Pilot/autopilot interface

As the autopilot is directly controlling the flight path of the aircraft it is extremely important that the pilot is fully aware of which modes are engaged and what those modes are directing the aircraft to do. The assessment of the interface examines the display of autopilot status, the method of mode selection/deselection, and the requirement on the crew to monitor the system operation. A particularly important issue is the warning provided to the pilot that a function has been deselected or has failed. The pilot must be able to assimilate with ease which aspects of the flight path are no longer under automatic control. Of equal importance is the ease with which the pilot can 'fly through' an autopilot mode if he or she elects to fly the aircraft manually rather than disengage the mode. This situation usually arises when there is a sudden requirement to manoeuvre the aircraft.

7.5.3.2 Autopilot performance

Testing the performance of the autopilot typically involves engaging each mode and assessing how accurately it is able to carry out its function. As autopilots rely on external signals for many functions such as VOR tracking or automatic ILS approaches, this type of testing has to be conducted within an assessment of the accuracy of the signals received. Only when it has been established that the external signals are being received accurately can the ability of the autopilot to manoeuvre the aircraft in response to these signals be evaluated. Testing of the autopilot is conducted initially in ideal conditions of low turbulence to establish the baseline performance and finally in operationally representative conditions such as high turbulence and large crosswinds.

Although there are too many autopilot modes to discuss in detail the tests used for the most common functions are detailed below:

- *Altitude holds.* These are tested by engaging the hold and measuring the accuracy with which it maintains the datum in still and turbulent conditions. As an example a recent UK military helicopter specification called for the barometric height hold to maintain the aircraft within 25 feet of the engagement height. Deliberate use of controls in the other axes is also made to determine if the hold is able to cope with the off-axis interference (for example, the angle of bank is increased, airspeed changes are made, or with a three-axis system which maintains altitude through the pitch channel, power changes are made). Large disturbances can cause the series actuators to saturate and the slower-acting parallel actuators may be unable to respond with sufficient speed thereby causing a large disturbance from the datum. Such testing may lead to restrictions being placed on the flight envelope with the hold engaged. The actuator activity is always examined closely to ensure that saturation has not occurred. For 3-channel systems that also employ trim follow-up the cyclic activity demanded to maintain the series actuators within authority is an important issue. The capture of the hold datum is also explored by, for example, engaging the hold with a rate of climb or descent. The ease with which the hold datum can be changed is also examined; for example, it should not be necessary to completely disengage the hold and then re-engage it when at a new altitude.

- *Radar altimeter height holds.* Because these holds are employed close to the surface and often in a degraded visual environment they require particularly careful evaluation. The accuracy requirements for this type of hold are always more stringent than for barometric height holds. To quote from the same UK military helicopter specification mentioned previously the radar height hold required the height to be maintained to the lesser of 10% or 10 feet. As radar height holds are usually employed in overwater operations testing is conducted with a variety of sea states at a variety of airspeeds and angles of bank. The testing takes place at heights that are operationally relevant; this can be as low as 40 feet in the case of anti-submarine helicopters. If a smoothing circuit is incorporated to deal with wave peaks and troughs then its effectiveness in operation is checked by comparing the performance of the hold when using the raw signal. Careful evaluation is made of the radar altimeter failure modes and the indications to the pilot if the altimeter unlocks.

- *Vertical speed holds.* These are assessed by selecting RoC and RoD and then timing the aircraft through an altitude band. In an autopilot where the mode operates through the collective channel then the danger of the system causing an excessive power demand is investigated and limitations may be imposed. Similarly a minimum airspeed is always imposed on three-axis systems which hold vertical speed through the pitch channel to prevent the hold operating on the wrong side of minimum power speed. The ease with which the system captures the selected vertical speed and its behaviour when the direction of the commanded speed is reversed are also investigated. In addition the ease of changing from the vertical speed mode to an altitude hold mode is assessed.

- *Airspeed holds.* Testing of an airspeed hold is approached in the same way as a height hold by noting the deviations from the datum in still and turbulent conditions. Power changes are also made to check the hold accuracy as the power is changed and when steady in climbs or descents. As some autopilot systems provide airspeed hold but rely on the pilot to select the datum speed the accuracy and workload associated with selecting a new airspeed can be of great interest. In addition the effect of pressure errors on the system are investigated by inducing deliberate sideslip.
- *Turn co-ordination.* The effectiveness of turn co-ordination features is tested by conducting turns at various rates without use of the yaw inceptor. Manoeuvres such as roll reversals are also used. A particular area of interest is the manner in which the turn co-ordination is mechanized and integrated with the heading hold. On some systems turn co-ordination will disengage and the heading hold will engage below set yaw rates, bank angles, or airspeeds. For example, in forward flight the SFIM 85T31 autopilot fitted to the AS 355 changes from heading hold to turn co-ordination as the angle of bank exceeds 5°. The heading hold will then automatically re-engage on completion of a turn as the heading rate falls below 1.5 degrees per second. The Boeing Chinook system on the other hand uses lateral cyclic displacement from trim to engage turn co-ordination. Clearly the implications of this automatic changeover requires investigation in a variety of flight conditions and role manoeuvres. This is particularly important if the logic changes above a certain airspeed to account for the need to turn the helicopter in forward flight as opposed to performing a lateral sidestep in the low speed regime.
- *Navigation modes.* Autopilots can offer a number of navigation features. In each case the testing examines the accuracy with which the autopilot can track the navigation feature being used as well as the performance of the system when capturing the datum. For example, a VOR mode is tested by intercepting the required radial at a variety of intercept angles and airspeeds. The feedback gain should be high enough to correct errors in tracking but not so high that it leads to uncomfortable and frequent roll inputs; this becomes particularly important as the aircraft approaches the navigational aid or waypoint. This type of assessment may highlight problems in the relationship between the roll and yaw channels in turning the aircraft and maintaining the heading. The action of the autopilot at turning points is often an area of great interest as the system logic has to decide when to start the turn onto the next leg and then has to capture this new track. A consideration is the matching of the demands of the autopilot with the capabilities of the aircraft. It should not be possible to programme a search pattern which the aircraft is physically incapable of following.
- *Automatic ILS approach mode.* Testing of this mode is conducted in a variety of wind strengths and directions to assess how well the localizer and glideslope indications are maintained. The full range of permitted airspeeds is also used. As with navigation mode testing the capture of the localizer is assessed at different approach angles. Three-axis systems are sensitive sometimes to power variations

during the approach therefore collective inputs are made once established on the localizer and glideslope and tracking accuracy is noted.

- *Automatic transitions.* A common feature on naval and SAR helicopters are automatic transitions to and from the hover. As this autopilot mode operates close to the surface, often in conditions of degraded visual cues, it is particularly important to test it thoroughly. Because these autopilot modes are required to control the aircraft from one extreme of speed to the other and climb and descend the aircraft, they are usually given considerable authority to be able to perform their task. Tests are conducted at a variety of weights, sea states, centres of gravity, wind speeds, turbulence levels and entry conditions. If automatic series actuator trim is not provided tests are made with different initial actuator positions to determine at what point saturation occurs. Considerations also include the ease of deselecting parts of the system, the monitoring requirement on the crew, and the safety of the flight path of the aircraft. The actions that the pilot would have to take in the event of an engine failure during a transition are an important part of the assessment.

7.5.3.3 Safety of operation of autopilot

As the autopilot is controlling the flight path of the aircraft there are a number of safety issues that are addressed in the test programme. For three-axis autopilots which control vertical modes through the cyclic pitch channel there is a potential problem if the pilot does not apply sufficient collective pitch to maintain airspeed, particularly when climbing in a vertical speed mode. In this case airspeed may fall below the minimum power speed resulting in a higher pitch attitude and a further fall in airspeed. This in turn may lead to the operational pilot becoming disorientated. For these types of systems there should be an airspeed cutout or, at minimum, a warning to the crew that airspeed is decreasing. For systems that control the vertical modes or altitude holds through the collective channel the possibility of the system making demands that result in an exceedence of a power limit is investigated. For example, above a certain airspeed or rate of airspeed increase a height hold may cause an excessive torque demand. In this case the testing may result in limitations being imposed on the maximum airspeed or the rate of airspeed change that is permitted with the hold engaged.

7.6 FAILURE TESTING

While the majority of testing involves the assessment of an aircraft with all its systems serviceable, a smaller but important portion involves evaluating the effect of system failures. The most important failures are those that affect the flight path of the aircraft and for these the test methodology employed always remains the same.

The initial stage of failure testing is to study and fully understand the system under evaluation. This is not simply a question of understanding what the individual parts of a system do but also understanding the way in which the system is used in the role. It is only with a detailed knowledge of the operational use of the aircraft that the implications of failures can be fully realized. As with all testing the incremental approach is fundamental when assessing failures. This is particularly important in the case of critical components such as the flight control system and powerplants where a failure can take an aircraft rapidly from a safe flight condition to a potentially hazardous one. Even failure testing of aircraft systems that are considered non-critical should be approached incrementally as unexpected results may follow. The golden rule for failure testing, particularly of critical systems, is *only vary one parameter at a time*. For example, when evaluating the height/velocity diagram of a helicopter the height at the moment of failure should be kept constant and the airspeed varied or vice versa. To vary two or more parameters simultaneously makes it difficult to predict the outcome and to determine trends. There is also a greater risk of unexpectedly meeting a 'cliff edge' change in failure characteristics if multiple parameters are varied simultaneously. The identification of trends through the analysis of results is an essential part of the approach to critical system failure testing.

7.6.1 Planning for failures

Before flight testing can take place a significant amount of preparation and planning is required. A failure modes, effects and criticality analysis (FMECA) will have been conducted to determine the probability of component failure and the effect of these failures on the system as a whole. It is then the job of the test team to evaluate the ultimate consequence of these system failures in the role. As part of this process, or in parallel to it, the test team will gain their detailed system knowledge. As a result of the FMECA it may be necessary to have specialized equipment designed and built to inject failures if the normal cockpit controls cannot be used to achieve this. It may also be necessary to approach the manufacturer for clearance to conduct certain tests and to provide more generous limitations. The difficulty of the anticipated recovery manoeuvre(s) will affect the requirement for crew training so these are defined at the planning stage and simulator time booked if appropriate.

7.6.2 Alerting the pilot

Specification documents (such as Ministry of Defence Standard 00-970 [7.2]) stipulate that for all failures that can affect the operation of an aircraft some means of alerting the crew must be provided. These documents also lay down the characteristics of the warnings, for example, the colours to be used and viewing arcs. Warnings can take the form of audio tones, flashing attention-getters and captions, either individually or in combinations. The first stage when assessing any warning is to determine if it performs its primary function, in other words, does it warn the pilot? Since the attention-getting quality of any warning is its most important feature it is assessed under an extensive range of environmental and mission conditions. The appropriateness of the level of warning provided is also considered for any failure state. Clearly the more serious the potential outcome of a failure the stronger the attention-getting qualities of the associated warning must be. Thus major emergencies such as engine fires or low hydraulic pressure are normally indicated to the crew with audio tones and red captions. An excess of these captions and audio signals is counter-productive and therefore the cockpit assessment will determine if they are only used when the crew must act immediately to ensure the aircraft's safety.

Of course warnings are not restricted to the alerting of failure states, they are also used to indicate to the pilot that the aircraft is approaching or exceeding the flight envelope; the most common parameters being structural limits, torque and rotor speed. The assessment of flight envelope warnings concentrates on the four key areas of *accuracy, clarity, utility* and *reliability*. Dealing first with accuracy it is obvious that an inaccurate warning is of little use, but it is sometimes not realized that an inaccurate warning can be worse than no warning at all. Take the example of a high rotor speed audio warning which is set at a level that is below the maximum permitted N_R. In this situation pilots may rely on the audio system and may not control rotor speed until the alert is heard. An inaccurate or unreliable warning could have serious consequences in this situation. Turning to clarity, any alerting system must provide an unambiguous message to the pilot that will direct him or her to react correctly. Audio warnings associated with low rotor speed or engine failure and high N_R are particularly important in this respect, as an incorrect reaction by the pilot will usually exacerbate the problem. Utility is a measure of the usefulness of an alerting system in helping the pilot to respect the flight envelope limitations and whilst permitting exploitation of the full potential of the aircraft. Whilst an alerting system that tells the pilot he has already exceeded a limit may be useful to the ground crew in directing their post-flight rectification activities, it has little utility as far as the pilot is concerned. Similarly a system which is triggered very close to the limiting value may not provide an adequate margin to prevent an exceedence during dynamic situations. Well-designed warning systems can have high utility, such as those that assist the pilot in maintaining the optimum rotor speed during a single engine flyaway without having to monitor the cockpit gauge. In other words an alerting system should provide cues to the pilot on the proximity of the limit and thus allow him to 'fly the buffet'. Although most warnings and alerts are visual or aural there are tactile systems in operation as well; the Bell 430, for example, employs a collective lever shaker to alert the pilot to high values of torque. Clearly a warning or alerting system must be very reliable and not give spurious alerts otherwise crews will quickly lose confidence in it and eventually choose to ignore it.

The assessment of alerting systems has obvious dangers particularly where it is necessary to go to the flight envelope limit to activate the warning. An incremental approach is vital and the use of telemetry with careful monitoring of trends is commonplace. This may be a situation where more generous limits can be sought from the manufacturer during the trials planning stage. Alternatively it may be possible to adjust the warning to activate at an artificial value although this is not without hazard as the modification may invalidate the assessment especially if it involves recompiling software.

Alerting the pilot to failures of non-critical systems is often an area that is poorly engineered and this can have serious consequences. Failures of sensors that feed the flight instruments, navigation system or flight control system fall into this category. During the assessment of the aircraft it will be necessary to determine if the crew is warned adequately about the degradation caused by such failures. For example, if the compass reverts to a directional gyro mode on failure which reduces its accuracy; this should be indicated clearly to the crew.

7.6.3 Reacting to the failure

There are a number of factors that will affect the time it takes a pilot to react to a failure. These include the characteristics of any warning system installed, the reaction of the aircraft itself, the arousal state of the pilot and the proximity of his or her hands to the relevant control. When testing the effect of failures, the establishment of realistic pilot intervention times is crucial and most assessing authorities have developed a set of rules regarding the way in which these times are constructed. The following definitions are taken from Ministry of Defence Standard 00-970 [7.2] which divides the intervention time into rotorcraft response time and pilot response time.

7.6.3.1 Rotorcraft response time

The rotorcraft response time covers the time between the failure occurring and the pilot becoming aware of it. This time will depend on the characteristics of the rotorcraft, on the nature of the failure and in particular on the characteristics of any associated warning. For example, in the case of an audio tone indicating engine failure the response time is equal to the time it takes this to activate. If there is no warning then the pilot may only become alerted to the failure by a change in the flight path. This is usually the case for runaways of the AFCS where the rotorcraft response time is determined by the time it takes for the aircraft to achieve an angular rate of 3 degrees per second about any axis or a change in acceleration of $0.2g$ along any axis [7.2]. The rate is increased to 5 degrees per second and the acceleration to $0.25g$ for passive flight phases where the pilot is less involved in controlling the aircraft. Although the pilot may be alerted by other changes, in noise or vibration levels for example, they are usually harder to define and are not often used as the basis for specification compliance. In situations where the failure cues are not easily defined it is often necessary to conduct no notice failures during simulated mission tasks to determine the rotorcraft response time.

7.6.3.2 Pilot response time

The pilot response time is defined as the time between the pilot being made aware that there has been a failure and recovery action being initiated via the controls. The pilot response time is further divided into the decision time and the reaction time. The decision time varies depending on the level of involvement in the flying task while the reaction time is dependent purely on whether or not the pilot's hands are on or off the flight controls. Table 7.1 shows the pilot response time criteria taken from 00-970 [7.2].

Table 7.1 Definition of pilot response times [7.2].

Flight segment (pilot attention level)	Decision time (s)	Reaction time (s)	Pilot response time (s)
Active	—	0.5	0.5
Attentive Hands-on	1.0	0.5	1.5
Attentive Hands-off	1.5	1.0	2.5
Passive Hands-on	2.0	0.5	2.5
Passive Hands-off	3.0	1.0	4.0

The flight phases associated with the decision time have precise definitions:

- *Active flight*. The pilot is using the flight controls continuously to maintain or change the flight path of the aircraft.
- *Attentive flight*. The pilot has to pay particular attention to flight control for short periods. This may involve making occasional adjustments to the flight path using the flight controls, for example, during instrument flight. Alternatively it may involve monitoring the actions of the flight control system closely, such as during an automatic approach to the hover.
- *Passive flight*. This covers long periods of flight during which the pilot need only give a minimal amount of attention to controlling the flight path or monitoring the AFCS; for example, during instrument flight with autopilot holds engaged.

The attentive and passive phases of flight may then be further divided into *hands-on* and *hands-off* depending on whether or not the pilot would normally be grasping the flight controls. Deciding which flight phase should be used when conducting failure testing is clearly very important, as it will directly affect the decision time and therefore the entire intervention time. From the point of view of the manufacturer it is preferable to have the specification compliance testing conducted against the shortest possible decision time, as the aircraft is more likely to be satisfactory. For the test pilot, however, the only consideration is to ensure that the decision time which is used accurately reflects the way that operational pilots would fly the aircraft. Thus if he knows from his operational experience that pilots normally engage the radar altimeter height hold when flying low level over the sea, leaving the flight controls unattended for substantial periods, then clearly the passive hands-off criterion should be used.

Once the appropriate flight phase has been chosen, testing is conducted to determine if the aircraft can be recovered safely at the end of the total intervention time.

7.6.3.3 *Applying a control parameter*

Employing a method that allows incremental increases in the intervention time ensures the safe conduct of this type of trial. The most obvious way to do this is to introduce a small time delay between the failure being initiated and the pilot taking recovery action, with the delay time being increased gradually until specification compliance is achieved or an aircraft limit (or test limitation) is approached. In practice, however, it is extremely difficult to use time as the means of controlling the test. This is due to the inevitable inaccuracies that are generated by the neuro-muscular lag inherent in an observer calling actions on a stopwatch and the pilot reacting. It should also be remembered that there might not be a linear relationship between time and aircraft reaction; a small increase in intervention time may have a dramatic effect on the aircraft flight path.

A better and safer approach is to use a controlling parameter which can be presented more easily to the pilot and which is more directly related to the state of the aircraft. Consider a trial to check that a single engine helicopter can successfully establish autorotation following an engine failure with a specified intervention time. In this case the test pilot will primarily be interested in the value and rate of change of rotor speed so this should be chosen as the controlling parameter. The intervention time can then be increased gradually by initiating recovery at incrementally lower values of N_R. Likewise for a trial evaluating AFCS pitch lane runaways, aircraft pitch attitude may be the most appropriate controlling parameter. Whichever parameter is used the methodology is the same; small increments are chosen initially with the aim of obtaining trend information. The test team need to establish the relationship between changes to the value of the controlling parameter at the point where the pilot takes recovery action using the flight controls and the value of any parameter which approaches a limiting condition during the recovery. Once a trend has been established informed predictions can be made about the effect of any increases in delay time and the test can progress, usually in reducing increments, to specification compliance or a limit. It is obvious from the above that the control parameter must be presented to the test pilot in such a way that precise values can be seen. This then may require the installation of specific test instrumentation.

7.6.3.4 *Living with the failure*

As part of the trials planning process it will be necessary to agree with the aircraft operator the requirements that the aircraft will have to meet following a system failure. For example, a failure which leaves an aircraft controllable but very difficult to fly may be acceptable if the aircraft will only be required to recover to base in these circumstances. However, if the requirement is to continue and complete the mission then the same post-failure characteristics may be unacceptable.

Where systems are multiplexed, a failure of one system may have no direct effect on the capabilities of the aircraft, however the loss of redundancy and the possibility of further failures is always considered. This often requires the definition of a post-failure operational flight envelope (OFE) that may be considerably more restricted than the normal OFE. The considerations here are the probability of a subsequent failure and the effect it would have on the aircraft or crew. Loss of the remaining hydraulic system in a duplex installation is an example of a subsequent failure that has such serious implications that the post-failure OFE is usually restricted to landing as soon as it is safe to do so. In many other cases restrictions on speeds and heights or a warning to aircrew on the effects of subsequent failures may suffice. With failures of an engine in multi-engine aircraft the performance available OEI is determined and published for aircrew use.

7.6.4 Total loss of power

For obvious reasons it is always necessary to test the characteristics of a helicopter following a complete loss of power. The amount of testing required depends on the likelihood of such a complete loss occurring and also on the role of the aircraft. Clearly single-engine helicopters are more likely to find themselves in this condition than multi-engine ones and require extensive testing, however, it is still necessary to demonstrate that any rotorcraft can enter autorotation and perform a power-off landing. It has been known for twin-engined helicopters to run out of fuel. Equally in the event of tail rotor drive failure it is common practice to shut down all the powerplants. If the aircraft is being procured for the training role that involves entries into autorotation and engine-off landings then its qualities during power-off manoeuvres are of particular importance.

7.6.4.1 *Conduct of the test programme*

The testing of flight regimes associated with total loss of power clearly carries with it a significant degree of risk. To mitigate this risk the test programme is usually structured in such a way that before an area of testing is attempted all the preparatory tests have been completed. Thus tests start by investigating flight idle glide before going on to autorotation. The aircraft characteristics in FIG are assessed before rapid entries into descending flight and recovery to climbing flight are made. Of course the completion of engine and rotor governing tests such as engine acceleration, transient overswing and combustion stability are a prerequisite. Once the full range of FIG and autorotative tests has been performed the engine-off landing tests are conducted before going on to look at in-flight relights.

7.6.4.2 *Autorotation and flight idle glide*

The behaviour of the aircraft in autorotation is the first area to investigate and this involves the two main areas of handling and performance. Turning first to handling tests the flight control positions in trimmed autorotative flight are documented at a range of airspeeds. Of particular note is the pedal margin available at low speed, which may not be adequate to ensure yaw control during the engine-off landing (EOL) particularly if it involves a gusty crosswind. The aft cyclic margin is another area that has proved unsatisfactory on some helicopters being insufficient to allow adequate manoeuvrability in the flare. The behaviour of the rotor is also investigated to determine how easy it is for the pilot to control rotor speed and also how often intervention by the pilot will be needed to respect limits. The cues to high and low N_R, together with the effectiveness of any warning devices are evaluated. Manoeuvres that are representative of those a pilot would make during the approach to a forced landing are conducted to ensure that adequate control is available and that the workload required to manage rotor speed is not excessive. This usually involves determining the effect on the rotor speed of a range of gentle manoeuvres such as turns, flares and bunts. Areas of particular interest are the handling qualities at low values of rotor speed where problems with control margins and effectiveness are more likely. Clearly all low rotor speed testing is approached with great care.

Typical autorotative performance testing involves a measurement of the rate of descent and distance covered with various combinations of airspeed and rotor speed. From these results recommendations are made to the operational pilot concerning the airspeed for minimum rate of descent and the combination of airspeed and rotor speed that gives the maximum range. Pressure errors are often significant during autorotation and are documented as they can significantly increase the difficulty of manoeuvring the aircraft during a forced landing. For example, a pilot will attempt to maintain a set airspeed such as V_{MP} when manoeuvring in autorotation, however, if he is unaware of the presence of large ASI PEs he may make unnecessarily large pitch attitude changes as he attempts to keep the indicated airspeed under control. Much of the performance information gained here is employed later in the trials programme for determining the initial airspeed for testing engine-off landings and for estimating the height loss during re-light tests.

7.6.4.3 *Recovery to powered flight*

Ultimately any autorotation that does not result in an engine-off landing will require a recovery to powered flight. The aircraft characteristics during this phase of flight will be heavily dependent on the engine and rotor governing system. Any tendency of the engine to surge as the throttle is advanced or as the collective is raised will clearly add to the difficulty of the manoeuvre. Of particular interest is the height loss involved during the recovery, as this will affect the usefulness of the aircraft in performing realistic practice forced landing (PFL) profiles. The cross-coupling associated with rapid recoveries to powered flight may be significant and will give an idea of the difficulties that may be encountered during the engine-off landing phase. Tests commence at a safe altitude with recoveries from flight idle glide. It should be remembered that the recovery from flight idle glide is a common operational manoeuvre following rapid descents and not just associated with a temporary power loss.

7.6.4.4 *Entry into autorotation*

Once the aircraft characteristics in steady autorotation have been investigated, rapid entries into autorotative flight are evaluated. These tests should start with gentle entries into flight idle glide (throttle(s) in the flight position) and then progress incrementally onto rapid entries into autorotation (throttle(s) retarded). Tests are conducted closed loop and are made initially at V_{MP} in level flight. Eventually the full speed range is checked in addition to climbing and descending flight. Information is gathered on the control movements required to counter the cross-coupling effects associated with rapid collective lever lowering. An assessment is also made of the adequacy of the remaining control margin to cater with more aggressive entries perhaps associated with a surprised operational pilot who is not expecting a total power failure! The margins available in steady autorotation will already be known but the dynamic situation of rapid entries can create problems with the aft cyclic margin (particularly at high speeds) and the pedal margin (particularly at low speeds). The transient overswing properties of the rotor governing system are also important for FIG entries especially if the pilot is required to raise the collective lever again to control a rapid increase in rotor speed. Once experience has been gained of rapid entries with the pitch attitude held constant further tests are conducted to investigate entries with a nose-up flare simulating the high-speed/low-level engine failure recovery actions.

7.6.4.5 *Entries into autorotation with delay times*

A vital part of the test programme is to determine the delay time (intervention time) that is possible between an engine failure occurring and the pilot lowering the collective. As explained earlier the intervention time is made up from the rotorcraft response time and the pilot reaction time. The pilot reaction time to be applied is usually specified by the procuring authority but traditionally a total intervention time of two seconds has been used for helicopters suffering a total power failure in level flight. This may be thought of as a 0.5 s rotorcraft response time and an attentive hands-on pilot response time. Thus it should be possible for the pilot to move the controls up to two seconds after the engine failure has occurred without the rotor speed reducing below the minimum permitted value at any stage during the recovery. This type of testing is clearly high risk and is normally monitored using telemetry.

The following terms are used to enumerate the rotor decay (see Fig. 2.26 for a graphical representation):

N_{R0} = rotor speed at which power loss occurs

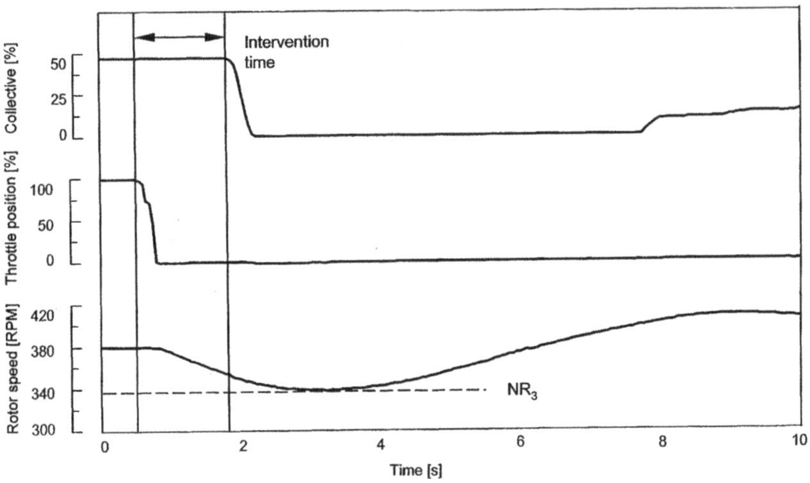

Fig. 7.12 Lever delay test data.

N_{R1} = rotor speed when pilot commences recovery action by lowering the collective
N_{R2} = rotor speed when collective lowered fully
N_{R3} = minimum N_R achieved
N_{R4} = a nominal value related to the normal autorotative rotor speed (once N_R has increased to this value it is assumed that the entry phase of the autorotation has been completed)

Tests start with a simultaneous 'throttle chop' and lever lowering from V_{MP} in level or descending flight. The minimum rotor speed achieved is noted and from this the total rotor speed drop is calculated ($N_{RO} - N_{R3}$). A delay time is then introduced by allowing the rotor speed to fall to a pre-determined value (N_{R1}) before the collective is lowered, as shown in Fig. 7.12.

The ΔN_R is calculated again ($\Delta N_R = N_{R1} - N_{R3}$) before a new N_{R1} is selected for the next point. This process is repeated using incremental reductions in N_{R1} until the minimum permitted rotor speed is achieved on the underswing, a handling difficulty is experienced, or the specified intervention time is met. As discussed in Section 2.12 the values of N_{R1} and N_{R3} are plotted to allow the trend to be monitored and thus the minimum rotor speed to be predicted for the next test point. It is important to note that a valid trend will only be established if a consistent technique is used every time; for example, yaw on engine failure is corrected immediately, the same rate of collective lever lowering is used and the airspeed is adjusted only after N_{R4} is reached. A check is made on the consistency of each lever lowering by noting and comparing the minimum value of normal acceleration achieved.

After the initial tests have been completed at V_{MP} the airspeed is varied to determine the effect that this has on the intervention delay time. At higher speeds the aircraft is flared as the collective is lowered to reduce the rotor speed loss. At lower airspeeds the aircraft is accelerated to the minimum rate of descent airspeed once rotor speed has recovered. In all cases information is gathered to assist the test team in conducting avoid area tests subsequently. Thus the height loss involved in regaining the nominal autorotative rotor speed (N_{R4}) and minimum RoD airspeed is noted. For the high-speed cases the height loss during the intervention time and the height gain possible during the flare are needed. On a full test programme of new rotorcraft further tests may be required to explore the effects of density altitude, AUM and initial power setting.

7.6.4.6 *Engine-off landing tests*

The aim of engine-off landing tests varies depending on the role of the aircraft. In the case of helicopters without a training role the aim is usually to prove that a safe landing is possible and to offer some advice in the aircrew manual. For aircraft with a training role where EOLs are likely to be commonplace a much more comprehensive series of tests is needed. The aim of these tests is usually to establish a box defined in terms of the airspeed and height at which the initial flare should be started. Within the box the handling qualities will be such that a student pilot can perform an EOL without undue skill; clearly the larger the box the more suitable the helicopter will be for teaching the skills required for engine-off landings.

The tests commence using the speed for minimum ROD in autorotation and a flare height determined from the powered recovery tests. Although the co-pilot makes a call at a pre-determined height on the radar altimeter, the responsibility for choosing when and how to perform the manoeuvre always rests with the handling pilot. The next point is then flown keeping the airspeed the same but changing the flare height. For each combination of speed and height the pilot rates the difficulty of the manoeuvre taking into account such things as predictability of the touchdown point, rotor speed control, proximity of the tail to the surface, touchdown speed and control margins. A series of these tests then allows construction of the box. At no stage is a point attempted that has a combination of an untried flare height and untried airspeed. Wind velocity, aircraft AUM and density altitude all affect the manoeuvre and therefore the power-off landings programme starts with the most favourable conditions and then moves incrementally to the least favourable.

7.6.4.7 *Avoid area testing*

The final stage of a power-off assessment programme is to define the *avoid curve*, avoid area or height/velocity diagram. An example of the type of presentation commonly seen in aircrew manuals is shown in Fig. 7.13. An avoid area diagram shows the operational pilot the combinations of airspeed and height above the surface from which a total loss of power is unlikely to be survivable. Armed with this information the pilot can plan the aircraft's flight path to minimize the time spent at risk inside the area. A number of assumptions are normally made in the construction of height/velocity diagrams:

- The aircraft is in level flight or the hover at the moment of power loss.
- There is a total intervention delay time of 2 seconds.
- There is no wind and the surface is suitable for a landing.

It is common practice to present different charts for sea level and a higher density altitude (in the region of 5000 feet); similarly several charts can be produced to account for variations in aircraft AUM.

The height/velocity diagram can be divided into four areas indicated as 1 to 4 on Fig. 7.13. Area 1 is dictated by the height loss required to achieve an autorotative

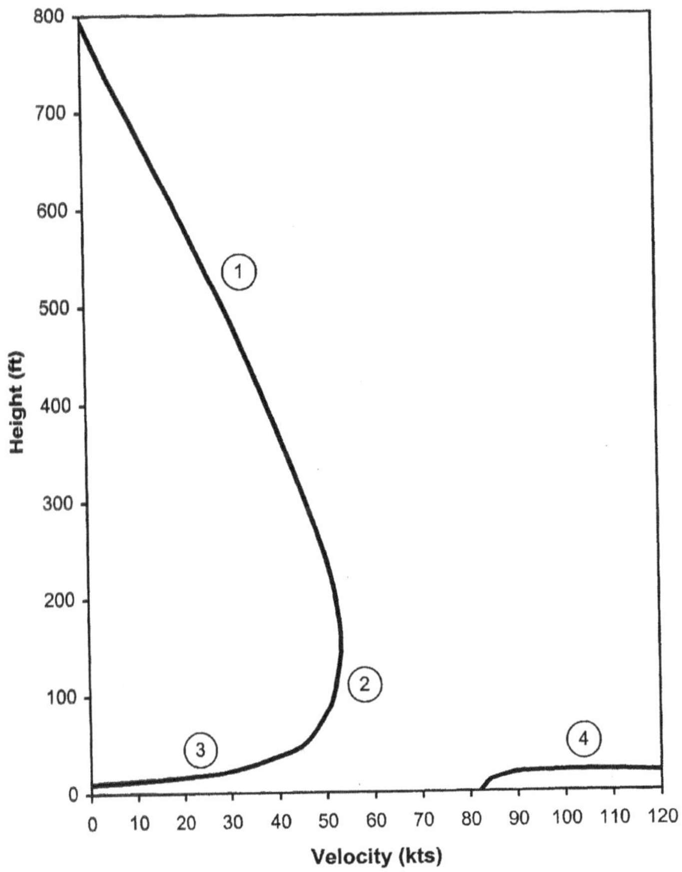

Fig. 7.13 Typical height/velocity diagram.

state and to accelerate the aircraft to an airspeed where an effective flare can be made. In Area 2 around the knee of the curve, the aircraft will not develop steady autorotation nor will a significant increase in airspeed be possible. Here the pilot will only be able to level the pitch attitude and use the remaining energy in the rotor to cushion the landing. Area 3 is dictated by the energy-absorbing qualities of the undercarriage and the rotor inertia. Area 4 needs to be avoided to allow sufficient height for the pilot to perform a flare to reduce groundspeed.

Performing avoid area testing carries with it a high level of risk which is mitigated by using test pilots who are fully familiar with the power-off landing characteristics of the aircraft being tested. Of course the standard risk reduction method of approaching each test point incrementally is applied rigorously in these trials. Testing starts from a known and benign point, for example, 500 ft and 60 KIAS; the height is then kept constant and the speed reduced by a small amount (5 to 10 kts) and another failure is simulated. After each landing the pilot awards a difficulty rating or an HQR. This process is repeated using small reductions in airspeed until the test pilot judges that the difficulty of performing the landing is such that following a sudden engine failure an operational pilot would not be able to achieve a survivable touchdown. For the testing of the low height points in Areas 3 and 4 the airspeed is kept constant and the height is incrementally increased or decreased as appropriate. It is a strict rule that only one parameter is varied between one test point and the next; these parameters include airspeed, height, wind velocity and direction, density altitude, AUM, aircraft configuration, intervention time and finally the individual test pilot.

7.6.5 Engine failures in multi-engine helicopters

7.6.5.1 Conduct of the test programme

A great deal of preparation is needed before the trials programme can start. A detailed knowledge of the engines and transmission system is essential. This should allow appropriate testing without unnecessary damage to the aircraft and powerplants. It may be necessary to approach the engine and airframe manufacturers for concessions to the normal limitations. Some thought has to be given to planning how the power failure is to be simulated and, most importantly, how the power output of the engine (or engines) which is not being 'failed' is restricted so that it is representative of a minimum specification engine. For hydro-mechanical fuel control systems this may be achieved by manually restricting the fuel flow using the normal engine controls in the cockpit or it may require modifications to the engine governing system itself. With FADEC systems it is often necessary for the manufacturer to provide a software 'patch' which causes the engine to operate at minimum specification power. Whichever way the required power output is achieved, trials risk is minimized if there is a method of rapidly restoring the power to its maximum value. It is vital to monitor the power output of the 'good' engine during the trial, particularly during the flyaway/vertical reject phase as it has been known for vibration to cause the cockpit controls to drift and reduce the available power.

The choice of which engine to 'fail' is an interesting question. During OEI testing on the triple engined Westland/Agusta EH 101 the choice was between the No. 1 and No. 3 engines as the central No. 2 engine suffered an extra 2% installation power loss in the hover under sea-level-ISA conditions. The No. 1 engine was eventually chosen when it was discovered that there was a further power loss on the No. 2 engine as the No. 1 ran down through ground idle and its exhaust gases were ingested through the No. 2 engine intake.

7.6.5.2 Post-failure performance and handling

The flight trials commence with an evaluation of the aircraft performance and handling with one engine inoperative (OEI). Initially one engine is throttled back in level flight at the speed for minimum power and the aircraft handling is checked. The next stage is to shut down one engine. This should take place at V_{MP} just in case there is any effect on the power available from the other engine(s). The minimum and maximum speeds in level flight that can be sustained on the good engine(s) are found together with the maximum sustainable angle of bank. It may be necessary to conduct a full level flight performance test to determine the power required and fuel flow for a range of airspeeds. This information is then often incorporated into the performance manual for use by operational crews. Once the performance has been established a series of gentle manoeuvres is made to determine if there are any handling problems. This whole process is then repeated at higher altitudes.

7.6.5.3 *Engine re-lights*

To achieve a successful engine start the mass flow of air through the compressor must be sufficient to achieve self-sustaining operation whilst preventing excessive turbine entry temperatures. At high altitudes the air density may be too low for a start to be achieved. This usually results in a re-light envelope that is smaller than the operational envelope of the helicopter. Since the air temperature affects density the re-light envelope is often expressed in terms of density altitude along with airspeed as this can affect conditions at the engine inlet. Testing of engine re-lights commences at low altitude and a mid-speed range and will then go on to expand the envelope in terms of density altitude and airspeed. In addition to establishing the envelope the testing also encompasses the effect that engine starting has on other aircraft systems. For example, the load on the electrical system may be such that the AFCS is affected or navigation equipment disrupted. Clearly if engine start is achieved using bleed air from another engine the effect of a start on the functioning engine requires assessment.

7.6.5.4 *Failure cueing*

Engineering an effective engine failure warning in a multi-engine installation can be a difficult task. Some systems are triggered by low oil pressure as the engine-driven pump slows down but these systems often suffer from significant delays. Another approach is to compare the power output of the engines and trigger the warning if the difference exceeds a threshold value. This type of system can be problematic if the power-sharing characteristics of the engine governors are not well matched.

7.6.5.5 *Sudden engine failures*

As is the case with single-engine machines the aim of sudden engine failure testing in forward flight is to establish the intervention time available and to check that there are no handling implications. Testing in the low speed flight regime is also required to establish the aircraft performance and to determine the optimum techniques for landing or achieving a safe flight condition. The intervention time is established using rotor speed as the controlling parameter in much the same manner as for total power loss testing. In a multi-engine installation the rotor speed decay rate will be clearly much slower, however, an incremental approach is equally important. The same range of parameters will affect this rate of decay as for the single-engine helicopter discussed in Section 2.12. In addition the acceleration characteristics of the functioning engine will have an effect as the faster the engine acceleration the slower the N_R decay will be. As described before sudden engine failures should be made in level flight starting at V_{MP} and then progressing to selected speeds throughout the flight envelope. It may also be necessary to conduct failures at the combination of speed and power normally used for climbing flight.

The most critical area for powerplant failures and hence for testing is in the hover and low speed regime. Here the testing is used to establish the techniques for conducting flyaways and vertical or forward rejects. Advice to operational crews is given in the form of a height/velocity diagram in the same way as is done for single-engine aircraft. Figure 7.14 shows an example of an avoid curve that has been produced as a result of this type of testing. The diagram is divided into four zones defined by height and airspeed. The important area is Zone 3: in this zone the aircraft is not able to perform a flyaway and a safe vertical or forward reject is also not possible. It should be remembered that the avoid curve may be sized on the assumption that at the boundary

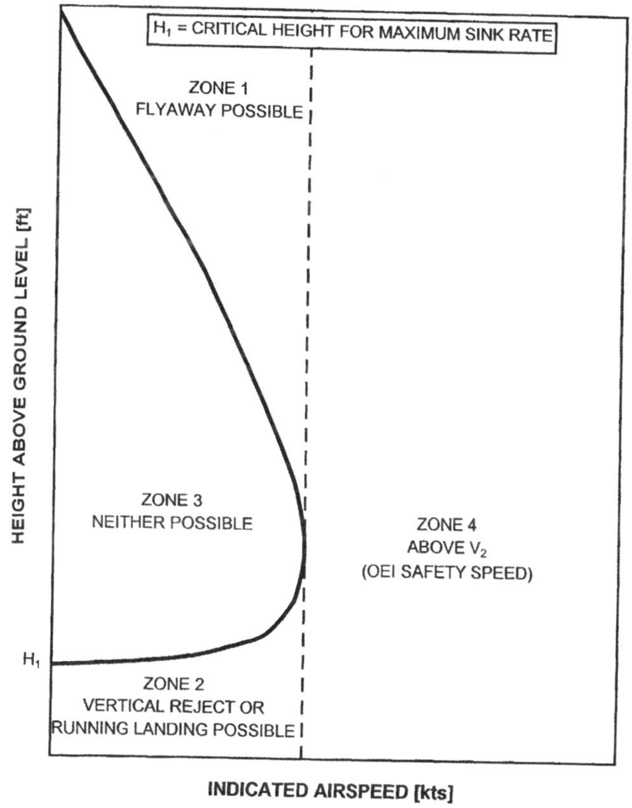

Fig. 7.14 Height/velocity diagram for a multi-engine helicopter.

a given level of aircraft damage may occur. The combination of height and speed from which a vertical or forward reject is possible depends mainly on the rotor decay characteristics for the test conditions and the energy-absorbing properties of the undercarriage. During the tests the rate of descent is kept within the undercarriage design limits and the minimum N_R is noted. The results are often compared with the predictions of any model used as part of the planning and risk assessment so that the model may be used to generate the advice given to operational crews.

For flyaway testing a model can be used to determine the optimum technique to be used to minimize height loss and one of the aims of the test programme is to establish if the technique is easy and repeatable to fly. If no model is available or the manufacturer fails to recommend a technique then a variety of flyaway profiles will have to be tried to establish the best method. The difficulty of flying any proposed technique must be borne in mind as the operational pilot will not be expecting a failure and will not be as adept or practised as the test pilot(s) conducting the trial. Accurate trends and therefore meaningful predictions can only be made if the chosen technique is flown consistently. At some point, however, it is often necessary to conduct testing to determine the sensitivity of the avoid curve data to a non-optimum technique.

Once the flyaway technique has been established testing starts by determining that the specified intervention time is achievable in the hover. For this the aircraft is hovered well clear of the ground and the flyaway profile practised. Initially the engine failure is coupled with a simultaneous flyaway. Then the intervention time is gradually increased by the pilot reacting to the failure at lower values of rotor speed (N_{R1}). Once the N_{R1} that equates to the required intervention time has been found it is used for the remainder of the hover flyaway testing. The total height loss suffered between the failure and the achievement of a safe flight condition is recorded as this can be used in defining the initial height AGL for the tests nearer the ground. It has been found from personal experience that the height loss during a flyaway is affected to a large extent by the initial hover height. The stronger visual cues to pitch attitude and RoD at lower heights assist the pilot in flying the manoeuvre and so help to reduce the height loss. To find the height loss near the ground the hover height used is reduced incrementally until consistent results are achieved with realistic visual cues. Typically a minimum height of 50 feet AGL is set for the recovery and the failure height set to ensure this limit is respected.

There are a number of risk reduction measures that are taken with this type of testing. As already stated an accurate predictive model reduces the chances of unpleasant surprises. Where possible two crew members are carried and the flight monitored by telemetry. It is worth noting again that for hydro-mechanical fuel control systems experience has shown that if cockpit controls are used to 'top' an engine, airframe vibration can cause the control to move. For this reason the power output of the 'good' (non-failed) engine requires careful monitoring to ensure that the settings remain unchanged. This can be achieved by reviewing the telemetry traces immediately after each test to check that when the maximum available power was demanded the torque, engine speed, and temperature figures did not drift from those seen on the previous test point. Risk to the aircraft and crew can also be mitigated if there is a procedure for rapidly restoring the engine under test and the 'failed' engine to full performance if needed. This may involve detailed discussions with the manufacturer and could lead to a non-standard engine handling methodology. Of equal importance is to have a *single* command such as 'Abort!' which can be issued by *any* member of the crew, *or* the observer in telemetry, to restore the engines and achieve a safe flight condition. Finally if cockpit controls have been used to limit an engine then the non-flying crew member keeps his or her hands on the relevant engine controls at all times.

OEI testing for civil procedures is in many respects the same as for military requirements, however, the aims of the test programme are slightly different and this is particularly true for the take-off and landing phases of flight. In military operations it is accepted that the aircraft may spend periods where an engine failure would place the aircraft at risk. The results of testing are therefore used to determine the level of risk for a typical mission. In addition information is produced which lets the operational pilot know where the risk lies (the height/velocity diagram) and which techniques can be used to avoid or minimize damage if a failure does occur. The military pilot can then make an informed judgement on how the aircraft should be operated. Civil operations have a different philosophy. When operating to JAR-29 Category A requirements [7.11] there is an intention that passengers are not placed at significant risk in the event of any failure including the powerplants. The overall aim of the civil OEI test programme then is to determine the boundaries of safe operation. This is achieved by establishing limiting combinations of AUM and density altitude at which passenger safety is assured *provided* the pilot performs the reject or flyaway in the prescribed manner. The estimation must include a suitable delay time to account for pilot recognition of a failure of the most critical engine (JAR-29.55 and 29.77). It

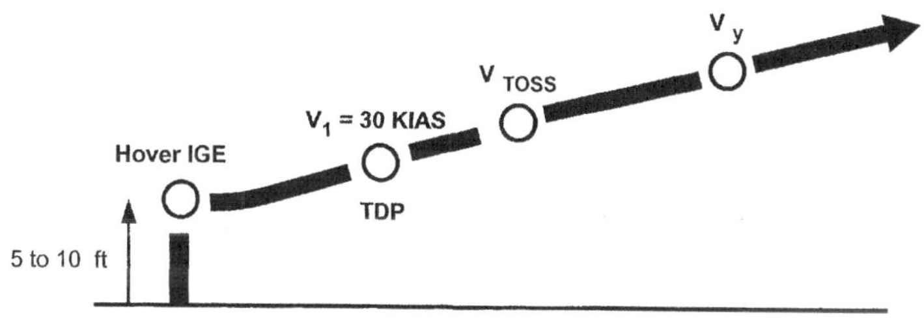

Fig. 7.15 Normal take-off – JAR Category A requirements.

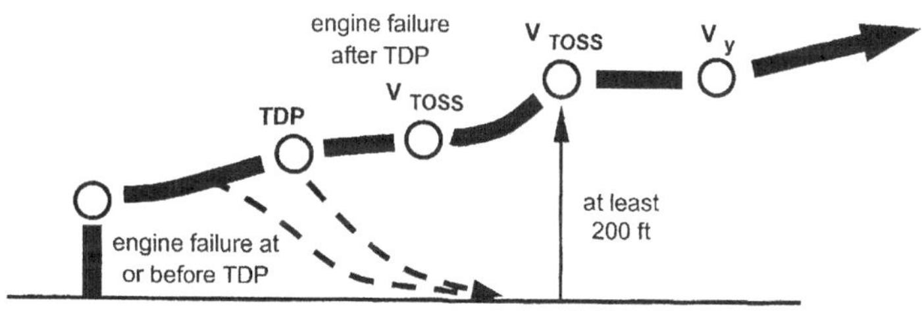

Fig. 7.16 Engine failure during take-off – JAR Category A procedures.

is worth examining the way that civil operators deal with the two most critical flight phases – take-off and landing.

Figure 7.15 shows a typical technique for conducting a normal take-off under Category A conditions. From a low IGE hover, with the height set such that following an engine failure the aircraft could descend vertically at a rate below the limit for the landing gear, the pilot applies take-off power and starts a climbing acceleration. The take-off decision point (TDP) is usually reached when the aircraft is slightly higher and has positive airspeed indications, typically 30 KIAS. The TDP is defined as the first point from which a continued take-off is assured and the last point from which a rejected take-off is assured within a specified distance (JAR-29.55). Different take-off techniques are often prescribed for different scenarios: short runways, confined areas or helipads and it is common practice for a manufacturer to specify a reduction in the maximum permitted AUM and H_D as the take-off area gets smaller and more congested. Once the TDP has been passed the aircraft is accelerated to take-off safety speed (V_{TOSS}) and then on to the speed for best rate of climb (V_y).

Figure 7.16 shows a typical flight path for an engine failure before or at the TDP and for a failure after the TDP in VFR conditions. In the first case an aborted take-off is performed by reducing speed and landing on. For a failure after the TDP the aircraft is quickly accelerated to achieve V_{TOSS}. This speed is then held until the height has increased to at least 200 ft (JAR-29.59) before the speed is increased to V_y.

The approach to landing is treated in a similar way. In Fig. 7.17 flight paths for an engine failure either side of the landing decision point (LDP) are depicted. The LDP is the last point in the approach and landing path from which a baulked landing can be performed (JAR-29.77). During the aborted approach the aircraft may not descend below 35 ft at any time.

For both take-off and landing under Category A flight the civilian pilot is required

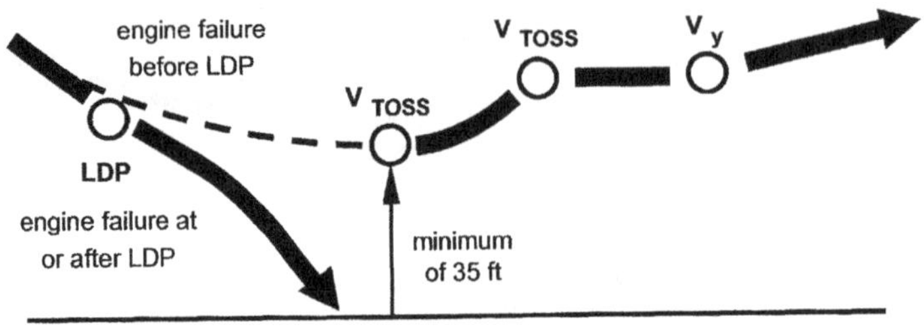

Fig. 7.17 Engine failure during landing – JAR Category A procedures.

to consult the aircraft flight manual to determine the aircraft performance. As indicated above the information is likely to include:

- maximum permissible take-off and landing weights;
- rate of climb with one engine inoperative;
- the acceleration/stop distance;
- distance to clear a 50 ft obstacle with an engine failure at the TDP;
- landing distance for an engine failure after the LDP.

The aim of civilian OEI testing is to determine the optimum technique and combination of heights AGL and speeds for conducting these take-offs and landings. In addition the helicopter performance will be noted so that the appropriate information regarding maximum weights can be presented in the flight manual. The test procedures used are closely related to those that have already been covered, however, a more detailed examination of the techniques used to determine the TDP is worthwhile.

For establishing the TDP the critical engine is 'failed' at a height and speed which allows the 35 ft minimum height to be respected easily. The height and speed are then incrementally reduced until either the TDP predicted by a model is reached or the minimum 35 ft ground clearance is just achieved. It is worth reiterating the 'golden rule' that only one parameter should be varied at a time. The next stage is to conduct a series of aborted take-offs starting at low height and speed and incrementally working towards the TDP established from the previous tests. It is mandated (JAR-29.55) that the TDP is easily recognizable by the pilot in terms of height and/or speed. Using speed alone can often be problematical due to the poor performance of most pitot systems at low speeds.

7.6.6 AFCS failures

The evaluation of failures within flight control systems is of particular importance because of the potential that these failures have to disturb the flight path of the helicopter. As flight control systems are fitted to provide additional stability and, in the case of autopilots, to control the aircraft they must have sufficient authority to make large enough inputs. Thus the AFCS designer is faced with balancing two conflicting requirements: high performance and benign failure characteristics. Of course one way of achieving both is to design the system with triplex or quadruplex redundancy and voting logic. This approach does suffer disadvantages in terms of complexity and cost that often make it unattractive for limited authority AFCSs. For these simpler systems the more usual solution is to employ simplex or duplex limited authority, fast-acting series actuators and high authority, slow-acting parallel actuators. Consequently the failure testing of these classes of AFCS is primarily concerned with evaluating the effect that failures may have on both series and parallel actuators and hence on the flight path of the aircraft.

There are a number of different types of failure that can occur within an AFCS and it is convenient to divide these into active and passive failures. The active failures are those that cause an uncommanded flight path disturbance while passive failures involve a loss of function but no disturbance. Active failures are clearly the more serious of the two and this category can be further sub-divided into oscillatory failures and runaways. Oscillatory failures are usually caused by problems with the feedback path that results in the actuator constantly reversing direction. Runaways may have many causes such as failures within the computation or erroneous sensor signals. It goes without saying that a series actuator runaway can be particularly hazardous should it occur in an agile aircraft that is flying close to the ground at high speed. Although passive failures are likely to be far less dramatic they can still lead to potentially catastrophic consequences. For example, if a radar altimeter hold drops out when operating at night, low level over the sea there is a considerable danger of the aircraft descending unnoticed and striking the surface.

7.6.6.1 Test technique

The main factors that influence the way the pilot reacts to the failure are the recognition that something has occurred and the level of involvement of the pilot in the flying task. From these two factors the desired intervention time is determined as has been discussed earlier. The test philosophy centres on injecting appropriate failures and incrementally increasing the intervention time until the specification requirements have been met or a limit is reached. The testing starts with practise of recoveries from extreme pitch and roll attitudes; this is approached cautiously as it is often the case that the highest stresses experienced by the airframe occur during the recovery and not during the runaway itself. If a simulator is available this can be useful in working up the crew. Failures are typically injected into the AFCS using a 'runaway box' starting at the most benign conditions of low speed, mid-CG and series actuators nulled (for dual lane systems). The testing then moves on to evaluate less benign conditions. In the case of actuator runaways the pilot is usually given an aircraft attitude at which to commence recovery: this is gradually increased until the required intervention time has been achieved or a test limit is reached. During the test plots of the intervention pitch (or roll) attitude versus maximum attitude and maximum load factor are maintained. For pitch runaways it is important to establish the height change that results from the full intervention time at each condition and in particular the height loss for forward runaways. During the test programme the utility of any AFCS cutout is assessed and the optimum strategy for recovery from runaways established. Cutouts may serve a variety of functions: to temporarily disengage part or all of the AFCS as and when a failure is recognized, to enable fault diagnosis and to re-engage serviceable parts of the system following a single failure.

7.6.6.2 Results of testing

Figure 7.18 shows a time history of a pitch runaway. Such traces are a typical way of presenting results. The results obtained during the test programme can have a number

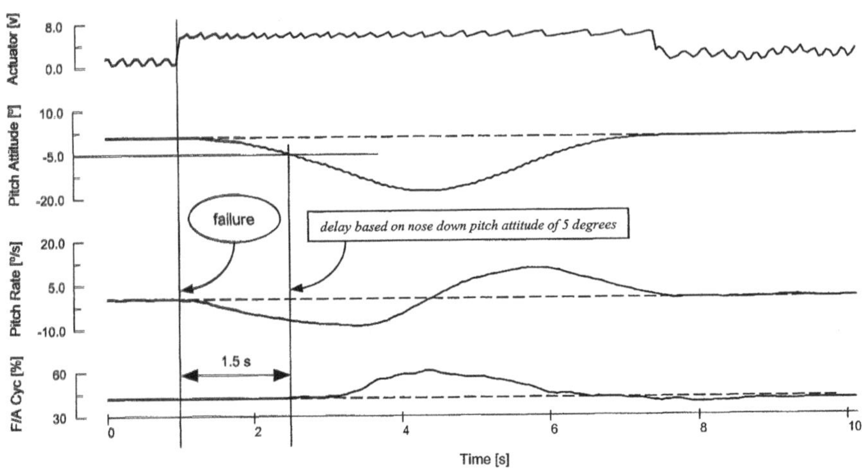

Fig. 7.18 Pitch series actuator runaway – test data.

of implications for the AFCS and for the way that the aircraft is operated. In the worst case it may require changes to the design of the system such as altering the speed or authority of the actuators. Making the system more fault tolerant by fitting a monitor channel which automatically disengages a faulty lane is another possible solution to unsatisfactory failure characteristics. More often the test programme will result in advice to aircrew and limitations on the use of the system. Such advice might stress the importance of timely and accurate nulling of the series actuators prior to NOE flight or operations close to the sea surface. Alternatively a deliberate forward bias of the series actuators may be obtained by trimming forward thereby reducing the severity of a nose-down runaway. Limitations that could be imposed include maximum speeds with the full AFCS and when operating with a degraded system. It might also be necessary to impose a minimum height for NOE flight if the height loss following a nose-down runaway is too great. The need for operational crews to practise recoveries from runaways may also be a recommendation from the testing. This may lead to a production version of the flight test 'runaway' box being incorporated into all fleet standard rotorcraft.

References

Chapter 1

1.1 Advisory Group for Aerospace Research & Development (AGARD) (1995) *Flight Test Techniques Series.* AGARD-AG-300. (In particular, Volume 14: *Introduction to Flight Test Engineering*, F. Stoliker (ed.), Sept.)

1.2 Advisory Group for Aerospace Research & Development (AGARD) (1980) *Flight Test Instrumentation Series.* AGARD-AG-160. (In particular, Volume 10: *Helicopter Flight Test Instrumentation*, K. Ferrell (ed.).)

Chapter 2

2.1 Dommasch, D. (1953) *Elements of propeller and helicopter aerodynamics.* Pitman & Sons, London.

2.2 Gessow, A. & Myers, G. (1967) *The Aerodynamics of the Helicopter.* Ungar Publishing, New York.

2.3 Glauert, H. (1948) *The Elements of Aerofoil and Airscrew Theory.* Cambridge University Press.

2.4 Layton, D. (1984) *Helicopter Performance.* Matrix Publishers, Beaverton, Oregon.

2.5 Newman, S. (1994) *The Foundations of Helicopter Flight.* Edward Arnold, London.

2.6 Prouty, R. (1986) *Helicopter Performance, Stability and Control.* PWS Engineering, Boston, Massachusetts.

2.7 Vogely, A. (1950) *Axial Momentum Theory for Propellers in Compressible Flow.* NACA TN 2164, Aug.

2.8 Wu, Sigman & Goorjian (1972) *Optimum Performance of Hovering Rotors.* NASA TMX 62138.

2.9 Lock (1934) *Tables for use in an improved method of airscrew strip theory calculations.* ARC R&M 1674, Oct.

2.10 Seddon, J. & Newman, S. (2001) *Basic Helicopter Aerodynamics*, (2nd edition) Oxford: Blackwell Science, Oxford.

2.11 Glauert, H. (1926) *A General Theory of the Autogyro.* ARC R&M 1111.

2.12 McCormick, B. (1979) *Aerodynamics, Aeronautics and Flight Mechanics.* J Wiley & Sons, New York.

2.13 Bennett, J. (1940) Rotary wing aircraft. In: *Aircraft Engineering*, March.

2.14 Stepniewski, W. (1973) *Basic aerodynamics and performance of a helicopter.* In: AGARD Lecture Series 63, March.

2.15 Coleman, R., Feingold, A. & Stempin, C. (1947) *Evaluation of the induced velocity field of an idealized helicopter rotor.* NACA ARR L5E10.

2.16 Mangler, K. & Squire, H. (1950) *The induced velocity field of a rotor.* ARC R&M 2642.

2.17 Bramwell, A. (1976) *Helicopter dynamics.* Edward Arnold, London.

2.18 Payne, P. (1959) *Helicopter dynamics and aerodynamics.* Pitman & Sons, London.

2.19 Gessow, A. & Crim, A. (1956) *A theoretical estimate of the effects of compressibility on the performance of a helicopter rotor in various flight conditions.* NACA TN 3798.

2.20 Castles, W. & New, N. (1952) *A blade-element analysis of lifting rotors that is applicable for large inflow and blade angles and any reasonable blade geometry.* NACA TN 2656.

2.21 McCormick, B. (1967) *Aerodynamics of V/STOL flight.* Academic Press, New York.

2.22 Peters, D. & Chen, S. (1982) Momentum theory, dynamic inflow and the vortex-ring state. *Journal of the American Helicopter Society,* July.

2.23 Wolkovitch, J. (1972) Analytical predictions of vortex-ring boundaries for helicopters in steep descents. *Journal of the American Helicopter Society,* **17**, No. 3, July.

2.24 Azuma, A. & Obata, A. (1968) Induced flow variation of the helicopter rotor operating in the vortex-ring state. *Journal of Aircraft,* **5**, No. 4, July.

2.25 McCormick, B. (1967) *Aerodynamics of V/STOL flight.* Academic Press, New York.

2.26 Heyson, H. (1975) *A momentum analysis of helicopters and autogyros in inclined descent, with comments on operational restrictions.* NASA TN D-7917.

2.27 Ferrel, Fredrickson, Shapley & Kyler. (1976) *A flight research investigation of autorotational performance and height-velocity testing of a single main rotor single engine helicopter.* USAAEFA Project 68–25.

2.28 Pegg, R. (1968) *An investigation of the helicopter height–velocity diagram showing effects of density altitude and gross weight.* NASA TN D-4536.

2.29 Condon, Bailes & Connor (1971) *Height–velocity test, AH-1G helicopter.* USAASTA Project 69–13.

Chapter 3

3.1 Engineering Science Data Unit (1977) *Introduction to non-dimensional methods for the measurement of performance of turbine-engined helicopters.* ESDU Item Number 73076, with Amendment A.

3.2 Knowles, P. (1967) *The application of non-dimensional methods to the planning of helicopter performance flight trials and the analysis of results.* ARC CP 927.

3.3 Engineering Science Data Unit (1977) *Equations for the calculation of International Standard Atmosphere and associated off-standard atmospheres.* ESDU Item Number 77022.

3.4 Engineering Science Data Unit (1974) *Non-dimensional methods for the measurement of level flight performance of turbine-engined helicopters.* ESDU Item Number 74042.

3.5 Langdon, G. (1963) *Determination of the range performance of a gas turbine-engined helicopter from flight test results.* Aeroplane and Armament Experimental Establishment, Boscombe Down, Report Number AAEE/Res/305, March.

3.6 Roots, M. & Blake, R. (2001) *Level flight performance: an engineer's guide.* Proceedings of the 32nd Annual Symposium of the Society of Flight Test Engineers, Seattle 12–14 Sept.

3.7 Langdon, G. (1962) *Measurement of the hovering performance of a helicopter by tethering.* Aeroplane and Armament Experimental Establishment, Boscombe Down, Report Number AAEE/Tech/217, May.

3.8 Stepniewski, W. & Keys, C. (1984) *Rotary-wing Aerodynamics.* Dover Publications Inc., New York.

Chapter 4

4.1 Bramwell, A. (1975) *Helicopter Dynamics.* Edward Arnold, London.

4.2 Babister, A. (1980) *Aircraft dynamic stability and response.* Pergamon Press, Oxford.

4.3 Prouty, R. (1985) *Helicopter performance stability and control.* PWS Engineering, Boston, Massachusetts.

Chapter 5

5.1 Ministry of Defence (1998) *Design and Airworthiness Requirements for Service Aircraft. Volume 2 – Rotorcraft.* Defence Standard 00–970/Volume 2/Issue 1 (Amendment List 12, dated March).

5.2 US Army Aviation and Missile Command (2000) *Handling Qualities Requirements for Military Rotorcraft.* Aeronautical Design Standard-33E, March.

5.3 Heffley, R., Curtiss, H., Hindson, W. & Hess, R. (1986) *Study of the Helicopter Roll Control Effectiveness Criteria.* NASA CR 177404, April.

5.4 Padfield, G. (1996) *Helicopter Flight Dynamics.* Blackwell Science, Oxford.

5.5 Hoh, R. H. (1988) Concepts and Criteria for Mission Orientated Flying Qualities Specification. In: *Advances in Flying Qualities.* AGARD LS 157.

5.6 US Department of Defense (1990) *Military Standard, Flying Qualities of Piloted Aircraft.* MIL-STD-1696A, 30 Jan.

5.7 Harper, R. & Cooper, G. (1966) *A revised Pilot Rating Scale for the Evaluation of Handling Qualities.* AGARD CP 16, Sept.

Chapter 6

6.1 Gregg, W. (1922) *Standard Atmosphere.* National Advisory Committee on Aeronautics (NACA) Rep. 147.

6.2 International Commission on Aerial Navigation (1924) *Standard Atmosphere.* ICAN AP 1275.

6.3 International Civil Aviation Organization (1954) *Manual of the ICAO Standard Atmosphere.* Document 7488.

6.4 US Committee on the Extension of the Standard Atmosphere (1962) *US Standard Atmosphere.* COESA.

6.5 International Civil Aviation Organization (1964) *Manual of the ICAO Standard Atmosphere extended to 32 km.* Document 7488/2.

6.6 International Organization for Standardization (1975) *Standard Atmosphere (Identical with the ICAO and WMO Standard Atmospheres from −2 to 32 km).* ISO-2533–1975(E).

6.7 Engineering Science Data Unit (1986) *Equations for calculation of International Standard Atmosphere and associated off-standard atmospheres.* ESDU 77022 with Amendments A and B.

6.8 Engineering Science Data Unit (1992) *Physical properties of design atmospheres.* ESDU 78008 with Amendments A to C.

6.9 Cohen, H., Rogers, G. & Saravanamuttoo, H. (1972) *Gas Turbine Theory,* (2nd edition). Longman Group Ltd, London.

6.10 Dixon, S. (1978) *Thermodynamics of Turbomachinery,* (3rd edition). Pergamon Press, Oxford.

Chapter 7

7.1 Pearson, F. (1990) *Map projections: theory and applications.* CRC Press Inc., Boca Raton, F1.

7.2 Ministry of Defence (1998) *Design and Airworthiness Requirements for Service Aircraft. Volume 2 – Rotorcraft.* Defence Standard 00-970/Volume 2/Issue 1 (Amendment List 12, dated March).

7.3 Royal Air Force – Institute of Aviation Medicine (1973) *An Anthropometric Survey of 2,000 Royal Air Force Aircrew 1970/71.* RAF/IAM Rep 531. (Also published as RAE Technical Report 73083, FPRC Report 1327 and HMSO R&M 3372.)

7.4 Aplin, J. & Nammari, H. (1995) *The 1995 Anthropometric Survey of UK Military Females.* DRA/AS/MMI/CR95076/1, April.

7.5 Ministry of Defence (1987) *Human Factors for Designers of Equipment.* Defence Standard 00-25/Issue 2, Sept.

7.6 Joint Aviation Authorities (1999) *Requirements for Large Rotorcraft.* JAR-29, change 1, dated 1 Dec.

7.7 Lewis, G. (1997) *A flight test technique using GPS for pressure error correction testing.* National Test Pilot School, Mojave, CA, July.

7.8 Olson, W. (1998) *Pitot-static calibrations using a GPS multi-track method.* Conference proceedings, 29th Annual Symposium, Society of Flight Test Engineers.

7.9 Fisher, I. (1977) *Instrument approach minima for helicopters making ground controlled approaches (GCAs) to runways equipped with precision aids.* Ministry of Defence, Procurement Executive, A&AEE Note 2073, June.

7.10 Ministry of Defence (1987) *General Specification for Aircraft Gas Turbine Engines.* Defence Standard 00-971, May.

7.11 Joint Aviation Authorities (1999) *Joint Aviation Requirements for Large Rotorcraft.* JAR-29, Change 1, Dec.

Index

Printed and bound by CPI Group (UK) Ltd, Croydon, CR0 4YY

16/09/2026

14963387-0008